Network Design Using EcoNets

Roshan L. Sharma

International Thomson Computer Press

I(T)P® An International Thomson Publishing Company

Boston • London • Bonn • Johannesburg • Madrid • Melbourne • Mexico City • New York • Paris
Singapore • Tokyo • Toronto • Albany, NY • Belmont, CA • Cincinnati, OH • Detroit, MI

For more information contact:

International Thomson Computer Press
20 Park Plaza, 13ᵗʰ Floor
Boston, MA 02116 USA

International Thomson Publishing GmbH
Königswinterer Straße 418
53227 Bonn, Germany

International Thomson Publishing Europe
Berkshire House
168-173 High Holborn
London WC1V 7AA England

International Thomson Publishing Asia
60 Albert Street #15-01
Albert Complex
Singapore 189969

Nelson International Thomson Publishing Australia
102 Dodds Street
South Melbourne, NSW
Victoria 3205 Australia

International Thomson Publishing Japan
Hirakawa-cho Kyowa Building, 3F
2-2-1 Hirakawa-cho
Chiyoda-ku, Tokyo 102 Japan

Nelson Canada
1120 Birchmount Road
Scarborough, Ontario
Canada M1K 5G4

International Thomson Editores
Seneca, 53
Colonia Polanco
11560 Mexico D.F. Mexico

International Thomson Publishing Southern Africa
Building 18, Constantia Park
240 Old Pretoria Road
P.O.Box 2459
Halfway House 1685 South Africa

International Thomson Publishing France
Tour Maine-Montparnasse
33 avenue du Maine
75755 Paris Cedex 15
France

Library of Congress Catalogin-in-Publication Data
Sharma, Roshan L.
    Introduction to network design using EcoNets / Roshan L. Sharma.
        p.   cm.
    Includes index.
    ISBN 1-85302-907-9
    1. Telecommunication systems--Design and construction--Data processing. 2. EcoNets. I. Title.
TK5101.S453   1997
621.382'15--dc21

Acquisition Editor: Simon Yates, ITCP/Boston
Publisher/Vice President: Jim DeWolf, ITCP/Boston
Marketing Manager: Christine Nagle, ITCP/Boston

Projects Director: Vivienne Toye, ITCP/Boston
Manufacturing Manager: Sandra Sabathy, ITCP/Boston

Production: Benchmark Productions, Inc., Boston, MA

Net
Usi

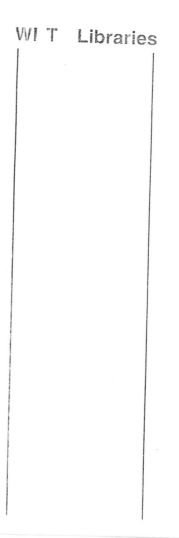

The Practical Networking Series

The Practical Networking Series is written by and for Network Designers, Administrators, and Managers. The books are professional guides to real-world problems at every stage of network maintenance, including planning, design, implementation, and operation.

Also available

Arthur B. Edmonds *ATM Network Planning and Implementation*, 1-85032-894-3
Charles E. Spurgeon *Practical Networking with Ethernet*, 1-85032-885-4

Contents

Preface

This book is a major revision of the author's previous book *Network Topology Optimization* published in 1990 by Van Nostrand Reinhold. Its use as a textbook for several network design related graduate courses throughout the country during the last six years has highlighted its strengths and weaknesses. To start with, the new interest in integrated broadband or multimedia networks, backbone networks, and advanced intelligent networks demanded an in-depth treatment of such new network types. The need for a more balanced treatment of theory and practice also became apparent quite early. Finally, its title sounded too formidable—somewhat related to the esoteric graph theory rather than network design. Although the need for a revised edition under a different title became acute, it also became apparent that nothing should be taken out of the original. Only the many new additions provide the salient new features to the this book.

The first feature of this book is the addition of five chapters dealing with the modeling of Automatic Call Distribution (ACD), integrated broadband, personal communication system (PCS), backbone, given topology and common channel signaling (CCS) networks with emphasis on the network design process based on the adage "topology is the soul of the network."

The second major feature of this book is the inclusion of internationally acclaimed network design software on an IBM PC formatted diskette. Although the software was described in the original book, it has seen two major upgrades requiring copyrights during 1991 and 1995. The immediate availability of the tool will now allow the reader to become acquainted with the actual network design process quickly and effortlessly.

The third feature of this book deals with the use of a consistent network design process applicable for all types of networks. The book constantly employs the "what if" approach for testing the validity of useful hypotheses during the network design process. The process quickly evolves into an iterative approach dictated by the designer's mind and facilitated by the friendly graphical user interface of the software package. This methodology is far superior to the closed-form solutions applicable to only small and purely theoretical networks. This book shows that even a small network presents a greater number of topological solutions than there are electrons in the universe. Only the design methodology as presented in this book can bring about some order.

The material of this book is also organized to provide either a one-semester or two-semester graduate course in network design. Each chapter provides enough material to be covered in three hours. The first nine chapters and Chapter 15, if followed in their natural sequence, can constitute an ideal syllabus for a one-semester introductory course. The remaining five chapters can be a basis for an excellent syllabus for a follow-up one-semester course.

Lastly, the book provides a large number of related exercises at the end of each chapter. These exercises can be the basis of self-motivated learning, regular homework assignments, mid-term and final exams, or term reports. Additionally, each chapter is now supplied with its own bibliography for compactness. Another interesting feature of this book is its inherent organization that allows the reader to read Chapters 1 and 6 only once, and then skip directly to the desired chapter(s) related to one's area of interest. A sizable number of specialists always prefer to learn about the theory through the actual process of solving meaningful exercises. They can make references to the material of Chapters 2, 3, 4, and 5 only when confronted with the need to understand the outputs. These four chapters provide the necessary theoretical groundwork for understanding all types of useful networks and interpreting the related output results.

The enclosed network design tool runs on any IBM-compatible PC equipped with a minimum of 16 megabytes of RAM and a hard disk space of about 10 megabytes reserved solely for network design. The user-friendly EcoNets software package employs proven analytical tools for optimization of network topologies very rapidly. It can be used to create an optimum network topology for any LAN/MAN/WAN application; a topology that can be fed, if desired, to a large mainframe-based network simulation tool for deriving detailed cost and performance attributes for the network. EcoNets' optimum network topology can also be the basis for computing accurate cost estimates of transmission and switching facilities by employing PC-based pricing tools available through many vendors.

The author is particularly indebted to a large number of graduate students who have, over a period of five years, taken a one-semester course on network design as a partial requirement for a Masters Degree in Electrical Engineering or Telecommunications obtained through the vehicles of both a live classroom and distant education. These students who were already working in a large variety of network design projects at major corporations, were kind enough to provide suggestions for improving not only the course material but also the graphical user interface for the enclosed network design tool. The author is particularly grateful to Robert Moussavi for reading the entire manuscript and making valuable comments.

The author can be reached at P.O. Box 822938, Dallas, TX 75382, or (214) 691-6790. Any questions or inquiries should be sent to this address.

1

Basic Networking Concepts

1.1 Introductory Remarks

This book is primarily concerned with the design of telecommunications networks. It is the purpose of this book to provide a balanced hands-on treatment of both traditional and modern telecommunications network design by using the principle of successive decomposition or iterative modeling as expounded in an earlier book, *Network Systems,* authored by Sharma, De Sousa, and Ingle (Sharma, De Sousa, and Ingle, 1982).

Telecommunications deals with the transmission, emission, or reception of signs, signals, written images, sounds, or intelligence of any nature by wire, radio, and optical or other electromagnetic means. A telecommunications network consists of two or more nodes that are interconnected through transmission links based on one or several of the many available technologies and network topologies. A hierarchy of nodes and/or links may characterize the network.

There are basically two types of telecommunications networks: switched and non-switched. The most common variety of a non-switched system involves communications between two or more humans. The field of human communications is a very complex subject (see Cherry, 1966). Another common variety of the non-switched system is a broadcasting network that primarily involves the transmission of only one type of information at any one time from a single location (or several locations repeating the transmission to different areas of reception), and the reception of the same information by a large number of listeners. Other types of non-switched networks deal with data processing and database management, and generally involve a single service node. This book does not cover non-switched telecommunication networks.

There are many examples of switched communication systems. The best example of a switched system is that of the public telephone network that allows subscribers to talk with one another on reserved paths. Public data networks also provide an excellent example of a switched system used for the interchange of messages or packets of data. Some of these data networks employ fully shared facilities. This book deals with the topological design of switched communication networks that provide a variety of services such as voice, data, image, and video separately or in an integrated fashion (multimedia). In order to accomplish that task, one must first introduce the necessary concepts such as classes of communication networks, network components (e.g., nodes and links), network traffic, and network architecture. See Figure 1.1 for an illustration of a generalized, standalone switched network that interconnects a set of local area networks (LANs) via several access subnetworks and a backbone network. Each LAN constitutes customer premise equipment (CPE) that consists of subscribers and a voice/data switch. Each access subnetwork consists of access lines (ALs) and lower level switches. Each backbone network consists of switches and trunks (TKs).

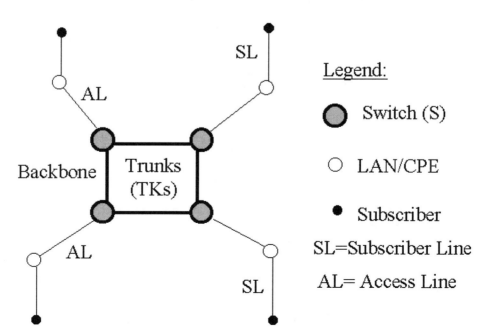

Figure 1.1 Basic components of a switched network.

1.2 Classification of Switched Communication Networks

Switched communication systems can be understood only in terms of their properties. These properties, when collectively considered, define what is known as the *network architecture*. One can classify any switched communication network by the following attributes:

1. Geographical coverage of the switched network

2. Method of accessing the switched network

3. Network topology

4. Transmission and multiplexing techniques employed to synthesize a switched network

5. Network management and control (NMC) techniques. These techniques are involved with the way intelligence is distributed throughout the network and how it is employed to switch/route traffic, control congestion and manage day-to-day operations. Therefore, it deals with the design of network hubs and switching nodes.

1.2.1 Geographical Coverage

There are four types of networks, characterized by their geographical coverage:

a. *Local Area Networks (LANs)*. About 60 percent of all networks fall into the category of LANs that span one or more buildings over a campus.

b. *Metropolitan Area Networks (MANs)*. About 20 percent of all networks cover distances of about 50 miles and fall into the category of MANs.

c. *Wide Area Networks (WANs)*. About 15 percent of all networks fall into the category of WANs. WANs covering a distance of about 500 miles account for 9 percent, and those covering longer distance account for 6 percent. Some authors have defined a WAN as simply an interconnection of LANs.

d. *Global Area Networks (GANs)*. About 5 percent of all networks fall into this category, which is bound to grow due to the global nature of many corporations. These networks present a unique design challenge due to the presence of many national boundaries and public policies.

1.2.2 Method of Accessing

Each subscriber node must be designed to access the switched network for service. Similarly, each network node must be designed to offer the advertised services to any qualified subscriber. Basically, there are three distinct access methods:

a. *Demand Access.* Demand access is generally employed in most voice and data communication networks. According to this interrupt-based technique, the subscriber goes off-hook to demand service. The service node provides a dial tone immediately after discovering the off-hook condition. The dial tone is an invitation for a further dialogue that ultimately results in a switched path between the calling and the called subscribers. The demand access is generally preferred by the users; however, such a scheme can present a challenge to the service provider during a period of a national disaster or a widespread failure. Some carriers stop providing a dial tone during such periods.

b. *Polled Access.* This is another popular method for allowing the computer-controlled switching node to interrogate each subscriber regarding the need for data interchange with another subscriber. The polled access method generally provides a very controlled environment, but only at the cost of a high overhead.

c. *Multiaccess.* The multiaccess technique is primarily employed by data LANs. A node transmits a message when some initial conditions are met (e.g., no carrier is sensed). If a collision occurs between two messages/packets transmitted by two sources, each node will retransmit after waiting for a randomly chosen period of time. A large fraction of data LANs employ the carrier-sense-multiaccess (CSMA) technique whereby each node senses the carrier on the shared bus before sending the message or packet. Many newer LANs employ the token access whereby a node can only transmit when it grabs a free token. Collisions are therefore prevented altogether since the need for a token is analogous to the poll. A few WANs employ the ALOHA multiaccess technique whereby a source node begins transmitting whenever it has a message or packet in its memory. A source will retransmit a message/packet if it does not receive a positive acknowledgment from the destination node within a certain period. Generally, the multiaccess method is an efficient scheme for cases where broadcasting a message (since every station is listening to the transmission over shared media) is required. However, it presents an unnatural boundary to WANs employing only point-to-point links. Complicated protocol conversion schemes become necessary.

1.2.3 Network Topologies

Network topology defines not only the manner in which the network nodes are interconnected with one another but also the types of links employed to interconnect each pair of nodes. Generally, there is a hierarchy associated with the network nodes and/or links. One can classify the topologies that have proven useful for designing switched telecommunications networks over the past 40-some years:

1. Mesh topology based on either a fully connected or partially connected backbone network

2. Multidrop topology based on a minimal spanning tree

3. Directed link topology based on directed graphs

4. Star topology based on either one star or many interconnected stars

5. Bus topology employing a bus shared by many stations

6. Ring topology employing a ring shared by many stations

Additional topologies can be obtained by mixing the basic topologies through the process of superposition and interconnection. Since most of the cost of a communication network is generally dictated by the cost of communication links, network topology must be optimized with great care for each application. The cost of hardware and network management, as it will be shown in later chapters, hardly influences the network optimization process. The goal of this book is to illustrate the process of network topology design in terms of both the underlying ideas and actual practice.

1.2.3.1 Mesh Topology

A fully connected mesh topology shown in Figure 1.2 (a), provides a direct link or path between every pair of network nodes of a given level of hierarchy. The mesh topology is economically feasible only when the number of switching nodes at a given network hierarchy is small. Since the cost of transmission in most existing networks is predominantly influenced by link costs and not switching, a fully connected mesh topology is only employed at a very high level of the network hierarchy (e.g., the backbone network). Even for those cases, the partially connected mesh topology shown in Figure 1.2 (b), will be found preferable over the fully connected topology considering either system costs or reliability considerations. A partially connected topology is recommended when the traffic flow between certain nodes is negligible and it can be switched on longer paths while resulting in a better economy of scale or better reliability.

1.2.3.2 Multidrop (MD) Topology Based on Minimal Spanning Tree (MST)

An MD topology based on MST is a subset of the generalized tree designed with the constraint that the sum of all link lengths in the network is minimal. Figure 1.3 (a) represents an MST topology. Assume that node 6 of Figure 1.3 (a) is a switching node. Using that information, one can create a four-level hierarchy of Figure 1.3 (b) according to the number of links between any pair of nodes.

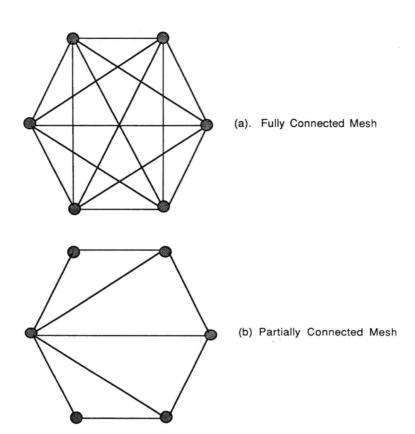

(a). Fully Connected Mesh

(b) Partially Connected Mesh

Figure 1.2 Mesh topology.

Chapter 5 will discuss algorithms for designing an economical MST topology. Some authors employ the concept of a multipoint (MP) topology instead of the multidrop (MD) topology. They apply the MD topology for only those situations where a common carrier's single Central Office serves more than two data terminals located at different addresses. Our definition of the MD topology assumes an MST consisting of all links with finite costs. Therefore, the MP topology can be considered as a subset of the MD topology.

1.2.3.3 Directed Link (DL) Topology

Several network architectures (e.g., Doelz's Virtual Fast Packet technology, also called the Extended Slotted Ring and frame relay technology), allow the synthesis of networks with only directed network links (also called the netlinks). These netlinks connect a subset of packet switching nodes in the manner similar to the way a salesman travels. See Figure 1.4 for an illustration of several netlinks, each based on the directed link topology.

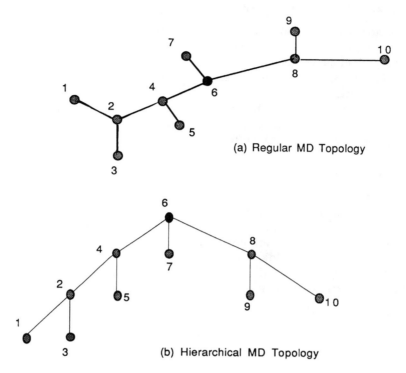

(a) Regular MD Topology

(b) Hierarchical MD Topology

Figure 1.3 Multidrop topology.

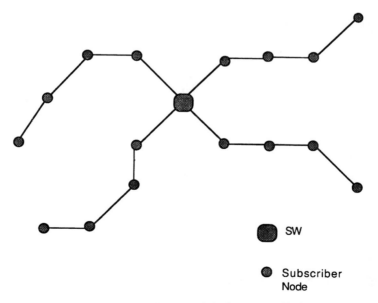

SW

Subscriber
Node

Figure 1.4 Directed link topology with four netlinks.

1.2.3.4 Star Topology

A network consisting of a single switching node and directly connected subscribers can be represented by a star topology (see Figure 1.5). For that case, each subscriber is connected to the switching node directly through a subscriber or user line. Such a topology is commonly used for connecting all the subscribers in a building or on a campus to a PABX, forming a voice LAN or a high-performance data switch forming a data LAN. The star topology is also used for connecting a very large number of local subscribers to a central office (CO) operated by the Bell Operating Companies (BOCs).

1.2.3.5 Bus Topology

The bus topology employs a fully shared broadcast transmission facility, called a bus, for exchanging information between nodes. Such a topology as shown in Figure 1.6 is generally the basis for a high-speed, distributed system used for interchange of information between computers and computer-controlled devices located within a building or a campus. The bus topology is commonly used for synthesizing high-speed LANs, which employ the CSMA access methods. Some users are lately choosing the new LANs based on the star topology for attaining higher performance or data rates or reliability.

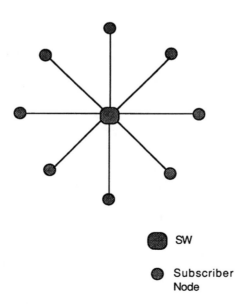

SW

Subscriber
Node

Figure 1.5 Star topology.

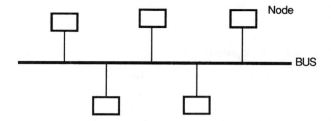

Figure 1.6 Bus topology.

1.2.3.6 Ring Topology

When the two ends of a shared transmission media of bus topology are closed to form a loop, a ring topology results (see Figure 1.7). Whereas Figure 1.7 (a) represents the S/F type ring topology, Figure 1.7 (b) represents broadcast type ring

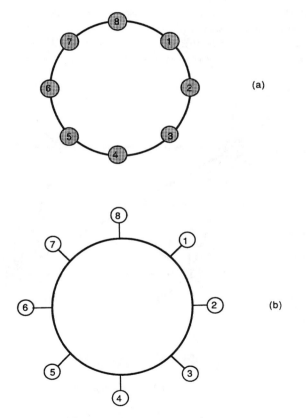

Figure 1.7 Ring topology.

topology. Many popular data LAN architectures employ the ring topology using the token access.

1.2.3.7 Mixed Topologies

Many mixed topologies can be derived from the previously mentioned basic topologies through the process of superposition or interconnection. For example, the multicenter, multistar topology of Figure 1.8 consists of the star and mesh topologies. This topology is commonly used for corporate voice and PS data networks. It will also be the basis for an integrated broadband network that handles several applications (e.g., an ATM network). Similarly, the multicenter, multidrop topology of Figure 1.9 consists of the mesh and the MST/MD topologies. This mixed topology is the basis for many SNA-based data communication networks. Such a mixed topology is finding some application in the design of a Personal Communication System (PCS) within a metropolitan area. The Doelz's Extended Slotted Ring (ESR) and Frame Relay Architectures generally employ the network shown in Figure 1.10, and it uses both the mesh and the DL topologies. The IBM Series/1 Chat Ring as based on the token ring architecture illustrates a mixture of the star and the ring topologies as shown in Figure 1.11. Whereas each Local Communication Controller (LCC) can serve a mixture of personal computers (PCs), printers and file servers, the LCCs on the ring provide

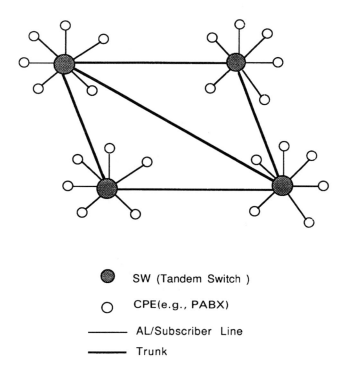

    ⬤   SW (Tandem Switch)

    ◯   CPE(e.g., PABX)

   ——   AL/Subscriber Line

   ▬▬   Trunk

Figure 1.8 Multicenter, multistar topology.

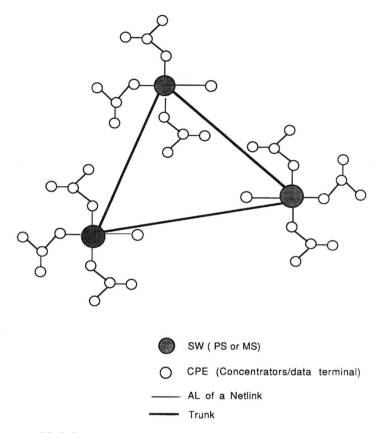

SW (PS or MS)

CPE (Concentrators/data terminal)

AL of a Netlink

Trunk

Figure 1.9 Multicenter, multidrop topology.

information exchange among a large number of computer-based terminals within a local area. A typical implementation of Datapoint's ARCnet LAN as shown in Figure 1.12 employs a mixture of bus and star topologies. Such a network topology is sometimes also called the starburst topology. It represents an older technology useful only for low traffic applications.

1.2.4 Transmission and Multiplexing Techniques

Transmission and multiplexing techniques define the modulation, encoding, transmission/reception, and multiplexing techniques employed to synthesize links for interconnecting the various network nodes. The issues such as analog or digital transmission, coaxial cable or optical fibers, and single or multiplexed streams of data are resolved.

A balanced design of a network system requires an integrated plan for transmitting both the user and control data from one node to another within the network. Knowledge of the transmission and multiplexing techniques is essential. Whereas, the transmission techniques usually involve the method of encoding

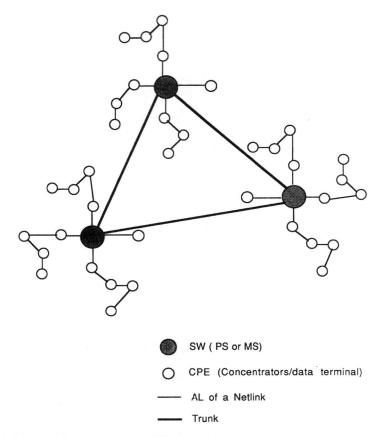

SW (PS or MS)

CPE (Concentrators/data terminal)

AL of a Netlink

Trunk

Figure 1.10 Multicenter, multi-DL topology.

each basic symbol prior to transmission over the channel, the multiplexing technique defines the way multiple streams of user and control data can be combined together to share a single physical link.

1.2.4.1 Transmission Techniques

Discrete information can be transmitted in the form of either analog or digital signals.

1.2.4.1.1 Analog Transmission Techniques

An analog signal is realized by multiplying a continuous modulating function (usually the information) with a high-frequency carrier waveform centered at F_c. Figure 1.13 illustrates the use of amplitude modulation (AM) and frequency modulation (FM) for producing continuous analog signals. Figure 1.14 illustrates the use of AM, FM, phase modulation (PM), pulse amplitude modulation (PAM), pulse width or pulse duration modulation (PWM or PDM), and pulse position

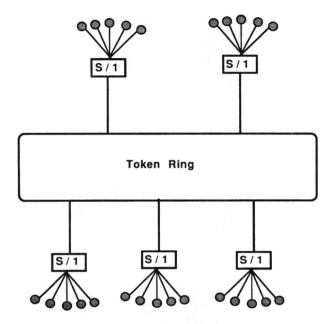

Figure 1.11 Mixed topology for IBM's Chat Ring.

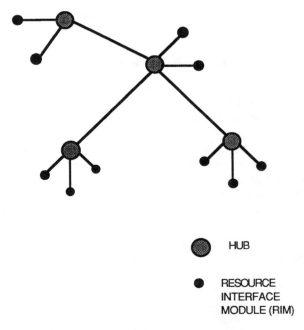

HUB

RESOURCE
INTERFACE
MODULE (RIM)

Figure 1.12 Mixed topology for Datapoint's ArcNet LAN.

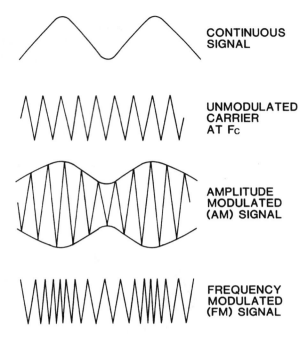

Figure 1.13 Modulated analog signals for a continuous message.

modulation (PPM) for producing discrete analog signals employed in data communications. These techniques are the basis of modulation and demodulation (MODEM) devices.

1.2.4.1.2 Digital Transmission Techniques

A digital signal is simply a base-band signal in the form of a digital pulse train that represents a coded version of the original information. Pulse code modulation (PCM) and delta modulation (DM) are two popular techniques for producing digitized voice signals.

Pulse code modulation employs the principle of sampling the continuous signal such as voice, quantizing the sampled value into a discrete level, and encoding this level into a fixed sequence of binary pulses. The larger the quantization levels, the higher the fidelity of the received signal. Figure 1.15 illustrates the manner in which a piece of the voice signal is converted into a sequence of binary pulses as a part of the entire PCM pulse train. According to Figure 1.15, eight quantization levels are employed, thus resulting in the total bandwidth (BW) requirement of 2 *signal BW*number of pulses per sample (=3). Assuming a typical voice signal of 4000Hz, the PCM bit rate required is therefore 2*4000*3=24000BPS. The telephone industry standard throughout the world is a PCM system with 256 quantization levels that requires a PCM bit rate of

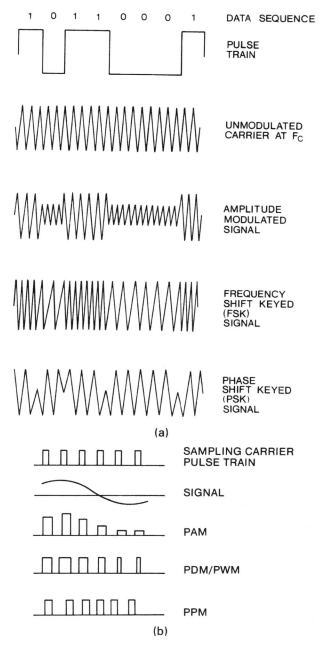

Figure 1.14 Modulated analog signals for analog (a) and discrete (b) messages.

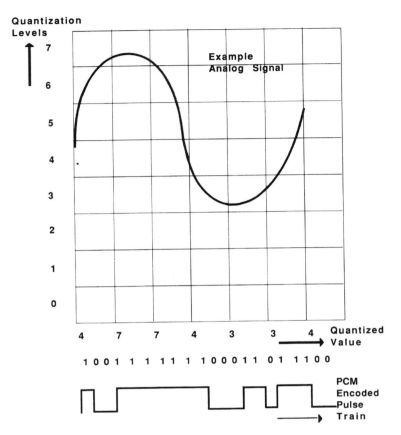

Figure 1.15 Pulse code modulated digital signal for an analog message.

2*4000*8=64000BPS. A new technique called the adaptive differential PCM (ADPCM) is being introduced that requires a PCM bit rate of only 32000BPS. It is based on the fact that a highly correlated signal (e.g., voice) changes slowly from one Nyquist sample to another and therefore it sends only the difference. The sent information is further reduced by adapting the step size (or delta) to provide increased dynamic signal ranges. Although the BW requirements of ADPCM look very attractive, one must note that 9600bps data calls cannot be handled by ADPCM. Furthermore, the cost of transcoders for interfacing ADPCM to existing network facilities must also be considered. In other respects, ADPCM is quite comparable to the traditional PCM.

Delta modulation (DM) is another form of the sample data technique that produces a 1 or 0 when the sampled value is higher or lower in amplitude than that for the previous sample. An alternating 1 and 0 represents a constant signal value. DM requires very little hardware complexity when compared to the PCM technology. For that reason, it is mainly used for mobile/military systems.

Although it requires less BW than a PCM system, it requires a higher sampling rate for achieving the same quality of received signal. Some new DM-based systems are based on the continuously variable slope DM (CVSDM) concept that is generally superior to the older DM and compares favorably with the PCM technique.

The pulse trains resulting from PCM or DM are transmitted in either the unipolar, polar, or bipolar form as shown in Figure 1.16. Figure 1.17 shows how the clear channel capability (CCC) is achieved for an extended super frame (ESF, as discussed later) using the bipolar, 8 zeroes-suppression (B8ZS) encoding scheme.

1.2.4.1.3 Why Digital?

Analog transmission was always beset with the effects of non-linear distortion, attenuation, and additive noise. Furthermore, the high cost of multiplexing and repeatering eventually caused its replacement by digital transmission techniques.

Digital transmission allows (1) simplicity in multiplexing and signaling, (2) lowering of costs due to the use of VLSI circuits, integration with digital switches and fewer repeaters, (3) higher fidelity of information due to low distortion and additive noise, (4) integration of many services with and without encryption, and (5) an integrated management and control of the entire network.

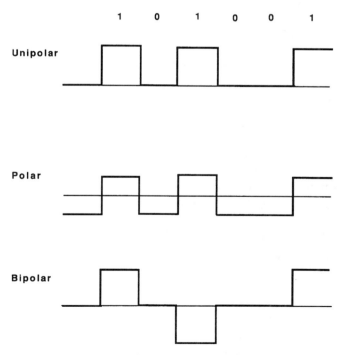

Figure 1.16 Digital signal transmission wave forms.

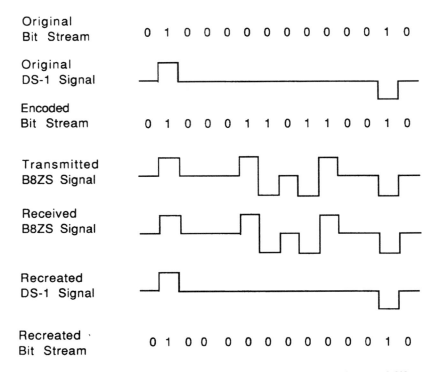

Original
Bit Stream   0 1 0 0 0 0 0 0 0 0 0 0 0 1 0

Original
DS-1 Signal

Encoded
Bit Stream   0 1 0 0 0 1 1 0 1 1 0 0 1 0

Transmitted
B8ZS Signal

Received
B8ZS Signal

Recreated
DS-1 Signal

Recreated
Bit Stream   0 1 0 0 0 0 0 0 0 0 0 0 1 0

Figure 1.17 Digital signal formation for clear channel capability.

At a public forum, it is hard to find anyone who will even dare to champion analog transmission these days. Before 1980, it was almost dangerous to talk in favor of digital transmission.

1.2.4.2 Multiplexing Techniques

Most transmission links employ some form of multiplexing of individual user and control channels for utilizing the inherent structure or bandwidth of the physical link and hence reducing the cost of per-channel transmission. Special multiplexing techniques are used for both the analog and digital transmission systems.

1.2.4.2.1 Multiplexing Techniques for Analog Transmission

The first multiplexing technique employed for analog transmission was called the space division multiplexing (SDM) that implies the bundling of individual voice grade channels in a physical/spatial manner. This method allows the sharing of the common conduit and hence the cost of cable laying. However, this scheme is beset with the effects of non-linear distortion and attenuation as the number of bundled channels or the distance increases. For those reasons, the SDM technique is mainly used for local loops and the LAN environments.

Frequency Division Multiplexing (FDM) combines the spectrums of individual channels to form a large spectrum if allowed by the physical link. It requires the use of guard bands at the transmitter to separate the individual channels and the use of filters to separate the individual channels at the receiver. FDM-based systems tend to be quite expensive and bulky due to an excessive amount of analog signal processing required. For a long time, FDM was a better method than the only other available technique of SDM. An interesting application of FDM was to send data under voice (DUV), a technique used for combining DDS channels with FDM voice channels.

1.2.4.2.2 Multiplexing Techniques for Digital Transmission

Time division multiplexing (TDM) is the most popular technique for combining several digital channels into a single channel. PCM-TDM is the most popular technique for combining 24 PCM channels to create a T1 carrier rated at 1.544Mbps. See Figure 1.18 for the T1 carrier structure that includes a framing bit for each 8*24(=192) bits of data. The 193-bit frame consists of user data, signaling data, and the 1 framing-bit used for synchronization. A better understanding of the PCM-TDM technique can be achieved by considering a set of 12

A : Structure of a PCM Frame

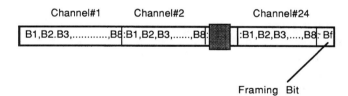

Framing Patterns (101010) for odd numbered and (001110) for even numbered frames

B: Structure of a PCM Multiframe Consisting of 12 Frames

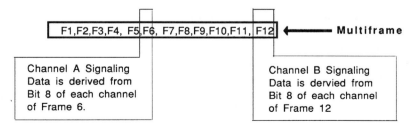

Figure 1.18 The PCM time division multiplexing technique.

frames. The odd-numbered frames (1,3,5,7,9,11) should yield the multiframe synchronization pattern of 101010, and the even-numbered (2,4,6,8,10,12) frames should provide a multiframe synchronization pattern of 001110. The signaling data (on-hook, off-hook, dial-digits, etc.) is derived from the eighth bit of each channel of the sixth and twelfth frames (Channels A and B). This is also known as the bit-robbing technique.

The bit-robbing technique just discussed adds a constraint to the 8 bits of user data for certain frames. Bit robbing always resulted in some voice quality degradation. To remove this constraint, a multiframe consisting of 24 PCM frames (instead of the 12 frames used traditionally) will be employed in conjunction with some new ways of encoding signaling data. Such a frame is called the extended super frame (ESF). The multiframe (ESF) synchronization will be now be derived from a single pattern "001011" derived from the framing bits from frames 4,8,12,16,20, and 24. A cyclic redundancy check (CRC) code will be imbedded in the framing bits of frames 2,6,10,14,18, and 21. This is a new scheme for providing an end-to-end checking of a link for diagnostics. A great deal of network management data dealing with items such as alarms, supervision, status, network configuration, performance indication, and maintenance will be imbedded in the framing bits of the remaining frames 1,3,5,7,9,11,13,15,17,19,21, and 23. The traditional signaling data dealing with on-hook, off-hook, dialing digits, and so forth, will be sent on a separate common channel signaling (CCS) network. With this approach, the users will be able to get the full 64Kbps capacity out of each PCM channel instead of the existing 56Kbps capacity. In order to implement the ESF concept and use the preferred bipolar digital transmission technique, one must employ the bipolar 8-zeroes suppression (B8ZS) technique to prevent the formation of long null sequences. See Figure 1.17 for an illustration of the way the B8ZS technique works to achieve ESF that provides the clear channel capability (CCC) in digital systems of the future for North America and Japan. The multiplexing techniques used in Europe and other countries never had the problems created by bit robbing.

T1-carrier is the first level of digital multiplexing hierarchy in North America and Japan. Figures 1.19 and 1.20 illustrate additional nomenclature associated with the PCM-TDM hierarchies in North America. Table 1.1 compares several generations of the well-known D-Type multiplexer.

ITU has another standard for Europe and elsewhere. That involves 30 channels of user data and 2 channels (channels A and B) of signaling channels to create a 2.048Mbps (32*64Kbps) carrier. This eliminates the need for bit robbing, thus resulting in simpler hardware.

The Synchronous Optical Network (SONET) as practiced in North America employs the base signal of STS-1 at 51.84Mbps. The Synchronous Digital Hierarchy (SDH) as practiced in Europe employs a base signal of STM-1 at about 156Mbps. See Figure 1.21 for the digital PCM, SONET, and SDH multiplexing hierarchies for the above standards.

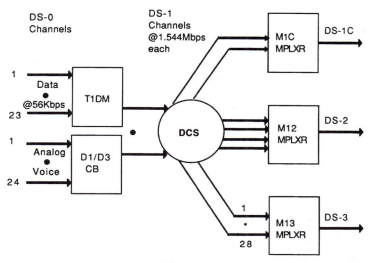

Notes: CB = Channel Bank
DCS = Digital Crossconnect System
MPLXR = Multiplexer
DS-i = ith Hierarchy Digital Signal

Figure 1.19 North American time division multiplexing hierarchy.

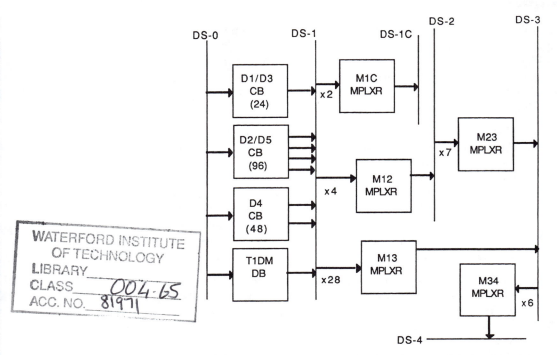

Figure 1.20 North American time division multiplexing hierarchy—second view.

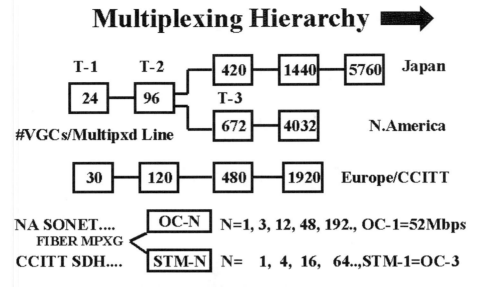

Figure 1.21 PCM, SONET, and SDH multiplexing hierarchies.

1.2.5 Network Management and Control (NMC) Techniques

Network Management and Control (NMC) technique is involved with the switching technology, traffic flow control, and methods of distributing and controlling intelligence in the network for day-to-day operation. Switching technology employed by the network determines the manner in which the network components or facilities are shared among the subscribers. Three types of switching techniques can be enumerated as follows:

a. *Circuit Switching (CS)*. Circuit switching provides a private, hardwired connection through the network, between two or more subscribers for the entire duration of the call. The flow of user information is completely transparent to the network nodes in the path of the call. The connection is brought down immediately after the calling party hangs up. The circuit switched networks have been mainly employed for voice and video teleconferencing services. The use of the public telephone network for data communications has also been increasing ever since the advent of personal computers (PCs) and low-cost modems.

b. *Packet Switching (PS)*. Packet switched networks permit the transfer of information between two subscribers through the routing of addressed packets of user data through the network. Unlike the circuit switched network, each PS node employs the store-and-forward technique for switching each packet for the purpose of routing it toward the destination on a virtual/permanent circuit established before the start of the transmission. In that sense, the PS node is not transparent to user data.

Table 1.1 Comparison of Popular PCM Channel Banks

Design parameter	D-Channel type				
	D1	**D2**	**D3**	**D4**	**D5**
#VG Ckts	24,72,96	24,48,96	24	24,48	24,48
#Bits/Sample	7	8	8	8	8
Signaling bit in Channel	Every 8th bit	8th bit of 6th frame	Same as D2	Same as D2	Same as D2
Companding parameter*	μ=100	μ=255	μ=255	μ=255	μ=255

Notes (*):

1. Uniform PCM is inherently inefficient since strong signals do not occur as infrequently as the weak ones. There are two companding techniques employed to enhance the weak signals and deemphasize the strong ones: μ-Law is used in Japan and North America, A-Law is employed in Europe.

2. The μ-Law companding transforms the input signal (x) into an output signal as follows:

   Output signal = sign(x) $\{$ (ln(1+ μ*[x]/V)) / (ln(1 + μ)) $\}$....0 $\leq$ [x] $\leq$ V

   where

   sign(x) is the polarity of x

   [x] is the absolute value of x

   ln(k) = natural logarithm of any real number k

   μ = μ-Law parameter value (usually = 255)

3. The A-Law companding transforms the input signal (x) into an output signal as follows:

   Output signal = sign(x) $\{$ (A* [x]) / (1 + ln(A) $\}$...........0 $\leq$ [x] $\leq$ 1/A

   $\qquad\qquad$ = sign(x) $\{$ (1 +ln(A*[x])) / (1 + ln(A))......1/A $\leq$ [x] $\leq$ 1

   where

   A = A-Law parameter value (usually =87.6)

Both the network links (virtual or permanent variety) and nodes are fully shared among all the subscribers. Most PS nodes use only the random access memory (RAM) to minimize the switching delays. Some recently introduced PS systems employ very small packets and the technique of nodal bypassing (when the destination is somewhere else on the virtual circuit) to minimize the delays even further. In general, all packets are switched according to the first-in-first-out (FIFO) scheme except when priorities are assigned to the packets.

c. *Message Switching (MS).* Message switched networks employ a long-term, store-and-forward (S/F) scheme for each received message. Each message is generally stored in full before forwarding it toward the destination. There are many variations of a message switched network depending upon the length of the message and/or the legal commitments related to the delivery of the message. Some message switched systems provide a disk-based storage for short-term retrievals and magnetic tape storage for a long-term retrieval. Most message switched networks segment the messages for high utilization of the storage media and fast retrievals from the storage media.

In general, a LAN, MAN, WAN, or GAN may employ any combination of the aforementioned access method or switching technology.

Another way to classify a network is by the amount of intelligence distributed throughout the network. Network links are assumed to possess only a temporary storage dealing with the information being transmitted. If a repeater performs any information processing to keep the link operating, it should be considered as an intelligent network node. Nodal intelligence is determined by the amount of random access memory (RAM) provided, the number of million instructions per second (MIPS) that can be executed by the nodal processor(s), the total number of instructions in all the software/firmware modules available, and the type/usefulness of the various software/firmware modules. The second way to classify networks is the extent to which the nodal intelligence is used to manage and control the performance and operation of the entire network. In some systems, each node may be very intelligent but it cannot communicate with the other nodes. Such a state can occur when different vendors manufacture the network nodes and when each vendor employs a proprietary scheme for internodal communication. This situation will prevent the realization of an integrated network management and control philosophy. NMC integration implies the extent to which network nodes communicate with one another to manage and control the entire network. The network is as intelligent as the amount of NMC integration employed by it. Advanced intelligent networks (AINs) are being installed to provide each user a direct control of its own network. See Appendix B for a description of such a network.

Basic functions of NMC can be classified into four distinct categories:

1. Routing of network traffic

2. Database related services

3. Maintenance and repair management

4. Network operation services

We will now describe each of these NMC functions in a little more detail.

1.2.5.1 Routing of Network Traffic

Allowing the traffic to flow on alternate routes results in a better use of network facilities and a better performance to subscribers. Three methods of traffic routing are generally employed by a system with alternate routing capabilities.

a. *Deterministic Method.* This is the simplest routing control involving the use of fixed primary (also called the high usage or HU) routes and a fixed number of alternate (also called the final) routes. Such a method was employed in the early versions of electromechanical switches, which stored the routing information in hardware. Since each switching node made its own decision as to the link it chose for the path, the originating or any other node did not know the exact path taken by the information. Such a method was also called the successive node route control (SNRC). The public voice networks have employed this method until recently when the common channel signaling (CCS) was introduced.

b. *Synoptic Method.* The network management center (NMC) samples the traffic flows in the entire network frequently and, based on the information collected, it decides which path the next call or message will follow. Sometimes, routing tables are dependent upon the time of day. In some systems, the cost of a call determines the route taken. This approach is similar to that in an ISDN or B-ISDN employing the CCS system. Using this scheme, the network becomes more intelligent in terms of the paths employed for routing traffic. Since most of the networks of the future will employ an NMC, the synoptic method will become commonplace.

c. *Adaptive Method.* Each network node gathers sufficient information from the neighboring nodes on a periodic basis and decides how calls or messages are to be dynamically routed toward their destination. Some control information must be collected (e.g., in the form of data stamps as it passes through different nodes) to gather information regarding the paths used. Additional care must be taken to avoid looping. This approach has been employed in fully distributed packet switching systems (PSSs).

During some holidays and periods of network outages, it may also be necessary to deny service to subscribers. A voice network achieves this goal by either not providing a dial tone or providing a busy tone to the calling subscriber. The data networks achieve this goal through flow control. This may involve a simple form of feedback to the subscriber or a complicated set of rules dealing with permission to send data.

1.2.5.2 Database Related Services

A properly designed NMC system must maintain a relational database that provides the following essential capabilities:

a. Network directory service

b. Network component inventory

c. Traffic/performance statistics gathering

d. Billing

1.2.5.3 Maintenance and Repair Management Services

A properly designed NMC system should provide the following capabilities required to maintain an almost fault-free system:

a. Collection of system alarms

b. Converting the system alarms into trouble tickets and sending these to qualified persons

c. Scheduling diagnostics and repairs

d. Predicting impending failures and taking evasive actions (self-healing)

1.2.5.4 Network Operations Services

A properly designed NMC system should provide the following functions to help the corporation run a smooth operation:

a. Scheduling training classes for the NMC-related personnel

b. Staffing of the NMC personnel

c. Ordering new network components or retiring other components for the purpose of maintaining an optimum network design on an ongoing basis. This effort will require the use of analysis and design tools described in this book.

1.3 Components of a Typical Network System

The smallest switched network system must employ at least one switching (or service) node and several communication facilities known as subscriber lines (SLs) for connecting the on-net subscribers to the switching node. An example of a small switched network may consist of only one Private Branch Exchange (PABX) serving a few subscriber (S) nodes (or telephones) via subscriber lines (SLs). On the other hand, a user may have many sites, each site fitted with customer premise equipment (CPE) consisting of either a voice LAN (e.g., PABX) or a data LAN (e.g., Ethernet). In that case, links known as access links (ALs) connect these CPEs to tandem switching (TS) nodes of a shared backbone network. The links that connect tandem switching nodes are called trunks (TKs). No useful network can afford to be an island. It must interface with the public networks to provide communication service to off-net subscribers (traveling employees or regular customers). To achieve that goal, some of its switching nodes must be connected to the Central Office (CO) switches of the public network via so-called off-net ALs (ONALs). See Figure 1.22 for an illustration of a useful switched communication network and its major components as discussed earlier. Again, a well-defined hierarchy related to nodes and links is quite apparent.

Links connecting higher-level network nodes are also called AL and TK bundles, each consisting of one or more individual links. Each link may be capable of either serving one or more voice conversations or carrying one or more messages or packets concurrently, depending on whether or not a multiplexing tech-

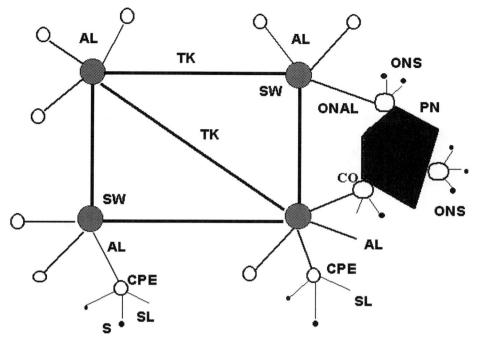

Figure 1.22 Components of a useful enterprise network.

nique is employed. This brings us to the concept of an individual server that is employed to handle one voice conversion or one message/packet at a time. In voice networks, an SL acts as a single server since only one telephone is generally served. In data networks, an SL may serve several data terminals or workstations (WSs) through the mechanism of packet switching or asynchronous time division multiplexing (ATDM). Despite that fact, a single data SL is still a single server since it is capable of carrying only one message or packet at the full capacity of the link.

1.4 Basic Traffic Attributes

A network system exists primarily to provide useful services to a set of subscribers. These services involve many forms of information exchange between network service nodes and subscriber nodes. Several traffic attributes characterize such services. These attributes and the underlying concepts are enumerated as follows.

A telephone/data call is an unstructured sequence of intelligence flowing in both directions between the calling and the called subscribers. For circuit switched voice networks, a call requires the establishment of a reserved, unshared path between the subscribers. A polite conversation between two persons in a voice network implies that intelligence flows in only one or the other direction. This results in a 50 percent utilization of most 4-wire communication servers in the form of ALs or TKs. This fact does not generally apply to the 2-wire subscriber links. A full duplex (FDX) data call may not suffer from this constraint since a 4-wire FDX server is generally used for two independent concurrent transmissions in both directions. Furthermore, in a packet switched network, shared virtual circuits are generally established in both directions before allowing independent data packets streams in any direction.

A data message represents the user's data directed from an originating subscriber toward a destination subscriber. Depending upon the switching technique, the message may or may not be transparent to the network node. Since a circuit switching node does not alter or analyze the contents of the information flowing through the network, the CS network is transparent to the users. Since a message switching or a packet switching node, on the other hand, must analyze a packet or a message before routing it to the destination, it therefore does not provide a transparent service to the users. An input message is generally called an inquiry and an output message is generally called a response. Many types of control messages flow between nodes to perform essential supervisory and signaling functions required by the various services. A message generally consists of many user-defined fields, each clearly demarcated to denote such items as originating and destination addresses, user data, and cyclic redundancy code (CRC) for error detection. A packet may also consist of similar fields, but a standard or a proprietary protocol generally defines their size and contents (except the user data). Several packets may be required to transmit a message.

A transaction generally involves a series of messages in both directions relative to the user and the network node. A transaction is simply a characteristic of the particular application or service. For example, a Binary Synchronous Communications (BSC) protocol involves a sequence of several message types between the data terminal and a host before receiving a response message from the host.

A session is almost identical to a transaction except that it may involve one or more transactions. A transaction or a session cannot generally commence until a permanent or a temporary virtual circuit has been established.

The *expected call duration* (ECD) is defined by the duration of a single voice session. Some analysts substitute call-holding time for ECD. A *message length* is generally determined by the number of bits or characters in the message. Some analysts simply use the letters ECD for the expected call-holding time. One can also define the *expected message duration* (EMD) as the message length divided by the link capacity denoted in terms of bits or characters per second. Call or message duration may be characterized by randomness. For that case, one must define a call/message duration distribution with at least a mean (μ) and a standard deviation (σ) before computing system resource utilization and response time. A random distribution of call/message duration also results when many types of fixed length messages travel randomly over a shared virtual circuit, each consisting of one or more physical links.

Call or message arrival rate (λ) defines the rate at which the calls or messages are presented to a network resource/server/system. The rate at which the calls or messages are successfully served is generally defined as the system throughput.

The concept of *traffic intensity* is generally employed to derive short-term statistical utilization of network resources. Traffic intensity is also employed for computing the number of servers such as access links (ALs) and trunks (TKs), which are required to provide a desired performance level. Chapters 3 and 4 describe the necessary relationships for both voice and data applications.

Offered *traffic intensity* (A) is generally defined as the product of arrival rate, λ and the expected call duration (ECD) for voice networks, or A = λ * ECD erlangs. Traffic intensity for voice networks is measured in terms of erlangs, named after the famous Danish mathematician. Traffic intensity when measured in erlangs determines the average number of servers (e.g., number of voice grade circuits) concurrently occupied. Classically, traffic intensity for voice networks was measured in hundred call seconds (CCSs); however, it does not have any physical meaning. For that reason alone, such a unit is not employed in this book. If one defines traffic intensity in terms of CCS, the reader can easily divide CCSs by 36 to compute erlangs.

The average utilization (ρ) of a AL or TK server in a voice network can be computed as the ratio of erlangs carried by the AL/TK bundle and number of servers in the AL/TK bundle. Erlangs carried is defined as a product A(1-B) where B is the fraction of calls blocked by the server group. See Chapter 3 for a methodology for computing blocking probability, B for given values of A and N servers.

Traffic intensity in data networks is generally measured in bits per second to account for the bursty nature of data traffic. One can compute its value as a product of λ and EML [= (packets per second)*(bits per packet)].

One can compute the short-term utilization (ρ) of a data trunk bundle by dividing the data traffic intensity (bps) by the total capacity of a data trunk bundle. In this case, if ρ is close to or larger than one, one must add ALs or trunks in order to reduce the average server utilization (ρ) below 1.0. Many experienced designers generally try to maintain the value ρ at about 0.667 (=2/3) to achieve a tolerable range of system response times. The value N*ρ can be defined as data erlangs only if some care is employed in defining data network server utilization. As an example, when a worker employs a circuit switched voice path to connect his or her WS to the company host for an entire day, and uses the connection to execute only one or a few transactions all day long, the link utilization approaches 1.0 for the working hours. In this example, the total number of data erlangs are as many as there are concurrently maintained in such connections. Furthermore, one should note that such CS servers are never shared while most packet switched data links are usually shared by many users. That is precisely the reason that most telecommunication managers do not tolerate the inefficient use of voice paths for data transmission. They would rather use a PS network for terminal to host communication.

Traffic flow is a measure of the traffic intensity on any directed path consisting of an ordered sequence of network nodes and links. A sequence of two nodes in a path implies a single circuit bundle connecting the two nodes. A sequence of N nodes in a virtual connection implies a directed path of (N-1) intervening links, each consisting of one or more physical circuits. Most well-designed network systems provide some traffic flow control to prevent congestion that may result in system failures and performance degradation. There are many ways to achieve flow control. Access limiting and dynamic routing of traffic are two of the most popular methods.

The concepts of switched telecommunications systems, network architecture, and the associated traffic attributes as introduced earlier can be employed to get a deeper understanding of all the switched networks in transition. To paraphrase the famous saying, one who doesn't understand the history is doomed to repeat all the mistakes of the past.

1.5 Networks in Transition

We will start by considering voice networks, followed by data networks, integrated broadband networks, and the always fully integrated digital network. Their evolution, architectures, and traffic attributes are described in a structured manner.

1.5.1 Voice Networks

The first example of switched telecommunication networks can be traced to public telephone networks (PTNs). Telephony began with Alexander Graham Bell's invention of the telephone in 1876. The first mechanical circuit switch was installed in 1878 in New Haven, Connecticut. The first automatic, step-by-step Strowger switch was installed in 1892 in La Porte, Indiana. The first crossbar type electromechanical switch was installed in 1938 in the United States. Direct distance dialing (DDD) as based on the 7-digit code was introduced in 1947 in the United States. Although the first PTNs were fully distributed in terms of switching, they employed only dumb switches. Consequently, the network didn't know the occurrence of component (links and nodes) failures. The users generally called the business offices (by using a neighbor's telephone) to help the telco to create a trouble ticket. Lack of intelligence prevented the network from employing any network management and control philosophy.

The first stored program controlled (SPC) No.1 ESS circuit switch (CS) was installed in 1965 in the United States. A deregulation of the customer-premise-equipment began in 1968 as a result of the Carterfone decision. This gave impetus to the development of SPC private automatic branch exchange (PABX) and privately owned tandem switches required for enterprise voice networks.

The first digital circuit switch employing pulse-code-modulation (PCM) and 255μ-Law was commercially announced by Collins Radio Company in 1971. That and many other digital switches from the vendor community became the building blocks of hundreds of enterprise networks for Fortune 500 companies. Ultimately, starting in the early 1980s, T1 circuits became the building blocks of backbone networks.

The divestiture of AT&T took place in 1984, creating about 250 Local Access and Transport Areas (LATAs), and resulted in seven Bell Operating Companies (BOCs) to handle local access and intra-LATA services, and many other common carriers (OCCs) such as the new AT&T, MCI, and US Sprint to handle inter-LATA and long distance services and leasing of long-haul analog/digital circuits. At present, a great deal of activity is related to the introduction of No. 5 ESS CO switches capable of handling ISDN-based services. See Appendix B for a description of the ISDN-related standards and network topologies.

Voice communication networks are generally classified by the type of switching nodes employed. The first generation of voice switches implied the use of dumb Strowger switches. The second generation of voice switches meant that SPC type circuit switches were employed. The third generation of voice switches generally implies the use of digital CSs. The fourth generation of voice switches could handle both voice and data. That, in turn, originally implied the capability to provide concurrent access to both voice and data terminals at each subscriber location and to share a modem pool for WAN applications. In addition to providing a traditional integrated voice and data (IVD) capability, an ISDN switch provides higher data rates and greater intelligence than the fourth-generation circuit switch. However, since ISDN has not yet evolved into a universal service, other switches based on the Asynchronous Transfer Mode (ATM) are now being introduced. The ATM-based switch should be called the fifth-generation switch. Since ATM is also related to the so-called broadband ISDN (or B-ISDN) standard, one may see the evolution of a truly integrated fifth-generation switch. In short, the various generations of voice switches can be characterized by an increasing amount of nodal and/or network intelligence required for cost-effective network management and technical control (NM/TC) in both private and public networks.

The types of transmission facilities employed can also classify voice communication networks. They originally employed analog transmission facilities with only space division multiplexing (SDM). With time, frequency division multiplexing (FDM) became the cost-effective method of carrying a large number of voice conversations over a coaxial cable. For example the L1, L3, L4, and L5 type carrier systems could carry 600, 1860, 3600, and 10800 voice-band channels and were introduced in 1946, 1953, 1967, and 1974, respectively. Later, analog and digital microwave systems (also called radio systems) became cost effective for long- and medium-haul routes for a medium number of voice channels. The T-carrier systems employing digital PCM-based multiplexing techniques became cost effective in early 1960s for short-haul routes. With the advent of fiber optics and longer repeaterless distances, the number of voice channels carried by the T-carriers has increased rapidly. One can now state without equivocation that digital transmission is now the only cost-effective means of carrying a large number of voice conversations and/or data traffic over all ranges of distance.

A subset of enterprise subscribers served by a PABX form a voice local area network (LAN) employing a star topology. Public subscribers served by a central office form a voice metropolitan area network (MAN), also based on the star topology. Voice WANs are invariably implemented by interconnecting several voice LANs and MANs according to a star topology.

At present, the availability of cost-effective digital switching and transmission techniques is quickening the pace toward a fully integrated broadband network. Since ISDN still employs a mixture of CSs (for voice and data) and PSs (for data), we are still far from realizing a multimedia network. But with the recent development of Frame Relay (using fast switching of variable length packets) and ATM (using small fixed packets and nodal-bypass mechanism) technologies, packetized voice is not far away. When packetized voice, image, and video becomes economically feasible, an integrated broadband network will finally become a reality, resulting in large economies of scale.

1.5.2 Data Communications

Concepts of mechanized data communications are much older than their voice counterpart despite notions to the contrary. Earliest versions involved coded messages using drum sounds, smoke signals, and semaphore devices. The modern variety of data communications began with the invention of telegraphy in 1835. In its original form, a telegraphy system represented a point-to-point communications system. The notion of a fully distributed system was absent.

The biggest impetus to modern data communications came from the development of the first commercial modem by Collins Radio Company in 1955. Such a modem soon became the foundation of simple two-node networks. Such point-to-point, coded-feedback communication systems were originally used for tape-to-tape and card-to-card data transport. A portion of the data were first divided into blocks before their transmission. Each block had an overhead code for detecting any errors caused by the transmission media. A number of data blocks could be outstanding until feedback information regarding their correctness was received and before continuing the transmission of additional blocks. The next epoch occurred with development of the first microprogrammable front-end (FE) communications processor in 1960. Later, Collins Radio Company made this processor the basis of the first SPC message switching system, ultimately replacing all of the then existing point-to-point, torn-tape Teletype systems by 1964. Collins Radio Company also developed in 1968 the C-System, which was a large LAN consisting of mainframes, minicomputers, and input/output devices, interconnected via several time-division multiplexed (TDM) loops, capable of providing both message and packet switching services to a large number of widely scattered subscribers. The first Advanced Research Project Agency (ARPA) sponsored packet switched network became operational in 1969 with only four nodes. The first public PS data network as operated by Telenet Corporation became operational in 1975.

Data communication systems can be classified by the switching technology employed. It should be emphasized at the outset that all data switches are of the digital variety since each switch only handles zeroes and ones. The first generation of data switches were those that provided interchange of data between the host and terminals using the mechanism of polled access. The second generation of data switches were the SPC message switches nodes. The third generation of data switches was based on CCITT X.25 packet switching. The fourth generation of data switches are based on fast (also virtual) packet switching employing the principle of fixed-length data blocks and nodal-bypass mechanism. The fifth generation of data switches will be based on broadband ISDN (B-ISDN) and ATM technologies. These fifth-generation data switches will be capable of integrating all forms of media such as voice, data, image, and video.

Data communication systems can also be classified by the type of communication facilities employed. Most of early data networks employed analog transmission facilities as dictated by the analog modems. Digital transmission facilities such as DDS and T1 facilities became the favorite choice of users increasingly due to use of availability of special hardware such as CSU and other digital interfaces for ISDN. The availability of fiber optical facilities is nowadays the foundation of all high-speed SONET (Synchronous Optical Net) networks.

1.5.3 Integrated Broadband Networks

During the early days, each enterprise network handled only one voice or data application, each characterized by a special set of performance requirements. This was basically influenced by the lack of distributed intelligence in the network. One bank had about 70 different data networks, each handling a unique application. In order to integrate several voice and/or data applications into all network resources, one must first understand the meaning of the word *integration*.

The first integration of voice and data occurred on backbone T1 trunks with the availability of T1 facilities and T1 multiplexers during the late 1970s. Voice accounted for about 70 percent of the need for transmission facilities. The data generally traveled on these backbone facilities for free. Many large customers achieved the same economies of scale on their large AL bundles by using T1 facilities and T1 multiplexers. Some large customers saved millions of dollars each month and helped usher in the so-called T1 network revolution.

The next upheaval came as a result of the deployment of SONET facilities in public networks and newer networking technologies such as Frame Relay, SMDS, and ATM. Frame Relay is a subset of the fast packet switching technology. Like X.25, it employs variable packet lengths, but it avoids delays in all of the nodes in the path by allowing only the two end nodes to manage data accuracy. SMDS and ATM employ fixed packet lengths of 53 bytes (with 48 bytes of user information). Only recently, Frame Relay has been utilized to transmit both data and voice. SMDS has always been used for data applications. As the ATM (or B-ISDN) standard matures, it will allow a full sharing of all switching and transmission facilities among all applications such as voice, data, image, and video. When that happens, a true multimedia network will be born.

1.5.4 Fully Integrated Digital Networks

Presently, the ATM technology is considered to be the ultimate one, but the ATM technology will not be the last one. There is almost an infinite capacity for improvement. Let's focus on the word *integration* alone. Integration can occur at the transmission, switching, access, data presentation (e.g., EDI), and control (or NMC) levels. Achievement of integration at all levels is several years away. When each location of the enterprise is served by multimedia sources (e.g., voice, data, image, and video), all of the traffic will be integrated through the use of fifth-generation switches employing ATM or B-ISDN technologies and transmitted over all levels of access lines (ALs) and trunks. Such full sharing of all switching and transmission facilities will provide significant savings to the enterprise. Integration at all other levels of the networks will be possible only through the use of fully intelligent and open systems. The AIN concept mentioned earlier represents such a direction.

A fully intelligent network will know (1) its topology well (which node is connected to which node and with what facility), (2) which nodes are malfunctioning and need maintenance/repair, (3) which paths are exhibiting traffic congestion and need some relief through traffic rerouting, (4) what reports and data are being presented to the right personnel at the appropriate time, (5) the level of cost effectiveness of the current NMC and network operations at all times, and (6) the level of service (both network planning and management) being provided to each customer. A truly open system will allow the integration of (1) multivendor hardware based on approved standards dealing with communication and NMC protocols, and (2) network services to all users.

A fully integrated digital network (FIDN) will eventually become a basis of an emerging new cyber utility that will provide unfettered sharing of information to all users with diverse media, bandwidth, and service requirements. This would be very similar to the existing electric and water utilities that provide any amount of utility through a simple wall outlet or a faucet and are managed through a very simple billing procedure. It is hoped that the reader is now aware of the unlimited possibilities emerging from the human imagination and needs of the human soul.

Exercises

1.1 Enumerate the basic components of a switched telecommunications network.

1.2 Give examples of nodes and links for three hierarchies in a voice network.

1.3 Give examples of nodes and links for three hierarchies in a data network.

1.4 Define two types of subscribers in a private voice network and write a line or two to justify their need.

1.5 Define two types of access lines (ALs) in an enterprise network.

1.6 Describe three methods of accessing a switched communication network.

1.7 Define traffic load (TL), traffic intensity (TI), and traffic flow for voice and data applications using generalized expressions only.

1.8   Two PABXs communicate with one another on a bundle of 64 voice-grade lines. During the peak hour, 12 calls are handled per minute (or $\lambda = 12$ calls per minute). The expected call duration(ECD) is 300 seconds. Express the traffic intensity (TI) in erlangs.

1.9   Describe the units to quantify traffic load (TL) and traffic intensity (TI or A), for both voice and data.

1.10 Four 64-kbps data circuits for data communications connect two nodes. The peak-hour traffic intensity is 192kbps. Express the average utilization (ρ) of each circuit and the overall traffic. intensity (TI) in equivalent data erlangs.

1.11 Define the three attributes of network architecture.

1.12 Define the concept of network topology and describe why it forms the soul of a network.

1.13 What are basic building blocks of North American SONET and CCITT Synchronous Digital Hierarchy?

1.14 Using some parts of Appendix B, describe the essentials of an Advanced Intelligent Network.

Bibliography

Cherry, Colin, 1966. *On Human Communication.* New York: Prentice Hall.

Sharma, R. L., P. T. De Sousa, and A. D. Ingle. 1982. *Network Systems.* New York: Van Nostrand Reinhold.

2

Generalized Tools for System Analysis and Design

2.1 Introductory Remarks

Any network system can be modeled at any level of detail. Even the largest network system can be modeled as a single node that handles a certain mix of user transactions per second. As the model gets complicated, one has to define the performance of each network system in terms of a larger number of transactions and different mixes. User transactions become mixed with other transactions required to maintain a given level of network system integrity while providing an acceptable level of services to the subscribers.

The task of modeling is simply that of representing the entire system in terms of a network of nodes belonging to certain levels of a hierarchy. The authors of *Network Systems* (Sharma, De Sousa, and Ingle, 1982) provide additional insights into the art of modeling network systems. It will suffice to say here that a successive decomposition (each resulting in a different model) is continued until no more useful understanding results.

In this chapter, we will introduce the concepts of systems analysis, design, and performance in a way that will be useful to readers for understanding the material of subsequent chapters.

2.2 Concepts of System Analysis and Design

Figure 2.1 illustrates the relationships between modeling, analysis, and design. It shows that it is very hard to separate the efforts generally associated with modeling, analysis, or design. Such efforts are completely intertwined. In general, an analysis effort should provide a prediction of system performance assuming correct distribution of transaction arrivals and their holding times. A system design effort, on the other hand, should yield a network system architecture and topology that will provide the desired performance. But in all cases, many passes will be needed to arrive at the correct system architecture and topology. The iterative process as illustrated in Figure 2.2 is probably the least understood by the telecommunications professionals. They still assume that the network design effort is a onetime effort executed on a large mainframe at the beginning of a network planning cycle. Many of them expect the use of detailed tariff-related databases for each resulting topology during the aforementioned iterative process. Such an approach will always slow down the design process even on a mainframe. Since the topology of a network is mainly affected by architectural/technological choices and never by small perturbations in tariffs (consult Chapters 7 through 14 for justification), one should only employ simple models of tariffs in order to make the design process of Figure 2.2 a truly interactive one. See Chapter 6 for a discussion of a software package that achieves this goal while providing a user-friendly interface. Since this package runs on a desktop workstation, it also encourages a continuous use for studying the effects of new tariffs and technologies during the entire life cycle of each network system.

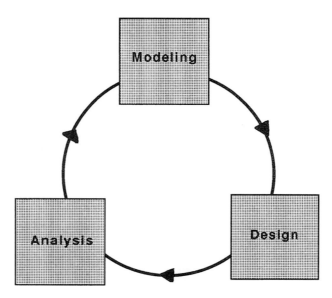

Figure 2.1 Relationships between modeling, analysis, and design techniques.

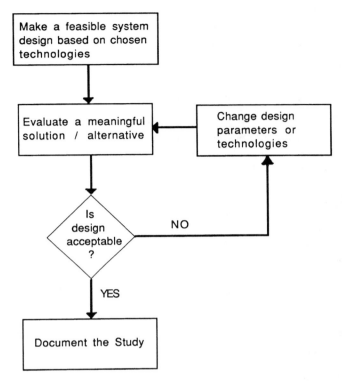

Figure 2.2 Iterative network design methodology.

Figure 2.3 illustrates a life cycle of a properly designed network system. Ten tasks are executed in series: modeling of existing environs, user requirement modeling (existing traffic intensities and performance), user requirements growth modeling, analysis and design of all the alternative system cutover models, return-on-investment (ROI) modeling, choosing of the most cost-effective system, system implementation, system performance measurements after the cutover, and documenting the results. Experience shows (1) that the task of network design takes the least time, and (2) the most important event is the ongoing collection of the analysis and design tools required to system engineer the present and the previous network systems. The database of all the tools represents the art and science of networking if these tasks are performed with diligence and patience for several complex network systems.

There are three basic methods of analyzing system performance:

1. Physical simulation, which requires the availability of the final system under investigation. Tests are performed under typical traffic loads to obtain all aspects of the performance. This is by far the most expensive and time-consuming approach.

2. Computer simulation, which requires software modeling of all operations and running the program until one achieves a statistical equilibrium. The

development of the simulation software may take a sizeable amount of time. Furthermore, the use of a mainframe for predicting the performance of every interesting configuration may cost a great deal of money.

3. Analytic simulation, which employs well-proven analytical techniques for modeling and predicting the performance of a network system or its elements. This method is by far the most elegant, the quickest, and the cheapest.

The author's experience shows that even the most complex system can be broken into smaller and smaller components until analytical techniques can be employed, and then the results are combined to predict the performance of the entire system. It must be emphasized that all methods yield approximate results since no one can simulate all possible environs. Large discrete systems are afflicted with too many design and traffic variables. The number of possibilities generally exceeds the number of electrons in the universe. In most cases, approximate analytical tools will generally yield more meaningful results than other methods considering cost and time constraints.

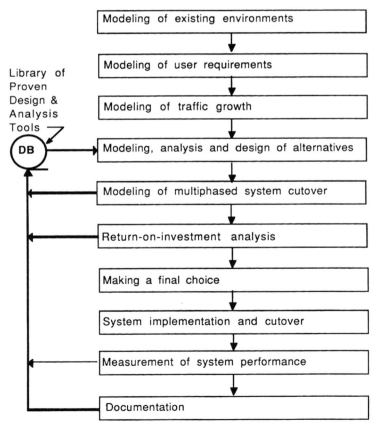

Figure 2.3 Life cycle of a network system.

2.3 Concepts of System Performance

It is a sad fact of life that very little concern to performance is being given to most of the network systems that are being implemented or for which standards are being developed. The simple answer lies in the complexity and high costs related to the time-consuming tasks of predicting the performance of any large network system. These observations apply both to public and private network systems. Performance of a network system generally deals with costs, throughputs, quality-of-service (QOS), and grade-of-service (GOS). These are described in the paragraphs that follow.

2.3.1 System Costs

This criterion is probably the most critical performance parameter since it determines the cost of service provided to the users. The cost of a call made on the public telephone network is a prime example of a system cost. Most system designers talk of system throughput in terms of the telephone call handling capacity as an example. Very little attention is given to the system costs and the efforts to lower those costs through network optimization. As a result, the users continue to pay higher rates for the services provided by the public networks.

Costs of a system should be divided into major components such as transmission, NMC, and hardware. Experience shows that the transmission costs are generally the predominant part of the total cost. For most private networks, the costs related to transmission may constitute as much as 70 percent. The costs incurred on a wide area network (WAN) management, operations and control may be 20 percent of the total costs. A common practice is to show the costs on a monthly basis since all carriers bill the users each month. The costs of NMC are also natural for a monthly exhibit. It is also easy to spread the costs of hardware for the entire life cycle of the system while considering the costs of money (e.g., interest, depreciation, etc.) and obtain an equivalent monthly cost. See subsection 2.4.4 of this chapter for a methodology to convert fixed costs (for hardware or fully owned transmission facilities) into equivalent monthly costs.

2.3.2 System Throughput

System throughput measures the overall capability of the system in terms of the maximum number of transactions that can be handled per unit of time.

The throughput of a circuit switched system is represented by the maximum number of call attempts and calls handled per average second of a busy hour by each network node and by the entire network system. The throughput of the entire system is also determined by the maximum number of calls that can be handled concurrently during the busy hour. The latter value is affected by the call duration and the number of nodes in the call path.

The throughput of a packet switched system is represented by the number of input packets or messages that can be handled per average second of a busy hour. Some vendors represent the throughput of their packet switch node by the maximum bit rate (input rate plus the output rate) handled during a second. Another useful measure of system throughput should be the maximum number of packets or messages handled by the entire network during a second or busy hour.

2.3.3 Quality-of-Service (QOS)

Quality-of-Service deals with performance issues such as the transmission quality, voice quality at the receiver, length of error-free periods, average bit-error-rate (BER), system reliability as determined by such measures as mean-time-between-failures (MTBFs) for each node and link and mean-time-between-system-failures (MTBSFs), and system connectivity (a measure of capability to provide service to all subscribers during component failures). In order to study system connectivity, one must first study the capability of each node to connect to a large number of subscribers and/or access lines or trunks.

2.3.4 Grade-of-Service (GOS)

Grade-of-Service criteria deal with the degradation in service caused by contention for critical resources when all those resources are functioning.

GOS for a circuit switched system deals with the statistical distributions of the following design parameters:

1. *System response time.* The elapsed time measured from the moment a user goes off-hook and the moment the user hears a dial tone.

2. *Call connection time.* The elapsed time measured between the moments the last digit is dialed and the connection is made between the calling and the called subscriber.

3. *Call set-up time.* The sum of two previous quantities and the time to dial all the digits.

GOS of a CS system is also determined by the probability of a call loss or blockage caused by the unavailability of a path through the network during the busy hour. GOS of a CS system can only be computed by taking into account the actual network topology and the blocking probability experienced by each network node-link bundle pair.

GOS for a packet switched system deals with the statistical distributions associated with the following design parameters:

1. *Nodal response time.* The elapsed time measured from the moment the last byte of the packet is received by the node and the moment the first byte of the same packet is transmitted.

2. *Call/session set-up time.* The elapsed time between the moment the last bit or byte of the call set-up packet is transmitted and the moment the first byte of the acknowledgment packet is received by the packet mode terminal.

GOS for a message switched system deals with the statistical distributions associated with the following design parameters:

1. *Nodal response time.* The elapsed time measured from the moment the last byte of the message is received by the node and the moment the first byte of the same message is transmitted by the node.

2. *Message retention time provided by the system.* The elapsed time measured from the moment the last byte of the message is received by a network node and the moment the system puts the message into a storage media requiring manual handling.

In most cases, the exact distributions for system performance measures are not available. But it is easy to obtain their average and 90 percentile values using the central limit theorem that is applicable for the case when a large number of random variables determine the system performance measure. The average and the 90 percentile values of any system performance measures are quite adequate for studying improvements in a system design or analyzing the performance of a given system. See Chapters 3 and 4 for the tools employed for analyzing the performance of a given or a new system.

The GOS for each network node must be studied before computing the GOS of a system. Vendor data and technical proposals are excellent sources for nodal GOS data in the form of a termination (via subscriber line [SL], access line [AL] or trunk) cost, maximum number of terminations allowed, the type of modularity associated with access line/trunk terminations, maximum traffic handling capacity in erlangs for each type of termination, nodal switching network architecture (e.g., number of stages and type of transmission interfaces), nodal blocking/delay characteristics, CPU type and capacity in MIPS, maximum size of RAM, type of secondary storage devices allowed, CPU word length, capacity to handle busy hour call attempts or nodal throughput, the number and type of services/features handled, NMC capability matrix and equipment (switching, air conditioning, power, etc.), footprints, and so forth.

2.4 General Purpose Analytical Tools

Experience shows that a large number of tools are often needed during the network design phase. Absence of such tools can make one's task quite difficult. An easy access to such tools can provide one with a keen insight into the operation of the network system or provide "what if" type solutions to many design alternatives. A set of generalized tools can be divided into the following categories:

1. Combinatorics and error analysis

2. Data stream analysis

3. System reliability analysis

4. Return-on-investment (ROI) analysis

The works of Sharma, DeSousa and Ingle (1982) and Schwartz (1987) may be consulted to get a deeper understanding of the actual tools described in this and the next two chapters. The EcoNets tool, as described in Chapter 6, employs many of these tools. Some of these tools are included in the Analysis Menu.

2.4.1 Combinatorics and Error Analysis

Combinatorics is a field dealing with possible combinations and permutations of limited items taken out of certain available items. As a result, this field deals with the set/group theory that deals with the enumeration of special patterns. It is not surprising that combinatorics is quite often used for applications dealing with computers and communications that involve 0's and 1's.

2.4.1.1 Number of Combinations or Permutations of x out of N Items

Consider N items, empty or occupied, as represented by 0's or 1's. It may be interesting to compute the number of combinations of x occupied items taken out of N items while ignoring the order into which the x items are occupied. It can be expressed as follows:

$$C(N, x) = N! / [x! (N-x)!] \tag{2.1}$$

where

$$N! = 1*2*3*....*N = \text{N-factorial}$$

The number of permutations of x occupied items taken out of N items requires the knowledge of order in which the items are occupied. One can express $P(N, x)$ as follows:

$$P(N, x) = N! / (N - x)! \tag{2.2}$$

The probability of a pattern {j, k, ...l} of occupied cells (see Figure 2.4 for an example pattern with L=6 and N=16) anywhere in the cellular array can be expressed as follows:

$$\text{Prob}\{j,k,...l\} = [C(L,j)*C(L,k)*...*C(L,l)] [P(N,x)] / [C(NL,m)] [u!v!...w!] \tag{2.3}$$

where

u, v, and w are the number of vertical columns that have identical number of occupied cells (e.g.,u, v, w are 1, 2, 1 for Figure 2.4):

j+k+....l = m (e.g. m=8 for the pattern 3, 2, 2, 1 of Figure 2.4)

u+v+w = x (e.g. u+v+w = 4 for Figure 2.4)

2.4.1.2 Some Applications of Combinatorics to Error and Throughput Analysis

Every system analyst is occasionally called upon to compute the output data error rates in a data communication system as a function of the raw error rate, and the error detection or error correction technique employed on the input data. One may also be interested in the optimum sizes of the transmitted block for maximizing the utilization of the limited storage or limited bandwidth of the output transmission channel. We will now describe some typical applications that require the use of several effective tools.

Application 1. A given binary array of information, of size N by L, consists of N bytes, each consisting of 7 information bits and 1 parity bit. This represents a one-dimensional parity scheme. Such a scheme will detect all odd number of errors (most probably only one error) occurring in each byte. It will generally allow a pattern of two errors in a byte to go undetected. One can add the last byte to provide a horizontal parity bit to each row of bits in the array consisting of N-1 bytes. This represents a two-dimensional parity scheme. Such a scheme will generally allow a pattern of four errors to go undetected. Given the raw probability of bit error, P_e, one can compute the undetected bit error rate (UBER). UBER can be expressed as the product of two probabilities: probability of a given pattern of errors occurring anywhere in the array as defined earlier, and the probability of exactly "x" errors occurring anywhere in the array for a given value bit error rate (BER) specified by the value P_e. Assuming only random errors and $P_e \ll 1$, one can obtain the value of UBER for patterns of errors defined as {2} and {2,2} occurring anywhere in the array can be obtained, and expressed as follows for the one- and two-dimensional parity schemes, respectively:

$$\text{UBER1} = \{[C(L,2)]*[P(N,1)] \, / \, C(NL,2)\}*\{C(NL,2) \, P_e^2(1\text{-}P_e)^{NL\text{-}2} \} \quad (2.4a)$$

$$\text{UBER2} = \{[C(L,2)*C(L,2)]*[P(N,2)] \, / \, 2C(NL,4)\}*\{C(NL,4) \, P_e^4(1\text{-}P_e)^{NL\text{-}4} \} \quad (2.4b)$$

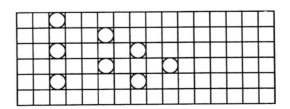

Figure 2.4 Cellular array pattern {3,2,2,1} of occupied cells.

where

C(N,x) and P(N,x) are the same as defined before.

Application 2. A block of data with N bits can employ a cyclic redundancy code (CRC) of R bits long that are imbedded in the N bits. Table 2.1 shows the relationship between the Hamming distance d or the number of errors that can be detected, d-1, to the information bits, k, and the block length n. The CRC code is determined for each information block of k bits and added to it just before it is transmitted. Experience shows that the Bose-Chaudhuri-Hocquenghem (BCH) codes that are a subset of the CRC codes are easier to implement than the one- or two-dimensional parity schemes. Furthermore, the BCH codes are more efficient than the one- or two-dimensional parity schemes since these also detect most of the bursts of 2 and higher bit errors. One can compute the UBER for a known raw bit error rate of P_e for the most likely undetectable number of errors (i.e., q=d where d is the Hamming distance) as follows:

$$UBER3 = C(N,q) \, P_e^q \, (1- Pe)^{N-q} \qquad (2.5)$$

See Figure 2.5 for some useful curves showing the benefits (in terms of UBER) and the penalties (e.g., in terms of redundancy) of the old one- and two-

Table 2.1 Number of Information Bits, k, Per Block of Length versus Hamming Distance d and Number of Detectable Errors, d-1, for Bose-Chaudhuri-Hocquenghem (BCH) Codes

Fixed integer	Block length	d (Hamming distance in bits)					
		3	5	7	9	11	13
m	n	k (Number of information bits)					
4	15	11	7	3	-	-	-
5	31	26	21	16	11	6	1
6	63	57	51	45	39	33	27
7	127	120	113	106	99	92	85
8	225	247	239	231	223	215	207
9	511	502	493	484	475	466	457
10	1023	1013	1003	993	983	973	963

Notes:

(1) $n = 2^m - 1$ = block length

(2) d = 2t + 1 where t is the number of errors that can be corrected

(3) d - 1 = number errors that can be detected

(4) BCH Codes are a subset of the Cyclic Redundancy Codes

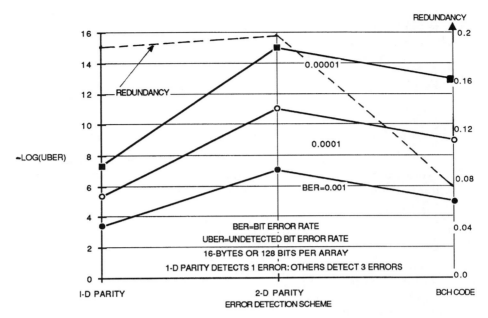

Figure 2.5 Probability of undetected error versus the error detection scheme employed.

dimensional error checking and the modern CRC codes. The values of UBER3 will be lower than those plotted for the reasons already discussed.

Application 3. Another interesting application is regarding the computation of an optimum block size for storing long messages with unknown lengths. Wolman (Wolman, 1965) has developed an interesting relationship that defines the optimum block length for a large number of message length distributions as follows:

$$L_{opt} = SQR\ [2*BR*L_m]\ bits \qquad (2.6)$$

where

BR = block redundancy due to overhead (in bits)

L_m = average message length in bits

Application 4. Equation 2.6 deals with a situation where an optimum use of the storage media is required for the case of no media errors. A closely associated problem deals with computing the optimum block size for a transmission channel characterized by random errors. Error control (implying almost no undetected errors) in transmission systems deals with blocked transmission and error correction through retransmission via feedback control. Each block contains a CRC code. A certain number (N) of blocks are transmitted before looking for a feedback (also called the Go-Back N scheme) message stating whether all blocks were received correctly or not. If one makes the block too long, each block may be received in error and must be retransmitted. If the block is too small, the overhead due to redundancy may be too high. Therefore, the designer is frequently called upon to solve for the optimum block length for a given raw binary error rate, P_e.

The optimum block length (see M. Schwartz, 1987) can be expressed as follows:

$$L_{opt} = SQR[BR / P_e] \dots\dots\dots for \ BR* P_e << 1 \tag{2.7}$$

where

BR= overhead redundancy per packet or block (= 48 for HDLC)

There is also some interest in computing the efficiency of transmission. It is expressed in terms of P_e and "a" which is the ratio Tt/Tb (Tt is the sum of block transmit time and the two-way turnaround propagation time, and Tb is block transmit time) as follows:

$$\zeta = \{BI/[BI+BR]\} *\{[1- p] / [1+(a-1) \ p]\} \tag{2.8}$$

where

BI = information bits per block
BR = overhead bits per block
$p = 1-(1- P_e)^{[BI+BR]}$

See Figure 2.6 for a plot of some useful curves relating ζ to block lengths for typical terrestrial and a satellite channels with P_e =0.00001, and for two channel capacities.

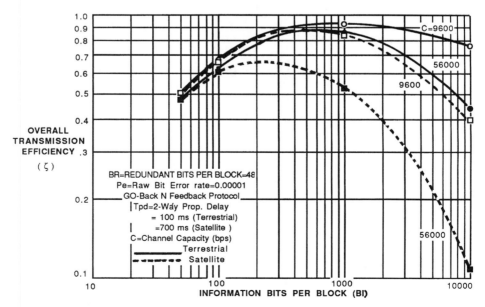

Figure 2.6 Transmission efficiency versus block size for typical terrestrial/satellite channels.

2.4.2 Data Stream Analysis

There are times when it is required to study the manner in which the packets of a message become distributed in time over a shared input/output transmission line when several local ports act as sources of message streams in a packet assembler and dissembler (PAD). See Figure 2.7.

Assuming a first-in-first-out (FIFO) mechanism, identical speeds at both the input and output channels, and effective output channel utilization less than unity, one can show that each message on the output link will experience some elongation in time. The average message expansion factor (AMEF) is related to the average number (N) of slots (or packets) per message and the number (S) of active sources (or local ports). An analytical solution was obtained through using computer simulation. The resulting expression for AMEF is as follows:

$$\text{AMEF} = 1 + [1 - 1/N - 1/S + 1/(NS)]*\rho \qquad (2.9)$$

where

　　N = average number of packets (or slots) per message
　　S = number of active sources of messages
　　ρ = utilization of the shared channel

Figure 2.8 plots several useful curves relating AMEF to N and S and for ρ equal to 1. It is interesting to observe that the maximum value of AMEF is 2 for an infinite number of active sources and large values of N. That is not intuitively obvious. Of course the value of AMEF is reduced to 1 for N=1 and any S. And this value is intuitively obvious. The result of Equation 2.9 can be used to estimate the delays in receiving the entire length of a multipacket message at the receiver when many sources are vying for the same destination over a large number of virtual circuits. There are many applications where the receipt time of the entire message is required. The 8th item of the Analysis menu of EcoNets can be employed to compute the value of AMEF for any combination values of N, S, ρ.

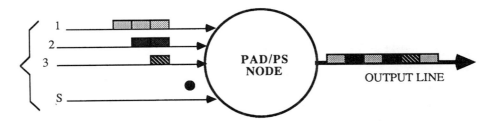

Figure 2.7 Merge operation involving several data streams.

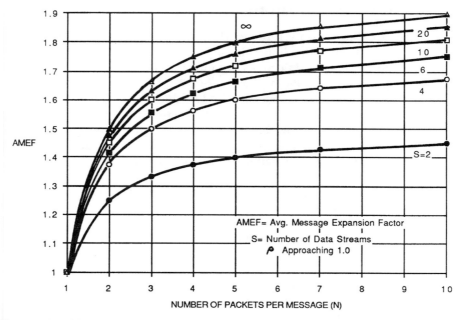

Figure 2.8 Message expansion factor versus the number of packets per message and the number of datastreams.

2.4.3 System Reliability Analysis

There are many occasions when it is necessary to compute the reliability of a nodal subsystem or the entire network system. In order to achieve that goal, we must define several quantities, since reliability can be expressed in several ways.

2.4.3.1 System Availability

System availability is defined as the percentage of time the system is available for its assigned mission. It can be expressed as follows:

$$A = [MTBF] / [MTBF + MTTR] \qquad (2.10)$$

where
MTBF = mean time between system failures
MTTR = mean time to repair a system failure

The system can possess any arbitrary scope or size. It could be a nodal subsystem or the entire network system. The value of MTBF for nodal products is generally supplied by the vendor. The vendor for a proposed system generally estimates MTBF of a network system, if required by the user. The value of MTBF of a well-defined product is generally computed by using one of many methodologies. Each large vendor has a reliability department that dictates the methodology. In every case, the product is modeled as a collection of unique indivisible parts, each part associated with a unique failure rate. Military standards exist to aid in the selection of such failure rates. Using such models, the MTBF of a product is determined. Some of the models are discussed in the following paragraphs. The values of MTTR are estimated using experimental data and some extrapolations based on product maturity.

The probability of a successful mission (or mission reliability) during an observed mission time T can be stated as follows:

$$R\ (T) = EXP\ [\ -\ T/MTBF\ ] \tag{2.11}$$

where

EXP $[x]$ = exponential function of $x = e^x$

Probability of K failures during a mission time T can be expressed as follows:

$$P[K,T] = EXP\ [-N_f]*[(N_f)^K\ ]\ /\ [K!] \tag{2.12}$$

where

N_f = average number of failures during a mission time T
= T / MTBF
K! = K factorial = 1*2*3*...*K

Each large system can be represented as a set of connected subsystems. The reliability of the entire system is a function of the way the subsystems are connected together and the manner in which the system functions together. There are three ways the subsystems can be connected together to allow the system to operate as a whole:

1. *Series connection.* The system requires all subsystems to function concurrently.

2. *Parallel connection.* At least one subsystem needs to function properly.

3. *K-out-of-N type connection.* At least K subsystems must operate for the entire system to provide a successful mission.

2.4.3.2 Reliability of a Series Type Redundant System

A large system can be modeled as a set of subsystems connected together in a series, as shown in Figure 2.9.

The reliability of the entire system can be expressed as follows:

$$A_s = \prod_{i=1}^{n} A_i = A_1*A_2*....*A_n \tag{2.13}$$

$$(MTBF)_s = \Big[\sum_{i=1}^{n} 1/MTBF_i\ \Big]^{-1} \tag{2.14}$$

$$(MTTR)_s = \{\ [1-A_s]\ /A_s\}*(MTBF)_s \tag{2.15}$$

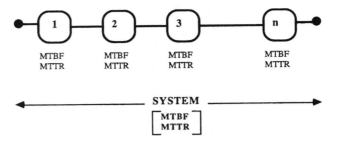

Figure 2.9 System with series type redundancy.

2.4.3.2 Reliability of a Parallel Type Redundant System

A system characterized by a parallel type redundancy is shown in Figure 2.10. The reliability of a system employing parallel type redundancy can be expressed as follows:

$$A_s = 1 - \prod_{i=1}^{n} (1 - A_i) \tag{2.16}$$

$$(MTTR)_s = \left[\sum_{i=1}^{n} 1/MTTR_i \right]^{-1} \tag{2.17}$$

$$(MTBF)_s = [A_s / (1 - A_s)] * (MTTR)_s \tag{2.18}$$

2.4.3.3 Reliability of a K-out-of-N Type Redundant System

See Figure 2.11 for an illustration of a system employing K-out-of-N type redundancy.

The reliability of such a system can be expressed as follows:

$$(MTTR)_s = (MTTR)_i / [N-K+1] \tag{2.19}$$

$$(MTBF)_s = (MTBF)_i [(MTBF)_i / (MTTR)_i]^{N-K} * [(N-K)!(K-1)!]/N! \tag{2.20}$$

$$A_s = (MTBF)_s / [(MTBF)_s + (MTTR)_s] \tag{2.21}$$

The models just shown can be used to compute the reliability of a composite nodal system or the entire network system. An experienced network designer generally reduces the computational models into easily accessible computer programs. Using this methodology, it is an easy task to compute the reliability of a large network system. The 11th item of the Analysis menu of EcoNets provides such a capability. The tool allows the designer to model up to two subsystems in series, each with up to five (5) elements arranged in either a series or parallel or

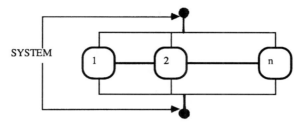

Figure 2.10 System with parallel type redundancy.

a K-out-of N manner. This can be continued until a critical path of a voice call or a data message is modeled for reliability. The major challenge lies in obtaining the correct values of MTBF and MTTR of individual subsystem (or products).

2.4.4 Return-on-Investment Analysis

A very cost-effective synthesis for a network system can be produced using the algorithmic methods described in this book. But if the designer cannot present ideas to the upper management of the user community in a manner to which they are accustomed to, the entire network design may come to naught. A common approach to making an effective presentation entails the following steps:

1. Make a model of the 5–10 year life cycle of the network system.

2. Determine the cost of deploying the current system during the life cycle.

3. Determine the cost of deploying the proposed system during the life cycle.

4. Determine the savings achieved during each year and during the life cycle.

5. Compute the payoff period for the new hardware.

In order to succeed in the effort, one must make economic models that consider the cost of money. A network system entails two types of costs: (1) transmission costs that are usually incurred each month, and (2) hardware costs that are usually incurred at the beginning of the life cycle. Of course there are exceptions.

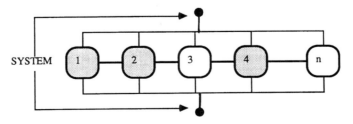

Figure 2.11 System with k-out-of-N type redundancy.

Some users own their own microwave facilities. Some users lease their hardware in order to avoid the onetime cost up front. For such cases, it is usually sufficient to complete Step 4 by showing the total savings achieved in each year of the life cycle and during the entire life cycle. But, in every case, one must present several alternatives to the management in order to avoid any surprises.

In order to compute the payoff period, one must first compute the life cycle costs of hardware by considering the cost of money. A simple technique is to compute the monthly payment (MP) for a principal value (PV) and a yearly interest rate (IR) and the total number of monthly payments (n). This value can be computed as a ratio of PV and a dividing factor (DF) as shown here:

$$MP = PV \; / \; DF \qquad\qquad (2.22)$$

where
$$DF = \{ \; [1 - 1/ \; (1 + i)^n \;] \; / \; i \; \}$$
$$i = IR \; /12 = \text{monthly interest rate}$$
$$n = 12*(\text{life cycle duration in years}) = \text{total number of payments}$$

The values of DF for IR=10% are listed in Table 2.2 for some typical life cycle values.

Using this approach, one can compute the total cost of the hardware for the life cycle as follows:

$$\text{Total life cycle cost} = MP*12*\text{Life cycle} = MP*n \qquad\qquad (2.23)$$

Several system alternatives can thus be compared easily by changing the products and/or varying the values of interest rates or life cycle. Using this approach, one can thus accomplish all of the five steps mentioned earlier. These results should then be reduced into effective graphs for final presentations to the upper management. The 12th item of the Analysis menu of EcoNets enables the designer to compute the values of MP and the total life cycle costs for any values of IR, PV, and life cycle duration.

Table 2.2 Values of DV versus Life Cycle Durations

Life cycle (years)	n (months)	DF
5	60	47.065
7	84	60.237
10	120	75.662
15	180	93.058

2.4.5 Computing Distance of a Transmission Line

The monthly cost of any transmission link is generally proportional to its length. Furthermore, the task of optimizing a network topology requires a large number of computations related to finding the closest and the farthest nodes from a given node. The execution time of a network design program will therefore be largely dependent on the distance and link-cost evaluation technique. Alternatively, one will always need an efficient technique for computing the distance between any two given points represented with vertical (V) and horizontal (H) coordinates. For North America, one can designate the location of any one given point by the readily available V&H coordinates. One can purchase databases that relate one's telephone number to the associated V&H coordinates. The three-digit NPA (commonly known as the area code) and the NNX (commonly known as the CO number) codes for a customer premise (CP) location determine the V&H coordinates of the associated wire center of a BOC. To illustrate, the NPA-NNX equaling 213–390 for a CPE located in the Southern California region corresponds to V&H coordinates of 9227 and 7920 for a wire center named as LAS. A database relating V&H coordinates to NPA-NNX is also a part of the PC-based private line pricers (PLPs). Such a database requires storage of millions of bytes. Furthermore, it generally requires many updates each year since NPAs are being constantly added or updated.

The V&H coordinate system is based on the Donald's projection of the North American continent that includes Canada, the United States, and Mexico. Such a projection system was derived to yield a simple method of computing straight-line distances between any two wire centers of the Bell System and also the associated costs of leasing a circuit. The distance between any two locations can be expressed as follows:

$$D12 = SQR \{0.1*[(V1 -V2)^2 + (H1 -H2)^2]\} \ ....miles \qquad (2.24)$$

where

V1, V2 are the V-coordinates of the two locations
H1, H2 are the H-coordinates of the two locations
D12 = straight-line distance between the two locations in miles

The distance D12 is never equal to the exact distance of the physical link employed to connect the two CPEs of the customer. Equation 2.24 is merely an aid to quickly compute the distance for pricing a direct distance dialed (DDD) telephone call or a leased line. The tariffs are always adjusted to provide a profit to the OCC.

For countries without such a coordinate system, the value of D12 can be expressed in terms of the readily available longitudes and latitudes as follows:

$$D12 = arccos[sin(LAT1)*sin(LAT2)+cos(LAT1)*cos(LAT2)$$

$$*cos(LNG2-LNG1)]*6371.256 \ Kmeters \qquad (2.25)$$

where
> LAT1 and LAT2 are latitudes of the two locations
> LNG1 and LNG2 are the longitudes of the two locations.

The author employed Equation 2.25 during the early 1960s until someone mentioned the availability of the V&H coordinates for North America. A comparison of Equations 2.24 and 2.25 should easily convince the reader that the use of longitudes and latitudes will consume an inordinate amount of computer time when designing a large network. See Chapter 6 for a recommended approximate method of obtaining equivalent V&H coordinates for countries situated in other continents. The EcoNets tool as described in Chapter 6 allows the use of such an approximate method.

One user of EcoNets has devised another interesting technique for computing the distance of a leased line. The user can now employ the Zip Codes of the two CPEs. The new technique avoids frequent updates to the (NPA-NNX, V&H) database. As we know, the ZIP Codes (ZIPCs) are more stable than the NPAs; however, one needs to create an approximate (ZIPC, V&H) database only once. The author has also learned that an Internet location exists that provides the latitude and longitude for any entered ZIPC. But such an approach will still require the use of the complex Equation 2.25.

2.4.6 Computing Cost of a Transmission Line

The monthly cost of a *leased* circuit will require the use of several tariff databases. The bandwidth of each leased line is always fixed. Therefore, the tariff is generally based on fixed costs for each mileage band and costs based on actual mileage. A *fully owned* transmission line can have any capacity. Onetime costs (or derived monthly costs valid for a given life cycle and cost of money) for a fully owned transmission line can be computed by constructing analytical cost models for the right-of-ways, wide-band (WB) modems and electronics, antennas/towers, media type/length/cross-section, repeaters, and trenching and media protection devices. Consequently, one can model an equivalent tariff by creating a model similar to that for a leased line and multiplying it by a factor, F_{pf}, defined as a system design factor. See Chapter 6 for a description of such models for the leased and fully owned transmission facilities.

For the pre-divestiture period, one required the knowledge of the high-density (HD) and low-density (LD) rate centers as defined by Tariff 260 for computing the monthly lease of a multipoint voice grade circuit. The customer had no choice as to the actual route of a leased line. The line-of-sight distance between the two CPE sites was just an arbitrary basis for computing the monthly rate. The monthly cost of a leased line depended on eight (8) mileage bands and on three schedules: Schedule A was applicable to a case of two HD rate centers, Schedule B applicable to the case of one HD and one LD rate center, and Schedule C applied to two LD rate centers at the two ends of the circuit. The monthly cost per mile for each of the schedules and for each of the applicable mileage band is listed in Table 2.3. Similar tariffs were available for other services such as DDS and for other common carriers. The pre-divestiture tariffs required much less

Table 2.3 Pre-Divestiture Multipoint Private Line (MPL) Tariffs

Mileage range	Monthly charges per mile		
	Schedule A	Schedule B	Schedule C
1–15	$1.89	$3.47	$4.63
16–25	1.58	3.26	3.99
26–40	1.18	2.10	2.94
41–60	1.18	1.42	2.21
61–80	1.18	1.42	1.68
81–100	1.18	1.42	1.42
101–900	0.69	0.69	0.71
over 900	0.42	0.42	0.42

description than their counterparts for the post-divestiture period. This can be attributed to the numerous service options that are now available to the users.

Since the 1984 divestiture of AT&T, the number of databases for all OCC (Other Common Carriers) has increased by leaps and bounds. The databases for IOCs (Inter-Office Channels) listed for each rate center the associated LATA (Local Access and Transport Area) number, name of the rate center, the State abbreviation, the CLLI code, point-of-presence (POP) rate center V&H coordinates, POP's wire-center V&H coordinates, and availability vector for the various services such as T-1, DDS, and VG applicable to each common carrier. Furthermore, one needed more than a single tariff for computing the monthly costs of leasing a circuit. For example, AT&T employed three tariffs, 9, 10, and 11 to compute the monthly costs of the various services listed as follows:

1. AT&T Tariff 9 defined the cost of the Inter-Office Channel (IOC) that connects the two POPs, CO connection between the POP and the CPE CO, CO functions such as CCSA/EPSCS/SCAN switching, M-24/M-44 multiplexing, customer controlled reconfiguring, and transfer and conferencing arrangements.

2. AT&T Tariff 10 contained no rate information but defined the databases necessary to evaluate the V&H coordinates of POPs and availability of various services. Such a tariff required more than 500 pages to describe it.

3. AT&T Tariff 11 contained all the rate and service data relative to local channels (LCs) provided by AT&T with Access Coordination Function (ACF). AT&T obtains such channels from the applicable local exchange carrier (LEC) within the two LATAs associated with the two ends of the circuit. The rate structure for AT&T's local channels is very similar to that for the LEC's special access tariffs and consequently it varies by each associated state. Such a tariff required about 1,000 pages to describe it.

Two important points need to be made here. First, AT&T did not guarantee the minimization of the private line cost for the user who can, in some cases, find alternate POPs to lower the cost by separately leasing the appropriate IOC and special LEC rates on local channels. The user could also purchase the ACF from AT&T to guarantee the end-to-end performance. Second, whereas the special access tariffs used by LECs employ the mileage between the wire centers, AT&T measures circuit mileage between rate centers. Similar considerations applied to other common carriers such as MCI and Sprint.

See Figures 2.12, 2.13, 2.14, and 2.15 for the simplified models of an entire leased line with its basic components, the intra-rate-center-LC and the inter-rate-center-LC, and the LEC bridged multipoint circuit configurations, respectively. See Table 2.4 for an illustration of AT&T's IOC Tariff 9 that was applicable during 1987. For designers interested in how tariffs change continuously, we will now illustrate AT&T's IOC Tariff 9 for the period 1991 to 1992 in Table 2.5. In order to show that private leased lines from other OCCs reflect similar costs, we will now list similar MCI tariffs (that became effective on December 1, 1996) in Table 2.6.

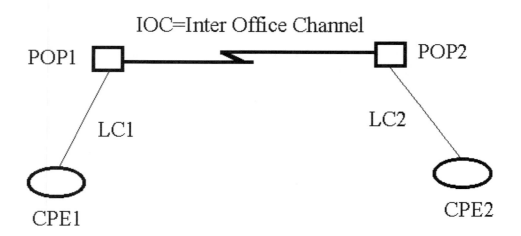

IOC=Inter Office Channel

POP1 POP2

LC1 LC2

CPE1 CPE2

Legend:
LC=Local Channel: CPE=Customer Premise Equipment
POP=Serving Point-Of-Presence Switch

Figure 2.12 Transmission components of a leased line.

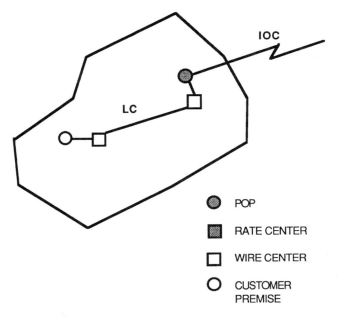

Figure 2.13 Topology of an intra-rate-center local channel.

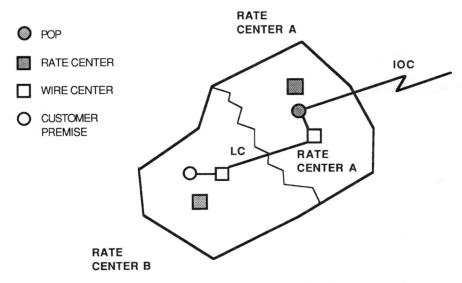

Figure 2.14 Topology of an inter-rate-center local channel.

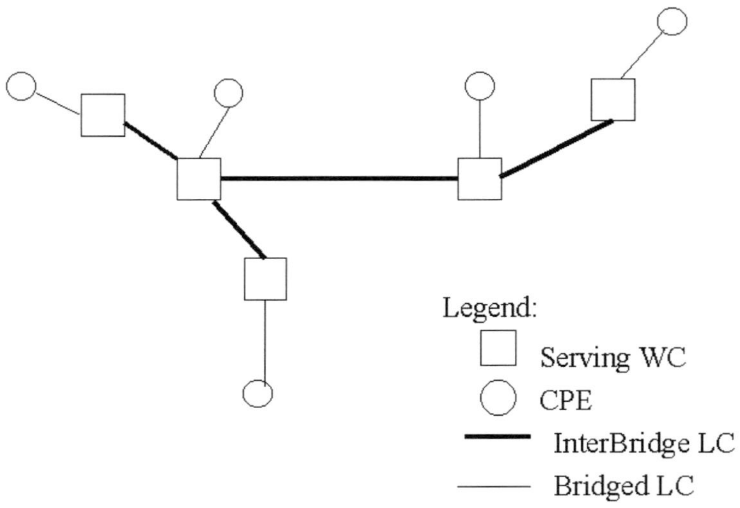

Legend:
□ Serving WC
○ CPE
▬ InterBridge LC
— Bridged LC

Figure 2.15 Topology of a LEC bridged multipoint circuit.

Based on the preceding discussion, we can now derive a generalized expression for computing the monthly cost of a leased line:

$$MC_1 = K + FC_i + CPM_i * D12 \tag{2.26}$$

where
  MC = monthly cost of the leased line in an appropriate currency
  K = average cost of the two local channels (LCs)
  FC_i = Fixed cost for the ith mileage band
  CPM_i = cost per mile for the ith mileage band

Table 2.4 AT&T's Inter-Office Channel (IOC) Tariff 9 (3/13/87)

Voice-grade mileage band	Fixed monthly cost	Per mile
1–50	$ 58.74	$ 2.33
51–100	122.24	1.06
101–500	164.24	0.64
Over 500	324.24	0.32
DDS (2.4–9.6Kbps) mileage band		
1–50	$ 79.34	$ 2.33
51–100	142.34	1.06
101–500	184.84	0.64
Over 500	344.84	0.32

Table 2.4 *Continued*

DDS (56Kbps) mileage band

1–50	$ 302.00	$ 10.07
51–100	566.50	4.78
101–500	742.50	3.02
Over 500	1407.50	1.69

Accunet T1.5 mileage band

1–Over 1000	$ 2600.00	$ 15.50 (1–2 years)
1–Over 1000	$ 2600.00	$ 14.85 (3 years)
1–Over 1000	$ 2600.00	$ 14.25 (5 years)

The equivalent monthly cost for a fully owned transmission facility can be expressed as follows:

$$MC_2 = MC_1 * F_{pf} \qquad (2.27)$$

where
   MC_1 = the monthly cost expression similar to that of Equation 2.26
   F_{pf} = a privately owned facility factor appropriate for BW

Depending upon the need for data rates or bandwidth, one can select the most appropriate private line service offered by different OCCs. Thanks to stiff competition, OCCs offer similar tariffs for fractional T-1, switched T-1, and virtual network services (e.g., MCI's VNET, AT&T's SDN, Sprint's VPN). CPEs with low traffic are generally connected to a private network through off-net ALs (ONALs) to receive OCC's low-cost virtual network services. Many OCCs offer special solutions for interconnecting data LANs to form wide area networks (WANs) to facilitate sharing of devices, software, and databases between different locations. Special interfaces are provided to interconnect popular brands of Routers employing diverse protocols on a variety of transmission rates. One should realize by now why one requires an iterative approach (directed only by the mind of an experienced designer) to study many viable "what if" solutions for finding an optimum one. Any hope of getting a fully automated tool during the next decade (or ever) is unlikely to be fulfilled.

At the moment of publishing this book, the tariffs for leased lines are going through another upheaval. Most experts believe that the tariffs for leased lines that we know today are about to disappear. In the near future, each customer will negotiate the tariffs with many OCCs in terms of traffic volumes and other prevailing rates, and select the best one considering special discounts offered. As a result, the competition for leasing communications facilities is bound to get even hotter during the coming years.

Table 2.5 AT&T's Inter-Office Channel (IOC) Tariff 9

Mileage band	Fixed monthly cost	Per mile
Analog private line service: **(Effective 3/21/91)**		
1–50	$ 72.98	$ 2.84
51–100	149.48	1.31
101–500	229.48	0.51
Over 500	324.48	0.32
DDS (56Kbps): **(Effective 10/1/92)**		
1–50	$ 340.80	$ 5.88
51–100	445.80	3.78
101–500	585.80	2.38
Over 500	1,105.80	1.34
Accunet T1.5: **(Effective 10/1/92)**		
1–50	$ 2,500.00	$ 3.50
51–100	2,500.00	3.50
Over 100	2,500.00	3.50
Accunet T45: **(Effective 10/1/92)**		
1–50	$16,000.00	$ 45.00
51–100	16,000.00	45.00
Over 100	16,000.00	45.00

It is also possible that large corporations may elect to purchase right-of-ways and own their own transmission lines. Many outfits already exist that will install privately owned fiber facilities, with end-to-end control. Such a developing scenario will have profound influence on the large network design packages, each of which boasts the largest and the latest tariff database. The ability to make models of tariffs at a moment's notice will become extremely important. Such an ability was always necessary to model enterprise networks for foreign countries where the U.S. tariffs don't apply. A modified model, but similar to the ones for leased lines, can be constructed for fully owned transmission facility. Equation 2.27 shows such a model. Such a model must include costs of the right-of-way, towers,

Table 2.6 MCI's Private Line Services: Tariffs Effective 12/1/96

Mileage band	Fixed monthly cost	Per mile
Analog VGPL		
1–50	$ 332.00	$ 0.27
51–100	332.00	0.27
101–500	332.00	0.27
Over 500	324.48	0.27
DDS (56Kbps):		
1–50	$ 413.00	$ 7.12
51–100	542.00	4.55
101–500	757.00	2.39
Over 500	1,408.00	1.09
TDS-1.5:		
1–50	$ 3,234.00	$ 3.87
51–100	3,234.00	3.87
Over 100	3,234.00	3.87
TDS-45:		
1–50	$22,236.00	$52.07
51–100	22,236.00	52.07
Over 100	22,236.00	52.07

installation, electronics, fibers, digital cross-connects, and well-managed operations. Chapter 6 shows how a simple model can be employed to compute the monthly cost of a privately owned transmission facility.

The succeeding chapters will emphasize a simplified methodology for synthesizing optimum network topologies for new and established networks using the EcoNets software package. Once the optimum topology has been synthesized using EcoNets, the commercially available line pricing packages (not the network design packages that are too expensive for that purpose) can be used to compute the monthly costs of leased transmission facilities to a high degree of accuracy. Some large customers simply supply the network topology to one or more OCCs and invite proposals for their network services. This is another creative way to obtain accurate estimates of monthly costs of transmission lines. Some well-known but high-priced network design packages can also be used to simulate the optimum network topology (as derived through the use of EcoNets) for computing some detailed performance attributes.

Cost may not be the only criterion for selecting a particular LEC or OCC for leasing a facility. The network reliability should also play an important part. If a node corresponds to a major corporate location, one should consider multiple access lines to that location by selecting alternate OCCs or leasing additional dial-up facilities or even some bypass facilities. This will guarantee some communications with the outside world in case an LEC's CO/WC or OCC's POP fails. In most cases, bypass of LECs (not a dirty concept anymore) will also reduce the transmission costs. The amount of "real" savings can be computed by evaluating the monetary loss occurring as a result of disaster or human error and comparing it with the recurring cost of additional facilities. In order to achieve the desired reliability objectives, the network designer must familiarize with the actual layouts of major switch and transmission facilities of an enterprise and take the necessary steps to guarantee the flow of at least the high priority traffic from every major location. These steps should influence the final network topology and overall network availability.

2.5 Traffic Engineering Techniques

2.5.1 Defining Time Consistent Averages (TCAs) of Traffic Intensities

Traffic plays an important factor in the network design process. The basic units of traffic intensity (TI) such as *erlang* have been defined in Chapter 1. If a network system already exists, one can measure the busy hour traffic intensity for each network link by reading the output of specially installed meters or commonly available call recording (e.g., Station Message Data Recording or Automated Message Recording) tapes and then deriving the system grade-of-service (GOS) in any desired form. In case the GOS is better or worse than that which is required, one may achieve cost effectiveness through better network dimensioning. If one has the resources, one may achieve an additional cost effectiveness through a new optimum network design based on end-to-end traffic flows.

Unfortunately, most networks are designed either from scratch or for environs that are quite different from the preceding ones. To illustrate, a corporation has just merged with another large corporation. Although the two existing networks maintained very good traffic data, each network operated in different areas with some overlaps and was based on different GOS requirements. Furthermore, experience shows that the combined network will generally create entirely new traffic loads on its resources as a result of different traffic flow patterns. Simple network tuning may not be adequate. For that reason, it may be necessary to redesign the network based on new traffic patterns.

Even when one obtains the busy hour (BHR) traffic statistics based on the actual observations or engineering assumptions, one will find that a network design based on peak hour traffic loads as too expensive. Alternatively, the GOS of a new system based on the peak traffic intensities will exceed those required by the new system. It should therefore be desirable to present a method for computing the traffic intensities that are finally used to dimension the network. Such values are generally called the time-consistent-averages (TCAs). Only the TCAs of traffic intensities should be used to dimension the network links and nodes. Analytical tools for dimensioning ALs and trunks are fully described in Chapters 3 and 4. The process of deriving the TCAs for traffic intensities is illustrated in the paragraphs that follow.

The following material presents some proven methodologies for estimating TCA of BHR traffic intensities for designing new networks or for redesigning existing networks. The act of "estimating" TI should be understood fully before one goes forward. Just because one has followed CCITT recommendations or derived the TI values from SMDR tapes, nothing guarantees that model days' busy hours will experience the expected traffic intensities. If all the assumptions (i.e., the number/types of sites and users and their operations) remain static, the theory states that one may have to observe an infinite number of model days before one can test the hypothesis that the expected traffic intensities were truly observed or not. Since the original assumptions never remain the same for a dynamic enterprise, one has to always remain pragmatic, vigilant, and patient.

There are basically two approaches to obtain the TCAs required for network dimensioning:

1. North American practice that requires the use of a typical month's traffic data

2. CCITT practice that requires the use of 30 peak hours from yearly traffic data

The major difference lies in the size of the database employed. For the North American practice, one needs only a single month's traffic tape. The CCITT practice requires the use of traffic tapes for each of the months of a typical year.

The first approach requires the development of a point-to-point traffic flow chart as shown in Figure 2.16. All 30 days and all 24 hours of each day are displayed in two dimensions (horizontal and vertical columns of the display), as done for a public network. Only the valid business days and valid business hours should be considered for designing a private network for a corporation.

The CCITT practice for computing TCAs is illustrated in Figure 2.17. According to this practice, only the first 30 highest values of yearly traffic flows are considered for computing TCAs for any given point-to-point traffic flow. The earlier remarks dealing with the computational complexities encountered in computing TCAs are still valid.

It should be emphasized that the methodologies of Figs. 2.16 and 2.17 are applicable to any point-to-point traffic flow. When there is no reference to a network topology, each point-to-point traffic flow refers to only the traffic flow between a given pair of subscribers. It is an enormous task to compute the TCA for each subscriber pair. Furthermore, this data is not directly useful for dimensioning the network without a topology. Therefore, TCAs for traffic flows or intensities are generally computed for either each network node (e.g., PABX or Data LAN) pair or each network. Such TCAs are then used to dimension the resources of an entire network topology being considered.

Hours	Day1	Day2	Day 30	TCAs
H1	A1,1	A1,2	A1,30	TCA1
H2	A2,1	A2,2	A2,30	TCA2
H3	A3,1	A3,2	A3,30	TCA3
.	.	.	.	.
.	.	.	.	.
H24	A24,1	A24,2	A24,30	TCA24

$$(1)\ \text{TCAh} = \text{AVERAGE } (\text{TCAh1},\text{TCAh2},\text{TCAh},3,...\text{TCAh},30)$$
$$(2)\ \text{TCA} = \text{MAXIMUM } (\text{TCA1},\text{TCA2},\text{TCA3},....\text{TCA24})$$

Figure 2.16 North American methodology for computing TCAs for each network node pair.

In order to compute the savings achieved through the use of time-of-day (TOD) routing schemes in the network, the designer must compute the TCAs for each hour and for each AL and trunk bundle of a given network topology. Such information helps the designer to dimension the circuit bundles as a function of the special TOD routing schemes. Experience shows that significant savings can be achieved for large corporate and public networks that encompass several time zones of large countries such as the United States and Canada.

2.5.2 Estimating Traffic Intensities for SLs in a New System

The International Telecommunications Union (ITU) provides a wealth of data regarding the erlangs associated with various types of subscriber lines (SLs) or terminals. Such data are based on the study of public networks of the world. Andrews (Andrews, 1967), AT&T (AT&T, 1966) and CCITT (CCITT, 1968) provide average values of busy hour (BHR) traffic intensities (as generated by both incoming and outgoing calls) as .058, .04, and .037 for a typical telephone subscriber line of an AT&T CO, AT&T rural CO, and CCITT CO, respectively. These

HValue1	HValue2	Hvalue30	
A1	A2	A30	TCA

$$\text{TCA} = \text{AVERAGE } (A1,A2,\ A3,......A30)$$

Figure 2.17 CCITT methodology for computing TCAs for each network node pair.

numbers must be modified if dealing with a doctor or lawyer's office or a business PABX line. A professional's line generally exhibits a busy hour traffic intensity of 0.25 erlangs. The use of personal computers (PCs) by many of the professional offices as a means of accessing data from host computers that manage commercial databases (DBs) is becoming quite commonplace. Since these PCs use the public telephone network (PTN), one must evaluate their effect on the traffic intensities. The values of BHR traffic intensities on such lines may vary from 0.2 to 0.9 per each CO business line. A business PABX line generally exhibits a TI of 0.1 erlang. The serious reader is advised to consult Eldin (Eldin, 1975) for an excellent discussion of CCITT recommendations on traffic engineering and network planning.

For data applications, there are no such things as the average traffic intensity figures for each terminal or port. Each application and each customer determines its own traffic intensities. In case there are no existing traffic models, one may try out several assumptions for a BHR. For example, one way to compute the BHR traffic intensity is to assume that each terminal produces a given (minimum) fraction of its capacity. Then. by using a traffic growth factor, one can thus easily vary the total traffic handled by the network. The software tool described in Chapter 6 allows the entry of fixed or variable amounts of traffic for each CPE in a site-related input file called the VHD file. It also allows the use of a traffic growth factor (TGF) that can be varied to study the effect of any combination of traffic intensities. The VHD entries can be a function of the number and types of users (data, voice, or video users) and compute the BHR traffic intensities (erlangs for voice, and bits per second for data) handled by each site and defined in the VHD input file for a new system.

In order to compute the total voice-related TI handled by a node, one must add the traffic flowing from the node and all of the traffic originating at all other nodes. For voice, a single 4-wire AL/TK can handle only one originated call. For data, a single 4-wire AL/TK can handle two independent traffic flows.

2.5.3 Estimating Traffic Intensities for AL Bundles in a New Network

To compute the total traffic, T_i handled by the ith AL bundle (that connects one CPE to one of the available switches as shown in Figure 1.1), one must study the CPE node in detail. If it is a voice LAN in the form of a PABX, then the total traffic handled by the particular AL bundle is equal to 0.1*No. of Voice Lines served by the PABX. If it is a data LAN, the total traffic handled by the particular AL bundle is equal to BHR data rate handled by the LAN-Associated Server for wide area application. If a CPE is a small office with a few voice lines and two or three workstations that communicate with the company host, the traffic intensity can be easily computed for voice or data or for multimedia application for dimensioning the AL bundle. The network design tool must be capable of computing the economic viability of either a leased AL or a virtual service for this small office.

2.5.4 Estimating Traffic Intensities for Trunks in a New Network

Since the exact traffic flows in a new network are not known to exactly compute the traffic on each trunk bundle, one can estimate the traffic intensities on trunk bundles by using an assumption that traffic from each source is uniformly distributed to all destinations as a function of their loadings. A symmetric traffic matrix can be defined for a backbone network if the following steps are executed:

Step 1. Compute the total traffic (TT_i) handled by the ith switch node = $\Sigma\, T_k$ = sum of traffic handled by all AL bundle (k=1, 2,...,N_{ali} where N_{ali} is equal to the number of AL bundles served the ith switch).

Step 2. Compute the grand total (GT) of all the traffic handled by all the switches.

Step 3. Compute the traffic (T_{ij}) flowing on the trunk from the ith switch to the jth switch as follows:

$$T_{ij} = [\ TT_i * TT_j]\ /GT \qquad\qquad (2.28)$$

$$= T_{ji}$$

(for a symmetric traffic matrix)

If the value i equals j, one obtains the value of intranodal traffic. The intranodal traffic implies that the source and the destination are terminated on the same backbone switch of the network node. As a result, this traffic does not influence the sizing of any trunk bundle. One can vary the amount of internodal traffic drastically by dividing the values Tij by the value of a design parameter called backbone trunking factor (BBTF) and adding the difference to the intranodal traffic. The use of BBTF is illustrated in Chapters 7 and 9 concerned with the network design process.

After computing the internodal traffic intensities for each direction of flow, the designer should then apply the most appropriate traffic analysis formula to determine the size of each trunk bundle and its length in miles. One can also compute the cost of trunks by using the most appropriate tariff. In most instances, the resulting traffic flows on the backbone trunk bundles do not imply a minimum cost trunking network.

The entire process of designing the AL and TK network is accomplished through the use of the EcoNets software package as described in Chapter 6. The TCA traffic is always defined by the 4th column of the VHD input file (a database for each site that also defines the nodal ID, and associated V & H coordinates). For each VHD input file data, EcoNets first computes the traffic handle by each AL and TK bundle for each assumed set of optimally located number of switches (Nsw = 1,2,3,..., S). EcoNets also enables the designer to either reduce the cost of a given BBNet or obtain a desired BBNet topology (e.g. ring or meshed rings) for obtaining high reliability. EcoNets also enables the designer to define all the routes between any two nodes of a given network topology for designing a system routing table. EcoNets should also be employed to vary the mix of AL and trunk-link types and repeat the preceding process to find the globally optimum network topology.

2.5.5 Estimating Traffic Intensities for ALs and Trunks in Existing Systems

Most network systems employing SPC switching nodes have the capability of providing a magnetic tape file with a recording of each voice or data call. There are two types of traffic tapes for voice networks. The SMDR (Station Message Detail Recording) tape is produced by the cognizant PABX on a daily or monthly basis. Since the SMDR tape records the details of all telephone calls (both the local and outgoing calls) switched by the PABX, its records can be used to compute the observed traffic intensities on each associated SL and AL bundle. The AMA (Automatic Message Accounting) tape is generally produced by the network management node on a monthly basis. Since the AMA tape records the details of only the inter-PABX calls, its data can be used to compute the observed traffic intensities on only the AL bundles of an entire voice network. Similarly, the data switching nodes have also the capability of creating traffic files with detailed call records. Each call record can be analyzed for the source-destination pair, the time when the call was made, and the call duration and/or the amount of data handled. A simple program can be written to sum the traffic to model TCAs of traffic intensities for each From-To nodal pair considering all hours. This data can then be used to create a From-To Data (FTD) input file which in turn can be used to design a network. In order to dimension the AL/TK facilities of an affordable network, one must be able to superimpose the TCA traffic intensities from the From-To Data (FTD) input file on each AL and TK bundle as a function of the network topology chosen. The software package as discussed in Chapter 6 accomplishes this later task. It simply superimposes the internodal TCA traffic intensity on all the en route AL and TK bundles (of a fully connected backbone network) of each assumed topology consisting of N_{sw} switches. The total cost of the AL and TK bundles can then be computed by choosing optimally located N_{sw} switches (see Chapter 5 for an algorithm for locating N_{sw} switches among the available CPE nodes). When several values of N_{sw} switches are considered iteratively and the total costs are plotted, the designer can easily compute the optimum network topology with the correct number of switches and/or types of communication facilities. The software discussed in Chapter 6 also enables one to optimize the backbone network topology according to either cost or reliability considerations. The EcoNets software as discussed in Chapter 6 also enables one to enumerate all the paths between any two nodes of any given network topology for the purpose of creating a large routing table for the switch design.

Sharma, De Sousa, and Ingle (see Sharma, De Sousa, and Ingle, 1982) provide an analytical method for dimensioning the backbone network using the HU and alternate hierarchical routes of classical hierarchical voice networks. Such a technique is not applicable to the emerging new systems employing the dynamic, non-hierarchical routing (DNHR) scheme associated with modern broadband networks. The older scheme is also not useful in designing modern corporate networks that employ T1 or T3 multiplexers. See Chapter 11 for several design methodologies for integrated broadband enterprise networks.

Exercises

2.1   Why does the network design process require an iterative approach?

2.2   Of all the tasks executed during the network life cycle, which task requires the least expense?

2.3   What are the criteria for defining the performance of a switched network system?

2.4   Enumerate the reasons for preferring the analytical simulation tools over the physical or computer simulation tools.

2.5   Compute the optimum block size for short-term buffering or long-term storing inside of a switching node when the average message length (L_m) is 10,000 bytes and a block redundancy (BR) of 32 bytes.

2.6   Compute the optimum block size for radio transmission for a block redundancy (BR) of 24 bytes and an expected bit error rate (p_e) of 0.001. Also compute the efficiency of media transmission for a message size of 1,000 bytes.

2.7   Enumerate the three transmission components of a leased line.

2.8   Describe the method of modeling the monthly cost of leasing these components.

2.9   Enumerate the two transmission component types of a multipoint circuit leased from a local exchange carrier (LEC).

2.10 Describe the main reason why the derived TCA BHR traffic intensity is used instead of the pure BHR traffic intensity.

2.11 Describe the method of estimating traffic intensities for a brand-new system.

2.12 Describe the method of estimating traffic intensities for an established (or existing) system.

Bibliography

Andrews, R. et al. 1967. "No. 2 ESS Service Features and Call Processing Plan." *BSTJ*, vol. 48, October 1967.

AT&T. 1966. "Rural Electrification Administration: Telephone Engineering and Coordination Manual," Section 325.

Bell Telephone Laboratories. 1977. "Engineering and Operations in the Bell System." New York.

"Economic Studies at the National Level in the Field of Telecommunications: (1964–1968)," Chapter VI, CCITT Books.

Eldin, A. 1975. "Basic Traffic Theory, Seminar on Traffic Engineering and Network Planning, Organized by ITU, New Delhi, India, November 24 to December 5, 1975."

Schwartz, M. 1987. *Telecommunications Networks.* New York: Addison-Wesley.

Sharma, R. L., P. T. De Sousa, and A. D. Ingle. 1982. *Network Systems.* New York: Van Nostrand Reinhold.

Wolman, E. 1965. "A Fixed Optimum Cell-Size for Records of Various Lengths." *JRNL. ACM*, vol. 12, no.1, pp. 53–70.

3

Analysis and Design Tools for Voice Networks

3.1 Traffic Analysis Tools

Traffic analysis is required to compute the performance of a CS voice network in terms of internodal blocking and the number of internodal circuits as a function of the allowed blocking probability B. There are generally two formulas employed to compute the blocking probability for a loss system in which calls are lost if service cannot be provided at the time of a call:

1. Poisson's formula

2. Erlang-B formula

Both of these formulas have been found useful; however, confusion still prevails regarding the related assumptions. One should consult the works of Syski (Syski, 1960) and Collins and Pederson (Collins and Pederson, 1973) for an in-depth study of the underlying theory of these formulas. One should also consult a handbook prepared by Siemens (Siemens, 1970) for a rich display of curves for computing both loss or blocking probabilities and response times.

3.1.1 Poisson's Loss Formula

The Poisson's formula assumes (1) infinite sources are responsible for a steady rate of random call arrivals, and (2) lost calls are held for a duration equal to call holding times. If A represents the call intensity measured in erlangs (see Section 1.4 for a definition), the probability of n calls in the system can be expressed as follows:

$$P_n = A^n * e^{-A} / n! \qquad\qquad (3.1)$$

The probability of blocking on a system with N outlet circuits can be expressed as follow:

$$B = \sum_{n=N}^{\infty} P_n = 1 - \sum_{n=0}^{N-1} P_n \qquad\qquad (3.2)$$

Relationship 3.2 is commonly known as the Molina formula. Figures 3.1, 3.2, and 3.3 illustrate the relationship between A and N for various values of B. Such curves can be useful in computing the exact number of circuits required to handle A erlangs of traffic intensity for a specified blocking probability B.

3.1.2 Erlang-B Loss Formula

The Erlang-B formula assumes (1) infinite sources are responsible for a steady rate of random call arrivals, and (2) the lost calls are cleared as soon as they are blocked and these calls never return. If A represents the offered call intensity

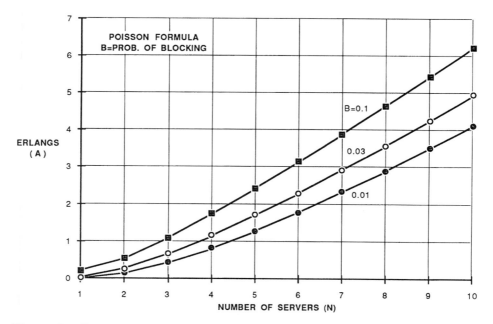

Figure 3.1 Erlangs (A) versus the number of servers (N) for some useful values of blocking (B): Poisson's formula for the range {1 ≤ N ≤ 10}.

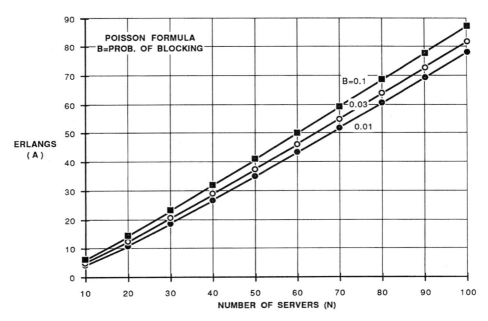

Figure 3.2 Erlangs (A) versus the number of servers (N) for some useful values of blocking (B): Poisson's formula for the range {10 ≤ N ≤ 100}.

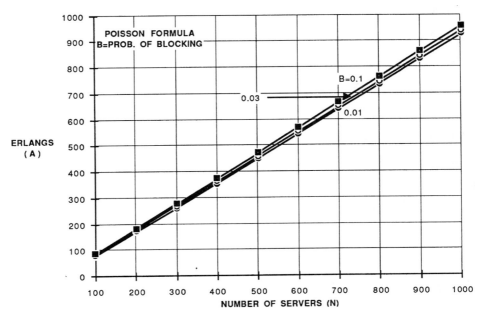

Figure 3.3 Erlangs (A) versus the number of servers (N) for some useful values of blocking (B): Poisson's formula for the range {100 ≤ N ≤ 1000}.

measured in erlangs and N is the number of servers, the blocking probability B(N) is given by the Erlang-B loss formula:

$$B(N) = \left[A^N/N!\right] \Big/ \left[\sum_{i=0}^{N} A^i / i!\right] \qquad (3.3)$$

The probability of blocking B(N+1) can be expressed as a function of B(N) in a recursive form as shown in Equation 3.4.

$$B(N+1) = A*B(N) \big/ \left[\, N+1+A*B(N)\right] \qquad (3.4)$$

This formula uses the fact that the probability of blocking B is equal to 1 if there are no servers; thus, B(0)=1. Similarly we can also get B(1)=A/[1+A] and so forth. The relation (3.4) can be transformed into a short software program to compute B for any given values of A and N. See Table 3.1 for a listing of such a program written in Basic language. See Figures 3.4, 3.5, and 3.6 for the curves

Table 3.1 A Software Program Written in Basic for Computing the Blocking Probability (B) Given A Erlangs and N Servers

```
1000 REM ERLANG B FORMULA GIVEN A AND N

1010 INPUT "ENTER OFFERED ERLANGS "; A

1020 IF A=0 THEN END

1030 INPUT "ENTER NUMBER OF SERVERS "; N

1040 IF N=0 THEN END

1050 GOSUB 5000

1060 PRINT "ERLANGS (A) = "; A

1070 PRINT "NO. OF SERVERS (N) = "; N

1080 PRINT "PRO. OF BLOCKING = "; B

5000 REM SUBR FOR COMPUTING B GIVEN A AND N

5010 B = 1.0

5020 FOR I = 1 TO N

5030 NUM = A * B

5040 DENOM = I + NUM

5050 B = NUM/DENOM

5060 NEXT I

5070 RETURN
```

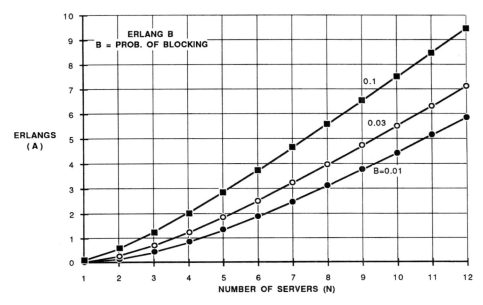

Figure 3.4 Erlangs (A) versus the number of servers (N) for some useful values of blocking (B): Erlang-B formula for the range {1 ≤ N ≤ 12}.

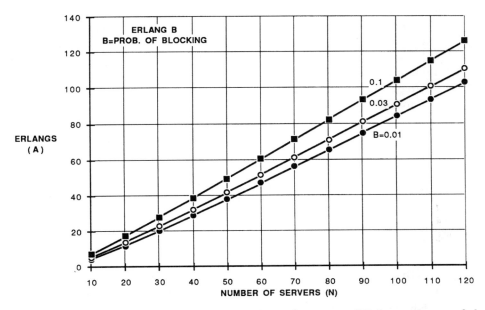

Figure 3.5 Erlangs (A) versus the number of servers (N) for some useful values of blocking (B): Erlang-B formula for the range {10 ≤ N ≤ 120}.

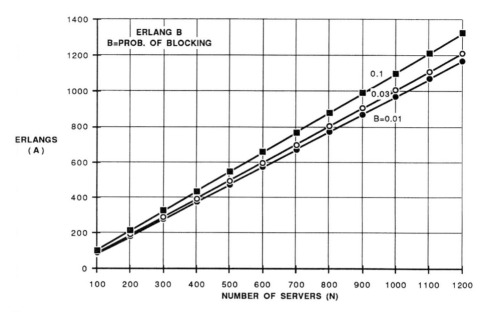

Figure 3.6 Erlangs (A) versus the number of servers (N) for some useful values of blocking (B): Erlang-B formula for the range {100 ≤ N ≤ 1200}.

relating N to A for some useful values of B. Although these curves can help one to compute the exact number of outlet circuits (ALs or TKs) required to handle A erlangs for the given values of B, one can also approximate the value of B for given values of A and N if enough values of B are used in the figures.

One can also compute the average utilization of the outlet AL or Trunk bundle that carries only A(1-B) erlangs. Such a quantity represents the efficiency of the network facilities. The average utilization of the AL or TK (or server) bundle is therefore as follows:

$$\rho = A(1\text{-}B) \ / \ N \qquad\qquad (3.5)$$

See Figure 3.7 for several curves relating average AL or TK (or server) utilization to the number of servers in the bundle and B. To obtain the value of ρ, one must first use the curves of Figures 3.4, 3.5, and 3.6 to compute N for a given A and B and then substitute these values in Equation 3.5. It should be emphasized that Equation 3.5 assumes an infinite number of traffic sources. The curves of Figure 3.7 clearly illustrate the desirability of larger bundles. Larger bundles always result in higher efficiency.

The curves of Figure 3.7 are very useful for interpreting the behavior of voice networks. A network with heavy traffic on its facilities will generally exhibit high efficiency. That is just fine for large users. A network with little traffic on its facilities will generally exhibit low efficiency. This penalty should not be acceptable to small users. We will present a set of relations (see Section 3.14) that can enable

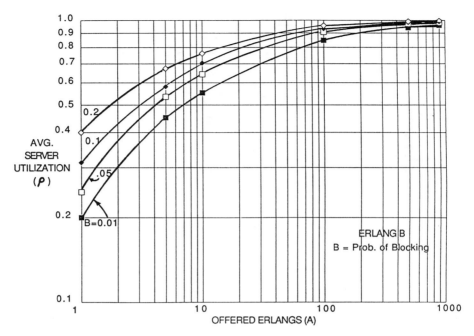

Figure 3.7 Average server utilization versus offered erlangs (A) for some useful values of blocking (B).

the designer to increase the average server efficiency. This can be accomplished by resizing the server bundle through the use of actual number of traffic sources, rather than an assumption of infinite sources.

3.1.3 Erlang-B Loss Formula with Retries

If the lost calls that were cleared from the system are allowed to return at random, the value of enhanced offered traffic intensity A' can be expressed (see Siemens, 1970) as follows:

$$A' = A(1+E+E^2+...) = A / (1 - E)\qquad(3.6)$$

where
  E = blocking probability computed for A' erlangs

It follows that A*E calls will be rejected on the first attempt. These calls will return with A*E2 calls being rejected again and so forth. See Figure 3.8 for a comparison between the preceding three formulas for the case of 10 trunks. The curves show that the difference between the Erlang-B formula with retries and the Poisson's formula becomes small for the useful range of B. The curves also show that the three formulas provide similar results for B smaller than 0.01 .

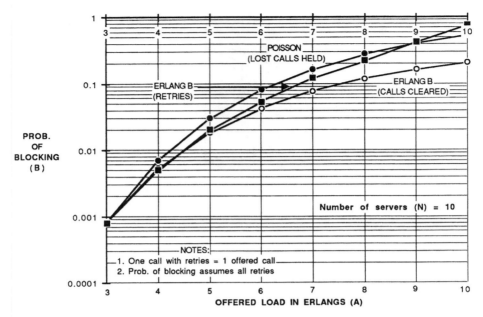

Figure 3.8 Probability of blocking (B) versus offered load in erlangs (A) and different teletraffic formulas.

3.1.4 Erlang-B Formula with Finite Number of Sources

The previous formulas assumed an infinite number of sources. It implies that there are occasions when a large number of active sources can result in a large number of call arrivals requiring a large number of servers. On the other hand, if the number of sources was very small, the worst-case situation will require a bounded number of servers. The value of N can be expressed for offered A erlangs, N servers, 1 call arrival rate per idle source, average call service time of T_s (= ECD), and M sources using the following relations:

$$B = \left[C(M\text{-}1, N)(\lambda T_s)^N \right] \Big/ \left[\sum_{i=0}^{n} C(M\text{-}1,i) \ (\lambda T_s)^i \right] \tag{3.7}$$

$$P_n = \left[C(M,n)(\lambda T_s)^N \right] \Big/ \left[\sum_{k=0}^{N} C(M,k) \ (\lambda T_s)^k \right] \tag{3.8}$$

$$Avg(n) \sum_{n=0}^{N} = n \ P_n \tag{3.9}$$

$$A = \lambda T_s \ [M - Avg(n)] \tag{3.10}$$

where

C(M,n) = no. of combinations of n items taken out of a total of M items (as defined by Equation 2.1 of Chapter 2)

Use Equation 3.7 to compute the value of (λT_s) for given values of B, M, and N. Then compute the value of P_n using Equation 3.8, the average value of n using Equation 3.9, and finally, the desired value of A using Equation 3.10, respectively.

Figure 3.9 shows plots of N for some useful values of A, B, and M. The results show that the infinite source case always requires more servers than the case of finite M. The plots also show that as M exceeds 500, the results for infinite sources apply. The plots also show that a case with few sources can result in a significantly lower number of servers. A designer may be intimidated by the mathematical complexity of Equations 3.7 through 3.10. One should not be intimidated anymore. The third item ("FindB-FiniteM") of the Analysis Menu of EcoNets software provides a capability of computing N for any combination of A, B, and M as long as N < M and N > A. The correct procedure is to select the values of M, N and then continue varying A (= real number) until the blocking probability is very close to the desired B. One has to repeat this for other values of N until a good curve (N vs. A) is obtained for a desired M. This capability can enable one to draw a large number of (N vs. A) curves for a sufficient number of M values and desired B. With the availability of these curves, the designer can obtain a significant saving for all of the AL/TK bundles that satisfy the finite M criterion.

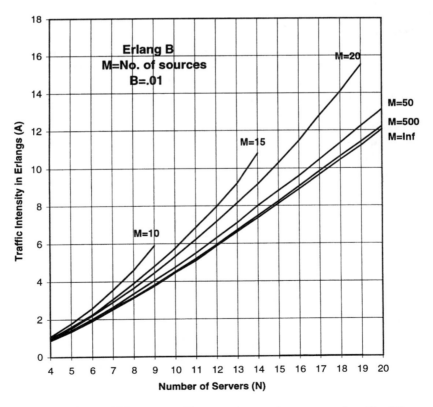

Figure 3.9 Erlangs (A) versus N servers and sources for B=.01.

3.1.5 Erlang-C Formula for Automatic Call Distribution (ACD) Systems

Automatic call distribution (ACD) systems provide uniform distribution of incoming telephone calls to P available agent positions in addition to very detailed and timely reports dealing with traffic statistics, ACD resource utilizations and agent performance levels.

See Figure 3.10 for a block diagram of an ACD system consisting of two subsystems: the trunk subsystem (with N trunks connecting the OCC's POP with the enterprise's ACD node to provide a holding place for incoming voice calls), and agent position subsystem (with P agents who answer the incoming calls on a first-in-first-out basis). A composite ACD system is a very complex entity to analyze as a whole. The sixth item ("CompositACD") of the Analysis Menu in EcoNets software provides a capability of evaluating the interdependence between the two subsystems. To illustrate, when an insufficient number of agents are provided, the trunks hold the calls longer than the average call holding time (=ECD), resulting in a higher blocking experienced by the calling public than expected by the Erlang-B formula. Alternatively, when more trunks are employed than what the trunking subsystem alone demands, the public will experience greater delay than the ACD subsystem was designed for. If the sixth item of the Analysis Menu is not to be used, the trunk subsystem of an ACD node can be designed by using the Erlang B formulae presented earlier, and the agent-position-subsystem can be designed by using the Erlang-C formulas described next.

As the call arrives, it is sent to the next available agent. If all the agents are busy when the call arrives, it must wait for the first agent that becomes available. The calls are handled on a first-in-first-out (FIFO) basis. The caller may wait for a long time until the call is answered; however, the caller may leave the system after waiting a certain amount of time. A well-designed ACD system minimizes the average wait time before the call is answered by scheduling a correct number of agent positions for each half-hourly periods as dictated by the expected daily

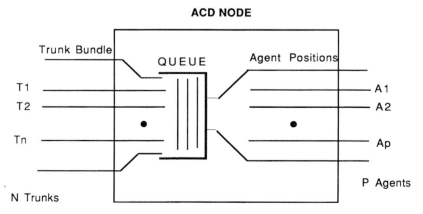

Figure 3.10 Block diagram for an ACD node.

traffic intensities. The probability of delay exceeding any given value of T(equal to DREQ in EcoNets) and average delay are expressed as follows:

$$\text{Prob(} \text{Delay} > T \text{)} = P_0 * \text{Exp} [- (P\text{-}A)T/ECD] = PEXD \qquad (3.11)$$

$$\text{Avg. Delay} = P_0 *[ECD/ (P\text{-}A)]........ \text{ for all calls} \qquad (3.12)$$

$$\text{Avg. Delay} = [ECD / (P\text{-}A)]........\text{for only the delayed calls} \qquad (3.13)$$

where
  A = offered traffic in erlangs = λ*ECD
  P = the number of agent positions that serve the incoming calls
  P_0 = Prob(Delay > 0) = (B*P) / [P - A(1 - B)]
  B = Prob. of blocking given A erlangs and P servers
  ECD= expected call duration
  Exp(x) = $\mathbf{e}^x$ = the exponential function

Some useful plots of computed agent positions P for useful values of A and three values of a ratio (T/ECD) are shown in Figure 3.11. The ratio (T/ECD) or (DREQ/ECD) is related to the GOS for an ACD. The fourth item of the Analysis menu can be used to numerically evaluate the set of Equations 3.11, 3.12, and 3.13. This item employs the terminology M/M/N of queuing theory as discussed in Chapter 4. This item of the Analysis menu allows the designer to specify the number of servers (or number of Agents P) and the offered erlangs A. After

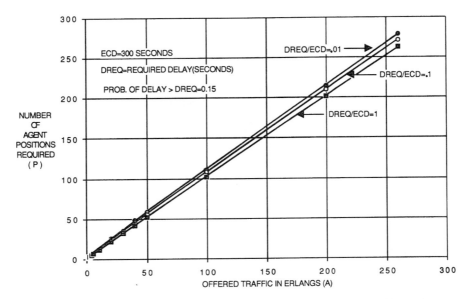

Figure 3.11 The number of agent positions (P) versus offered erlangs (A) for some values of required delay (DREQ) and probability of delays exceeding DREQ = 0.15.

entering these values and clicking on the Compute button, one gets five PEXD values for DREQ=0, DREQ=ECD/4, DREQ=ECD/2, DREQ=ECD*0.75, and DREQ=ECD. These can be plotted to obtain a set of useful charts. The program also provides the values of Avg. Delay (as defined by Equation 3.12) and StdDev(Delay or Wait).

A typical airlines reservation system requires that the probability of delay exceeding 10 seconds should not exceed 0.15. This is equivalent to specifying DREQ=10 seconds and PEXD=0.15. Equation 3.13 is used for computing the average delay per delayed call in case that becomes the GOS criterion. See Figure 3.12 for some useful curves relating Avg(T_w) to A and P. The set of Equations 3.11, 3.12, and 3.13 are generally useful for those ACD applications where the hourly traffic patterns for each day of the week are predictable. For applications where traffic is unpredictable (e.g., televised home merchandising), it may be impossible to predict a GOS for all hours. The number of manned agent positions is generally determined by marketing considerations alone.

3.2 Nodal Performance Analysis Tools

There are many types of network nodes that handle voice calls. See Figures 3.13, 3.14, and 3.15 for illustrations of a typical private/corporate network, an older public telephone network (PTN), and a modern public network (PN) of a typical OCC, respectively. A typical private network consists of many PABXs and a few

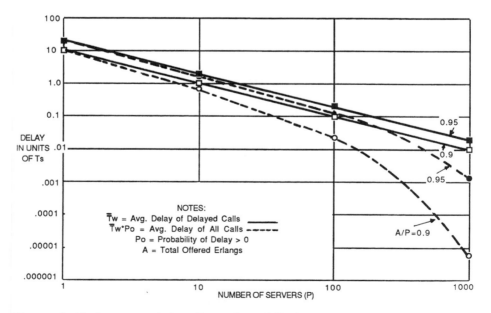

Figure 3.12 Average delay (in units of $T_{,s}$) versus erlangs (A) for some useful values of A/P ratios.

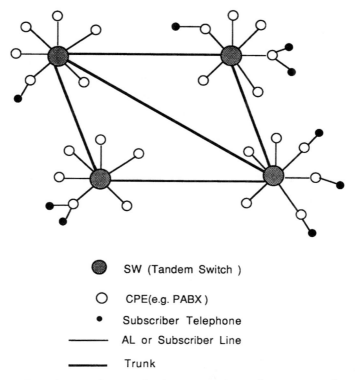

Figure 3.13 Topology of a typical corporate voice network.

tandem-switching nodes. The older PTN consists of a large number of COs interconnected via four levels of tandem toll offices. For a larger metropolitan city, many of the local COs may also be interconnected by a number of local tandems to provide toll-free service to a segment of the public. A modern PN consists of a large number of POPs interconnected via a single level of tandem switches. It employs the common channel signaling (CCS) system-based Dynamic Non-Hierarchical Routing (DNHR) to derive a significant reduction in transmission costs as compared to the deterministic routing scheme (as exemplified by HU and Final AL Routes) employed by the older PTNs. Some levels of toll offices in the older types of PTNs may now be unnecessary if one uses the new signaling system 7 (SS7, also called the CCS) and modern switching nodes with large capacities.

Some care must be taken in interpreting the network topology of Figure 3.15, which illustrates only a sparsely connected backbone network; however, this could be very deceiving. Some nodal pairs, although not connected in Figure 3.15, may be actually connected together by a large fiber trunk bundle that bypasses many switching nodes en route. This is a common practice for connecting two metropolitan cities with very heavy traffic between the two. Although the route may be very long compared to the line-of-sight distance, it avoids the cost of additional right-of-way routes. However, the reader must always be careful

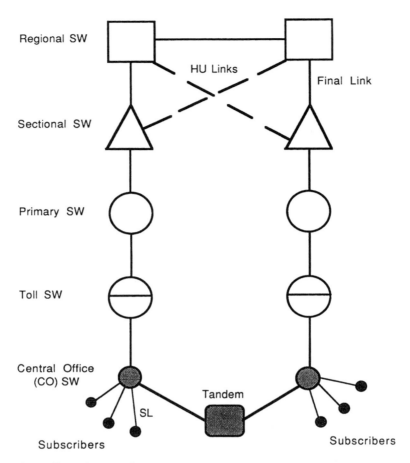

Figure 3.14 Topology of a conventional public telephone network.

when interpreting the network topology of Figure 3.15. *There are times when even a good picture can be very deceiving.*

A typical voice network must perform the following applications:

1. Request analysis (for class-of-service or COS, etc.) and provision of the dial tone

2. Called-party address analysis

3. Pathfinding and routing for single- and multiaddressed calls

4. Circuit switching function

5. NMC functions

Some of the applications (e.g., PABX, Data LAN, CO, and NMC) may be distributed over a subset of network nodes. Other applications may be handled by

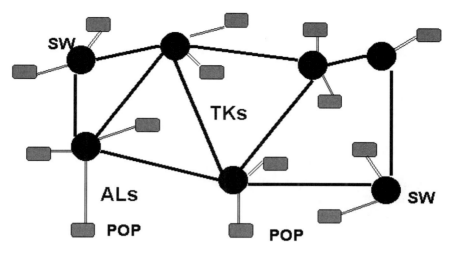

Figure 3.15 Topology of a modern public network.

several subnodes within a network node. For example, the subscriber-line scanning function may be handled by several line modules, each controlled by either a powerful microprocessor or a minicomputer. Each subnode utilization should be modeled independently as a function of the applications handled. Each subnode requires an operating system for allocating the central processing unit (CPU) or input/output device time to the various functions handled. The time spent by the applicable operating system for each subnode or resource must be accounted for, along with the time spent on all the functions handled. Using this approach, find the subnode or resource with the highest utilization factor applicable for the required peak period traffic. If it is higher than 60 percent, then redesign the resource to satisfy the constraint. Some of the critical resources are CPU time, and hard disk or RAM space. A simple design criterion to be followed is: "The lower the utilization factor of the busiest subnode, the higher the robustness of the network node."

3.2.1 Nodal Throughput Analysis

A throughput analysis deals with computing the number of calls that can be handled by a network node. For example, if the resource is the subnode's CPU, the highest utilization is computed as a product of the number of instructions related to the call functions and the average instruction execution time. The number of instructions is computed either by simply adding all the instructions actually executed by a typical call processing environment or by measuring actual program executions. The average instruction execution time is measured by computing the execution time of each instruction and the probability of instruction use. An alternate way for computing the average instruction time is to actually measure the times spent on typical sets of instructions.

The throughput or capacity of a network node can then be computed as follows:

$$C = 3600 \ / \ T_{cp} \qquad\qquad (3.14)$$

where

C = nodal capacity in calls per hour

3600 = number of seconds in an hour

T_{cp} = time in seconds spent on call processing by the busiest subnode CPU

The throughput is commonly determined in terms of either call attempts or calls in progress. Each call attempt generally requires only the servicing of requests and finding a path through the network. Each call in progress also involves the handling of post-call functions such as billing, call-data recording, and so forth. This procedure can be applied to other critical resources such as the hard disk and RAM. By comparing the results, one can discover the performance limiting resource.

3.2.2 Nodal Response Time Analysis

There are two quantities of interest for voice networks: (1) the average time for receiving a dial tone after going off-hook, and (2) the average time required to make a connection. For a local call handled by a PABX, the average time required to make a connection is equal to the time it takes to make a connection between the caller and the called parties. For a call requiring the help of another network node, the average time required to make a connection is simply a sum of the times required to find the best route and make the hardwired connection between the inlet and the outlet ports through the circuit switch (CS).

A CS network node employs P registers as part of common control equipment shared among all the users. The network node scans all the inlet ports for off-hook condition. As soon as it discovers a request for service, it connects a CC register to the inlet port. The first function of the register is to provide a dial tone to the user. The second function of the register is to receive all the dial digits. The third function is to hold the dial digits while the connection is made. The register is then released and it is put back into the register pool. The average time of register occupancy will be called as the call connect time.

The average time to receive a dial tone after going off-hook can be computed using the Erlang-C Equations 3.12 and 3.13, where the mean service time of $\text{Avg}(T_s)$ is the same as the call connect time and the P servers are equal to the number of dial-digit receiving registers. Equation 3.11 provides the probability that delay will exceed any given time t. Since most designers employ the criterion that delay will exceed 1 second only 15 percent of the times the user goes off-hook, Equation 3.11 is quite appropriate. There are times the designer must know the average wait defined as the average elapsed time between the moments the user goes off-hook and the user hears a dial tone. The average wait time can be expressed by Equations 3.12 and 3.13. Figure 3.12 illustrates several useful curves relating $\text{Avg}(T_w)$ to P and A.

The value of $\text{Avg}(T_c)$ or the average connect time is simply the sum of the route-finding and CS connection times. The routing time is the time spent in analyzing the routing tables for computing the best route. The CS connection time is the time spent in making the hardwired connection through the circuit switch. The value of applicable connection time is generally supplied by the vendor.

3.3 Network Performance Analysis Tools

There are occasions when several CS network systems must be compared with one another in terms of their performance. This will require the computation of network throughput and the GOS parameters described as follows:

1. Effective throughput of the entire CS network system

2. Total end-to-end connection time measured as the elapsed time between the moment the last digit is dialed and the moment the destination terminal is made busy

3. End-to-end blocking probability experienced by a call request

3.3.1 System Throughput

One way to compare a CS network system with another is in terms of the total number of calls handled during a busy period. If each CO type network node handled only local calls, the throughput of the entire system will be the sum of all CO nodal throughputs ($C_s = C_1 + C_2 + C_3 + ...$). This will also imply no need of tandem switches or trunks. In the other extreme, if every CO type node handled only tandem calls, the total network throughput Cs must be divided by a factor k that represents the number of identical CS nodes in the path of the call. The actual throughput of a practical CS network system will be somewhere in between the two limits.

3.3.2 End-to-End Connection Time

Another way to compare two CS network systems is to study their end-to-end connection time distribution. The total connect time can be expressed as the sum of (1) processing time delays experienced by each network node in the path of the call, and (2) sum of the delays experienced by the signaling data on each inter-nodal link. Since the delay encountered in each node and on each internodal link is a random variable characterized by an average and a variance, the end-to-end connection time will also exhibit randomness. Its average and variance can be expressed as follows:

$$\text{Avg}(\,T_c\,) = \sum_{i=1}^{N} \text{Avg}(\,T_{in}\,) + \sum_{j=1}^{L} \text{Avg}(\,T_{jl}\,) \tag{3.15}$$

$$\text{Var}(\,T_c\,) = \sum_{i=1}^{N} \text{Var}(\,T_{in}\,) + \sum_{j=1}^{L} \text{Var}(\,T_{jl}\,) \tag{3.16}$$

where
  T_c = total connection time
  Avg($\,T_c\,$) and Var($\,T_c\,$) are the average and variance of T_c
  T_{in} = delay in the ith node
  T_{jl} = delay on the jth internodal link
  N = total number of nodes in the path of the voice call
  L = total number of links in the path of the voice call

In case no single node determines the final value of T_c, the Center Limit Theorem (CLT) should apply (see Feller, 1957). Using the CLT, one can derive several useful expressions for 90 percentile, 95 percentile, 99 percentile, and the 99.9 percentile values of Tc, as follows:

$$T_c\,(90\%) \;= \text{Avg}(\,T_c\,) + 1.28\;\text{SQR}[\,\text{Var}(\,T_c\,)\,] \tag{3.17}$$

$$T_c\,(95\%) \;= \text{Avg}(\,T_c\,) + 1.65\;\text{SQR}[\,\text{Var}(\,T_c\,)\,] \tag{3.18}$$

$$T_c\,(99\%) \;= \text{Avg}(\,T_c\,) + 2.33\;\text{SQR}[\,\text{Var}(\,T_c\,)\,] \tag{3.19}$$

$$T_c\,(99.9\%) = \text{Avg}(\,T_c\,) + 3.09\;\text{SQR}[\,\text{Var}(\,T_c\,)\,] \tag{3.20}$$

The preceding methodology provides only approximate answers for a large network system during the planning phase. The exact solution is always unavailable either due to the unpredictability of exact traffic intensities or due to lack of closed-form mathematical solutions. Consequently, the approximate methods should be more than adequate for most cases.

3.3.3 End-to-End Blocking Probability

Two CS network systems can also be compared in terms of their end-to-end blocking probability. The blocking probabilities Bs experienced in the N network nodes in the path of a call can be computed as follows:

$$B_s = 1 - \left[\prod_{i=1}^{N} B_i \right] = 1 - \left[\, B_1 * B_2 * .... * B_N \,\right] \tag{3.21}$$

Or one can get a simpler relation for $N*B_i < 1$:

$$B_s \approx B_1 + B_2 + ... + B_N = \sum_{i=1}^{N} B_i \tag{3.22}$$

Exercises

3.1 Given the traffic load of 10 calls per minute, compute the number of servers required for the call holding times of 120, 240, and 360, and for the required blocking factor of 0.01.

3.2 For A=1, 10, and 60, find the number of servers required for the blocking factors of 0.1, 0.05, and 0.01 (a total of nine answers).

3.3 Compute the efficiency of circuit utilization for A=1, 10, and 100 for the blocking factors of 0.1, 0.05, and 0.01 (a total of nine answers).

3.4 Plot the curves relating ρ to N for B=0.01, .03, .05 and 0.1 for the range 1<N<1000.

3.5 Using the "FindB-FiniteM" item of EcoNets's Analysis Menu, plot (N vs. A) curves for M=5, 15, 25, 35, 400 for B=0.01.

3.6 Using the "FindB-FiniteM" item of EcoNets's Analysis Menu, plot (N vs. A) curves for M=5, 15, 25, 35, 400 for B=0.03.

Bibliography

Feller, W. 1957. *An Introduction to Probability Theory and Its Applications.* New York: John Wiley.

Collins, A. A., and R. D. Pederson. 1973. *TELECOMMUNICATIONS: A Time for Innovation.* Dallas: Merle Collins Foundation.

Siemens Aktiengesellschaft. 1970. *Telephone Traffic Theory.* Munich.

Syski, R. 1960. *Introduction to Congestion Theory in Telephone Systems.* Edinburgh and London: Oliver and Boyd.

4

Analysis and Design Tools for Data Networks

4.1 Traffic Analysis Tools

Whereas traffic analysis tools for voice traffic are required to compute the values of nodal/system blocking or connection times, traffic analysis tools for data traffic are required to compute the nodal/system response times. Whereas traffic analysis for voice traffic is used to size AL and trunk bundles in a voice network, traffic analysis for data must also be used to size the AL and trunk bundles in a data network. In order to develop meaningful analysis tools for data, one must first understand the behavior of data traffic.

Data traffic differs from voice traffic in many significant ways. Whereas each voice call is characterized by its well-behaved service time, T_s, data traffic is generally very bursty in nature. Data traffic generally involves the transmission of messages, both small and large, as required by the associated protocol defined as the set of rules for interchanging data between two communication devices. See Figure 4.1 for an illustration of the traffic flows between the terminal controller and a host computer for the asynchronous (or ASYNC) and Binary Synchronous Communication (BSC) protocols, respectively. In many cases, some of the messages have to be created by human operators while the connection (virtual or permanent) is being maintained. Some other messages may have to be retrieved by a host computer from its memory before transmission. In some applications,

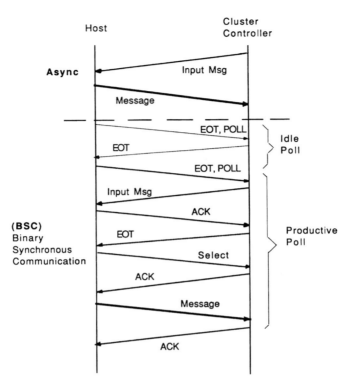

Figure 4.1 ASYNC and Binary synchronous communication protocols.

large files are transmitted continuously for a long time. Some voice networks also allow the transmission of data on circuit switched paths.

Almost all data applications employ store and forward techniques involving delays. Consequently, data traffic analysis tools can only be expressed in terms of queuing theory. See Figure 4.2 for a representation of two queues within a classical PS/MS node. One queue represents the input process and the other represents the output process. The Open Systems Interconnection (OSI) may require up to six queues in series in a single PS node. See Appendix B for the discussion of OSI. Most designers are afraid of queuing theory. They should not be. The methodology presented in following sections is actually simpler to apply than that developed for the voice network.

Basically two types of queuing models are of interest here. Each model assumes a unique traffic model. The first type is characterized by random arrivals, N servers, and negative exponential distributions of the service times, Ts. It is defined as the M/M/N type traffic with infinite sources. The second type of data traffic is characterized by random arrivals, N servers, and constant service times. It is defined as the M/D/N type traffic with infinite sources. These two types of traffic determine delays in the queue within a PS/MS nodes and data networks. Textbooks of Feller (Feller, 1957), Martin (Martin, 1972), Syski (Syski, 1960), and

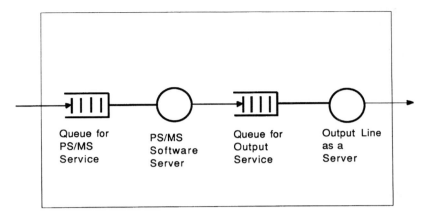

Figure 4.2 Queuing model of a typical PS/MS with two-stage service.

Schwartz (Schwartz, 1987) should be consulted for an in-depth treatment of the concepts used in this chapter. We will start with a queuing model with N = 1 since it is easy to comprehend.

4.1.1 The M/M/1 Queue Analysis

Considering that time in queue (T_q) is the sum of waiting time and the service time, and using ρ for server utilization (or loading) and Avg (T_s) for the average value of service time—it is also sometimes expressed as E (T_s)—one can express the average value of T_q for a single server as follows:

$$\text{Avg } (T_q) = \text{E } (Tq) = \text{Avg } (T_s) \, / \, (1\text{-}\rho) = 1 \, / \, (\mu - \lambda) \tag{4.1}$$

where
$\rho = \lambda * \text{Avg } (T_s) = \lambda * \text{E } (T_s) = \lambda * \text{EMD}$
λ = message arrival rate
$\mu = \lambda / \text{Avg } (T_s)$

The standard deviation of T_q, equal to Std.Dev.(T_q), is also equal to Avg(T_q) or

$$\text{Std.Dev. } (T_q) = \text{Avg } (T_s) \, / \, (1\text{-}\rho) = 1 \, / \, (\mu - \lambda) \tag{4.2}$$

The average number of items in the queue (equal to L) and the average wait time (equal to T_w) are also of interest.

$$\text{Avg } (L) = \rho \, / \, (1 \text{-} \rho) \tag{4.3}$$

$$\text{Avg } (T_w) = \rho \, / \, (\mu - \lambda) = \rho (\text{Avg } (T_s)) / \, (1\text{-}\rho) \tag{4.4}$$

4.1.2 The M/D/1 Queue Analysis

The time spent in the queue is the sum of time in waiting and getting served. Using ρ for the server utilization and T_s for fixed service time, one can express the average time spent in queue as follows:

$$\text{Avg}\ (\ T_q\ ) = E\ (\ T_q\ ) = T_s\ (\ 2 - \rho) \ / \ 2(\ 1 - \rho) \tag{4.5a}$$

The standard deviation of T_q can be expressed as

$$\text{Std. Dev.}\ (\ T_q\ ) = [T_s \ / \ (\ 1 - \rho)]*\text{SQR}[\rho/3 - \rho^2/12] \tag{4.5b}$$

where

SQR (x) = Square root of x

The average number of items in the queue and average wait can be expressed as follows:

$$\text{Avg}\ (\ L\ ) = [\rho^2 \ / \ 2(1 - \rho)\ ] + \rho \tag{4.6}$$

$$\text{Avg}\ (\ T_w\ ) = (\ \rho * T_s)\ \ / \ [2\ (1 - \rho)] \tag{4.7}$$

See Figure 4.3 for a plot of $E(\ T_q\ )$ and Std. Dev.$(\ T_q\ )$ for both the M/M/1 and M/D/1 traffic and queuing models. It should be obvious that the M/M/1 and M/D/1 yield the upper and lower bounds of delays, respectively. It should also be clear that there is no great difference between the two values of T_q for these queue disciplines in the range of $0 < \rho < 0.6$. Since actual queuing disciplines are never known exactly for data networks, the amount of attention given to high accuracy is unwarranted. The use of the aforementioned two queuing disciplines for either computing or interpolating an approximate value of delay should be adequate for most engineering practices.

4.1.3 The M/M/N Queue Analysis

Most queues in a large network system employ more than one server (i.e., $N > 1$) to either handle a large number of requests for data transmission or minimize delays experienced within the network nodes. For the M/M/N type queue characterized by random arrivals, negative exponential service times and N servers, one can define the performance of such a system by specifying the probability of delay exceeding a given value or by requiring the average wait time in the queue. Denoting the traffic intensity by A (equal to $N*\rho$ where ρ is average utilization of each server) and average service time by T_s, one can express several useful relations as follows:

$$\text{Prob.}(\text{delay} > T) = (P_0) * \text{Exp}[- (N - A)T / T_s] \tag{4.8a}$$

$$\text{Avg}(T_w) = T_s / (N - A) \text{ ...delayed transactions only} \tag{4.8b}$$

$$\text{Avg}(T_w) = T_s * P_0 / (N - A) \text{....all transactions} \tag{4.8c}$$

$$\text{Var}(T_q) = [T_s^2 / (N - A)^2] * [P_0 (2 - P_0) + (N - A)^2] \tag{4.8d}$$

where

T_s =Average service time, and

P_0 is the probability that delay is greater than zero. Its value can be derived by using Equation 3.11 for given values of N, A, and T_s.

Equation 4.9 is the same as used for an ACD systems where N agents handle the incoming traffic intensity represented by A. Figure 3.11 illustrates some useful curves relating N (or P) to A for a typical design constraint of delay exceeding 10 seconds for only 15 percent of the time. See Figure 3.12 for some curves relating Avg(delay) to A for some useful values of server utilizations. Figure 4.4 illustrates another set of useful curves that relate Prob.(delay > T) to T expressed as a multiple of T_s and for several values of N and average server utilization (as equal to A/N) of 0.667.

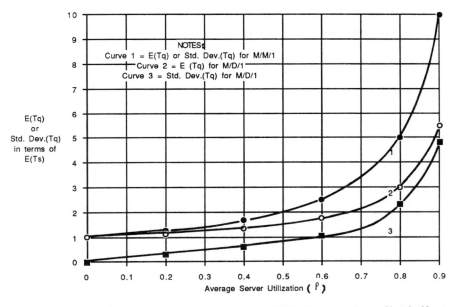

Figure 4.3 Useful curves for M/M/1 and M/D/1 queuing disciplines.

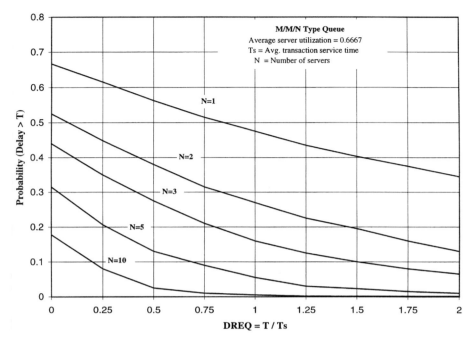

Figure 4.4 Probability of delay exceeding required delay (DREQ) versus DREQ and some useful values of the number of servers (N) for the M/M/N queuing discipline and average server utilization of 0.66667.

4.1.4 The M/D/N Queue Analysis

The analysis of an M/D/N type queuing system characterized by random arrivals, fixed service time, and N servers is quite complex. See Appendix A for a computer solution to derive the probability of delay exceeding zero (equal to P_0) and any value T. The published literature dealing with traffic theory does not provide simple expressions for the first two moments of the time spent in the M/D/N type queue. The author has simulated the M/D/N type queue for a large number server loads (ρ) and N. He has discovered that the following expressions agree very well with the simulation results.

$$\text{Prob.}(\text{delay} > T) = (P_0) * \text{Exp}[-2(N-A)T / T_s] \qquad (4.9a)$$

$$\text{Avg}(T_q) = \{1 + [P_o / 2(N-A)]\} T_s \qquad (4.9b)$$

$$\text{Var}(T_q) = \{[1 / (N-A)^2]*[P_o / 3 - P_o^2 /12]\} T_s^2 \qquad (4.9c)$$

The plots of Figure A.1 of Appendix A can be useful for computing P_0. If the plots of Figure A.1 are not handy, one can approximate P_0 for M/D/N as equal to about 6 percent lower than P_0 for M/M/N for N > 2. Figure 4.5 illustrates several useful curves that relate Prob. (delay >T) to T as a multiple of T_s and several values of N and an average server utilization (as equal to A/N) of 0.667. The reader may like to observe some similarities between Equations 4.9b, 4.9c and Equations 4.5a, 4.5b, respectively. For N=1, both A and P_0 reduce to ρ.

The curves of Figures 4.4 and 4.5 can be useful for designing systems with the desired response time constraints. In case the traffic characteristics fall in between the worst case (i.e., M/M/N) and the best case (i.e., M/D/N), it should be adequate to interpolate the results for design guidance.

The EcoNets software allows the designer to model both the M/M/N and M/D/N queues for any pair of values, A and N. The ratio A/N defines the average utilization (ρ) of each server. Items MMN and MDN of the Analysis menu can be used to derive the values of Prob.(delay > T) for T=0, $T_s/4$, $T_s/2$, $0.75*T_s$ and T_s. These Analysis items also provide the values of two moments of T_q—i.e., Avg (T_q) and Std. Dev.(T_q)—for the two queue disciplines.

4.1.5 The Multipriority Queue Analysis

No priorities were considered in the queuing formulations presented in the preceding subsections. All items, which are serviced according to the first-in-first-out queuing discipline, are assumed to have the same priority. There are applications

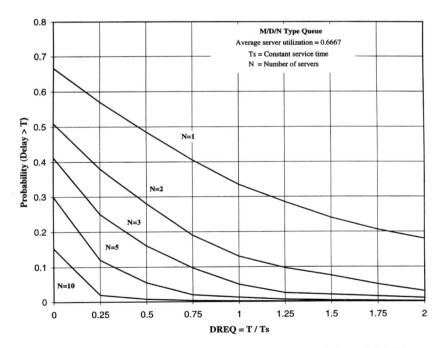

Figure 4.5 Probability of delay exceeding required delay (DREQ) versus DREQ and some useful values of the number of servers (N) for the M/D/N queuing discipline and average server utilization of 0.66667.

that require the computation of average wait times for items characterized with R (where R is greater than one) priorities. Generally, two cases are of interest:

1. No interrupts are allowed for lower priority items when a higher priority item arrives. Only after an item is fully serviced can an item of higher priority in queue be serviced.

2. Interrupts are allowed for lower priority items when a higher priority item arrives. The so-called preemptive-resume priority queuing discipline is employed.

The queuing analysis becomes quite complex for most practical situations. Martin (Martin, 1972) has enumerated a limited number of solutions for the M/M/1 type queuing model. One can also use the MultiPriority Queue item of the Analysis menu of EcoNets as described in Chapter 6 to study the average wait times in a queuing system characterized by up to five priorities. The model is valid only for the non-interrupt, first-in-first-out service discipline.

4.2 *Nodal Performance Analysis Tools*

There is a definite need to compare data communication systems. It can be done in terms of system performance. But that cannot be done until one computes the performance of each node. A network system consists of several types of network nodes. Some provide users with an access to the network services. Some others provide transport to the user and/or supervisory traffic. One can characterize any network node in terms of two types of performance parameters:

1. Throughput in terms of packets or messages or bits handled per second by the node

2. Delays experienced by each packet or message while passing through the node

4.2.1 Nodal Throughput Analysis

The throughput of a network node can be described in terms of the bits or packets or messages handled per unit time. The contents of Section 3.2.1 dealing with nodal throughput for a CS network also apply to the PS or MS nodes. The values of nodal throughputs are generally provided by the vendor. System analysts working for an equipment manufacturer must have a methodology for computing the nodal throughput that appears in the product brochures. The following discussion should help one achieve that purpose.

In order to compute the throughput of a network node, one must first model all the relevant subnodes within a network node and the work assignment for each. Whereas some subnodes may only involve microcomputers or minicomputers, other subnodes may involve both computers and peripheral devices. The subnode showing the highest utilization (ρ) should determine the nodal throughput for a peak hour. If the most active subnode reflects an excessive utilization (i.e., $\rho > 0.60$), such a subnode must be redesigned by using a faster computer and/or a peripheral device, and/or a better software module. While computing the nodal throughput, one must never forget the response time requirements. Only the highest throughput that guarantees a given response time should be considered. The task of computing nodal throughput is generally tedious and time consuming.

The nodal throughput for a PS node is generally dependent upon either the CPU speed or the limited speed of the input/output links. The availability of random access memory (RAM) used to be a critical factor in the past. Cheap VLSI devices have changed all that. Modern vendors provide PS nodes with a capacity to switch 1,000 to 100,000 packets per second. Assuming 1,000 bits per packet, a throughput rate of 1Mbps to 100Mbps can be reached if and only if input/output links have an adequate total capacity. ATM switches provide even higher throughputs.

The nodal throughput of a message switch is generally determined by the disk capacity or speed. Each message is divided into blocks. A header block is prepared for each message and chained to the first message block and other message header blocks. The remaining message blocks are also chained together. The addresses of each header block are generally preassigned when the last block of message is received on a given link. Each disk read/write request can be handled according to the FIFO scheme. This will result in a low nodal throughput, since each disk access will encounter full disk latency (equal to half the disk rotation time). The throughput of an MS node can be increased by batching disk read/write requests and ordering these according to the rotational latency before servicing. While a previous batch is being served, a new batch is being created. A steady state is reached when the size of the previous batch and the next batch of disk requests become equal during a peak traffic period. See Figure 4.6 for an illustration of such a state. The larger the batch, the higher the throughput and the greater the delay through the node. Sharma (Sharma, 1968) has derived an analytical solution for the batching and ordering approach based on the combinatorial models developed in Chapter 2. His paper also considers the effects of arm-position mechanism of the disk. See Figure 4.7 for some useful curves relating the number of disk rotations to the batch size, and some useful values of cells per disk track (or cylinder). These curves imply a significant increase in throughput. For example, 100 items will normally require about 50 disk rotations for the case of no batching. For the case when there are 32 blocks in a sector, 100 items will require only 6 disk rotations when batched. That represents an eightfold increase in throughput. To handle larger batches, one should consult Beizer (Beizer, 1978) who presents additional but approximate expressions for very large batches.

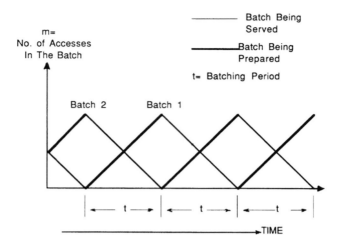

Figure 4.6 Batching, ordering, and servicing of disk accesses.

4.2.2 Nodal Response Time Analysis

One can compare the performance of two network nodes in terms of their response times. In order to analyze the average delay experienced in a network node, one must first model the queue as shown in Figure 4.2. The queuing discipline, number of servers, and average service time of each message or packet must be known before computing the response times. There are basically four different techniques for sharing transmission facilities by the traffic generated by subscribers.

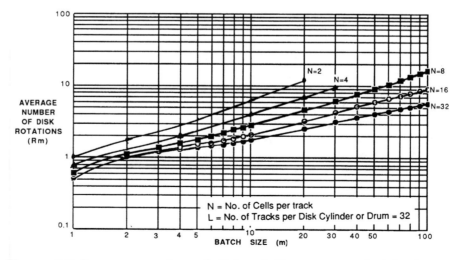

Figure 4.7 Average number of disk rotations versus batch size, and some useful disk parameters.

1. A traditional multipoint network consists of a host, a front end (FE), and several control units (CUs) (also called controllers) sharing a single transmission line. The host and the FE act as one store-and-forward message switching node. Even the shared link handles only one conversation at a time. No packetization of messages is allowed. The nodal response time analysis is generally computed for the host node only since it does the polling. The controller is a slave to the host.

2. Several types of PS node are available that can be connected to one another according to either a pure multidrop or star topology, or a mixed multidrop/star topology. However, these configurations yield very high utilizations of communication facilities since every packet is subjected to the store-and-forward mechanism using the FIFO (first-in-first-out) queue discipline. The asynchronous time division multiplexing (ATDM) technique is used for the output process in all PS nodes. This results in full link sharing and traffic concentration. The local ports and long-haul trunks are generally connected to the node according to the star topology. All PS nodes employing the X.25 protocol obey these rules. Figure 4.8 illustrates such a nodal topology.

3. An interesting example of a virtual packet switching system is AT&T's ISN/Datakit system. Each node serves several local ports, access links, and long-haul trunks for information exchange, link/trunk sharing, and trunk concentration. The concepts of a shared (or virtual) packet and a shared contention bus are employed to reduce the nodal response time for interactive traffic. The topology of its PS node is illustrated in Figure 4.9, involving several communication modules for serving CPUs, local async ports, access links to concentrators, and trunks to other PS nodes. Most modules transmit on a shared bus (with contention) and receive on the broadcast bus. The switch listens on the contention bus and transmits on the broadcast bus. Fraser (Fraser, 1983) describes the operation of such a node in some detail, but without developing a generalized equation for computing the nodal response time.

4. A modern PS node generally employs the mechanism of fast packet switching. Each node serves several local ports and one fully shared trunk for concentration and information exchange. See Figure 4.10 for an illustration of the nodal topology forming a basis for the Doelz network architecture as described by Doelz and Sharma (Doelz and Sharma, 1984). This topology also applies to other technologies (e.g., Frame Relay and ATM) based on fast or virtual packet switching. The time on the shared ESR trunk is divided into very small slots. Each time slot equals a *virtual packet* transmit time. The PAD unit enclosed within each netlink node creates protocol independent virtual packets from each active source. The PAD performs the reverse function for the virtual packets arriving for a local port. The en route traffic on the trunk has priority over the local port traffic queued for the trunk. The locally generated packets are queued and sent on empty slots. An en route

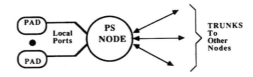

Figure 4.8 Simplified topology of a PS/MS node.

packet not destined to a local port is not subject to the S/F mechanism and therefore bypasses the node. Such a capability provides a significant improvement over the traditional PS technology based on the X.25 protocol. Doelz's ESR, Frame Relay, and ATM technologies employ such a capability.

Now we will study the nodal response times for each of the described nodal designs.

4.2.2.1 Response Time Analysis for a Traditional Multipoint Node

A typical multipoint network consists of a system node (consisting of a host and a FE), a network of modems, and cluster controller units, each serving a subset of subscriber terminals scattered over a large area. The delays experienced by a subscriber are effectively determined by the system node that controls all the polls. Although the delays seem to obey the M/M/1 model, the solution is rather complex. See Section 4.3.2.1 for a simplified solution due to Schwartz (Schwartz, 1987). The delays experienced in each controller unit are kept at a minimum level by employing the correct hardware and serving a correct number of terminals.

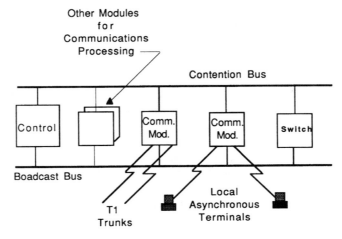

Figure 4.9 Simplified topology of AT&T's ISN PS node.

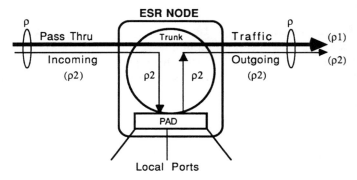

Figure 4.10 Simplified topology of Doelz's ESR/PS node.

4.2.2.2 Response Time Analysis for a Conventional PS Node

A study of Figure 4.8 shows that a typical PS node receives packets from several input channels and puts these packets in an input queue. The PS node then analyzes each packet for addresses and other functions. The processed packets are then put into an output queue. As soon as the output link(s) is available, these packets are transmitted. The queuing discipline and the number of servers should then be determined for each of the two queues. To compute the total delay through the network node, one must add the delays encountered in each of the input and output queues according to the analysis tools provided in Sections 4.1 and 3.3.2. Simple addition assumes that these two queues are stochastically independent. Some 20 years ago, such independence was very hard to achieve due to the lack of cheap RAM storage and powerful microprocessors. For both M/M/N and M/D/N types of queues, it should be easy to predict the expected value of delay (or the first moment) or estimate the 95, 99, 99.9 percentile values of delay. By choosing the correct values of delays in the two nodal queues and assuming independence, it is possible to estimate the probability of exceeding the sum of two values. One can express two useful results as follows:

$$E\ (\ T_q) = \ E\ (\ T_{q1})\ + E\ (\ T_{q2}) \tag{4.10a}$$

or

$$[\ P\ (\ T_{q1})\ +\ P\ (\ T_{q2})\ ]\ \text{is}\ >\ \{\text{a given value T}\} \tag{4.10b}$$

where
$E\ (x)$ = Expected value of x
$P\ (x)$ = Probability of x
$(\ T_{q1})$ = Delay in the input nodal processing queue
$(\ T_{q2})$ = Delay for the output transmission queue

4.2.2.3 Response Time Analysis for the ISN Node

According to Fraser (Fraser, 1983), each packet assembled for the output trunk may be composed of several segments of interactive datastreams from local ports or data from other trunks. This technique is used to minimize the delay associated with assembling a packet/message for slow interactive traffic. Furthermore, the high-capacity shared contention bus and the switch provide very high-speed packet switching.

No generalized equations for computing the nodal response times on each VC can be derived due to their dependence on the exact LAN topology (e.g., number and mix of local line speeds). Fraser (Fraser, 1983) shows that the switching time should be about 250 microseconds per packet. For a 40-byte output packet size, the average delay for the output transmission is about 10 ms, and 0.5 ms for a 60 percent loaded 56Kbps and 1.35Mbps trunk, respectively.

4.2.2.4 Response Time Analysis for a Fast Packet Switching Node

We will derive some useful results for an ESR node since published data is easily available. The results should be applicable to other nodes based on fast/virtual packet switching. Response time of an Extended Slotted Ring (ESR) node can be analyzed for two designs: (1) asynchronous point-to-point whereby each user queues the inquiry message for transmission by pressing the Send button, and (2) virtual multipoint (VMPT) that permits a controlled sharing of the transmission facility among several virtual multipoint circuits through the mechanism of polling. Several conventional multipoint (MPT) circuits can be superimposed on a netlink controlled by the fast packet switching mechanism (i.e., ESR). One should remember that a typical multipoint link in an SNA link is not utilized efficiently. In contrast, the utilization of an ESR netlink can go as high as 90 percent without incurring a severe response time penalty.

Asynchronous point-to-point (or simply ASYNC). Following the nomenclature of Figure 4.10, one can compute the nodal delay as follows:

$$E(T_q) = \text{.........pass-through packet]}$$

$$E(T_q) = [\ \rho_2(1+\rho_1)\ /\ 2(1-\rho)]*T_s \text{......local packet]} \tag{4.11}$$

where

$\rho = \rho_1 + \rho_2$ =total input/output loading of ESR
T_s = fixed slot time on the ESR
ρ_1, ρ_2 are the pass-through ESR and locally generated loads, respectively

The messages are removed from the shared media by the destination nodes. Equation 4.11 was derived jointly by Doelz and Sharma (Doelz and Sharma, 1985). The expression has some similarities to the expressions for queuing delay for an M/D/1 queue such as Equation 4.5. This expression was arrived by studying the results of computer simulations and fitting the results to an analytical formula. The precedence of the incoming line prevented a quick intuitive solution. It should also be mentioned that model is applicable to traffic flows over a Frame Relay or ATM virtual circuit (VC).

Virtual Multipoint (VMPT). The delays for a virtual multipoint (VMPT) circuit with N drop nodes are also of interest. In fact, the ESR architecture allows the sharing of a single transmission media among several virtual circuits, each based on the conventional MPT operation. For a VMPT operation that allows broadcasts, the incoming load is not reduced by the amount of traffic destined for the local ports of a given drop node. Furthermore, the traffic, ρ_2, originated from the local ports of a given drop node adds to the incoming load. In other words, the output load is related to the input loads as follows:

$$\rho_o = \rho_i + \rho_2 \tag{4.12}$$

The relationship between the input traffic and output traffic loads is illustrated in Figure 4.11. The average queuing delay in the jth node for VMPT operation has been expressed by Doelz (Doelz, 1985) as follows:

$$\text{Avg} (T_{qj}) = [1/2N][\rho_i \rho_0][NUM_1 / DEN_1]*T_s \tag{4.13}$$

where

ρ_i, ρ_o are as illustrated in Figure 4.11
T_s = fixed slot time on ESR
N = number of drop nodes on the netlink
$NUM_1 = [\rho_i + \rho_0 + j\rho_i / N]$
$DEN_1 = [1 - (j-1)\rho_i / N][j - \rho_o - j\rho_i / N] - \rho_i \rho_o / N$

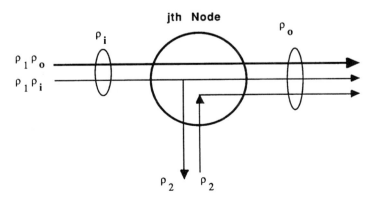

Figure 4.11 Illustration of a VMPT nodal traffic flow in a Doelz's ESR PAD/PS node.

4.3 Network Performance Analysis Tools

There are times when several network systems must be compared with one another. This can be done by comparing their performance in terms of system throughputs and end-to-end response times as defined in Chapter 2.

4.3.1 Network System Throughput Analysis

The nodes of a wide area network (WAN) are generally interconnected by non-shared links according to a well-defined topology. If every node of a WAN handled only the local traffic, the throughput of the system will be equal to the product of nodal throughput and the number of PS central offices. Such an extreme case will require no internodal trunks for user data. If each PS node handled only the tandem traffic, then the system throughput of a WAN can be approximately expressed as follows:

$$C_s = N * C_s / L_p \qquad (4.14)$$

where
C_s = system capacity in packets or messages per second
N = number of identical PS/MS nodes in the network system
L_p = average number of nodes in the path of a packet or message

The reader should note that Equation 4.14 applies only to data networks with identical PS/MS nodes. The data networks usually employ two types of nodes: (1) concentrator nodes (i.e., PADs) that serve subscriber lines only, and (2) large packet/message switching nodes. Equation 4.14 also assumes that a sufficient number of concentrator nodes are available to handle the traffic intensities required to achieve the network capacity, Cs. The voice networks, on the other hand, employ two types of switching nodes: (1) CO type nodes for handling the originating and destination traffic, and (2) tandem nodes for handling only the tandem traffic.

The nodes of a local area network (LAN) are generally connected by means of a shared media. In case LAN nodes are connected by separate links, the LAN throughput can be computed using the same methodology as used for WANs. The throughput of a typical LAN depends upon the shared media capacity (R_0), number (N) of stations, interface delay T_{int}, and type of access employed. Stuck (Stuck, 1983) describes a simplified methodology for computing the throughput of a LAN based on the three well-known access techniques and for two cases of interest: Case A, when only one out N stations are active, and Case B, when all stations are equally active. The peak throughput rates for Token Ring, Token Bus, and CSMA/CD LANs can be expressed for two special cases of interest as follows:

1. Token Ring: $R_p = L_m / [T_{msg} + Tp + N*T_{int}]$ bps......CASE A (4.15)

   $R_p = L_m / [T_{msg} + Tint + T_p /N]$ bps......CASE B (4.16)

2. Token Bus: $R_p = L_m / [T_{msg} + N*(T_{int} + T_p)]$ bps.......CASE A (4.17)

$R_p = L_m / [T_{msg} + T_{int} + T_p]$ bps...............CASE B (4.18)

3. CSMA/CD: $R_p = L_m / [T_{msg} + T_{ifg}]$ bps....................CASE A (4.19)

$R_p = L_m / [T_{msg} + T_{ifg} + (2e-1)*(T_s + T_j)]$bps CASE B (4.20)

where

R_p = peak average throughput rate in bps

T_s = slot/ frame time = $(2T_p + T_{int})$

T_j = jam time for CSMA/CD = $48/R_o$

T_p = Propagation time on the shared media

T_{msg} = message transmit time over the shared media = $(Lm+96)/R_o$

T_{ifg} = interframe gap time for CSMA/CD =9.6 microseconds

T_{int} = interface delay time = 4 microseconds (token bus)

=$1/R_o$(for a LAN based on token ring)

L_m = bits per message

e = 2.71828

R_0 = transmission capacity of the shared media in bps

The reader should consult the works of Schwartz (Schwartz, 1987), Stuck (Stuck, 1983), and Stallings (Stallings, 1984) for additional insight into the LAN operations and related design parameters. See Figures 4.12 through 4.15 for the plots relating net output data rates to transmission capacity, R_o, in bps for all three LAN architectures, 100 stations, two message lengths L_m, and cases A, B of interest. The capacity of the shared medium (R_o) is varied between 0 and 24Mbps. The value of 2e represents a crude approximation for the number of retransmissions required to achieve a correct transmission. The curves of Figures 4.12 to 4.15 were obtained by converting Equations 4.13 to 4.18 into a spreadsheet and varying some of the design parameters. The results show that a token ring is generally the most efficient among the cases considered. The reader should realize that LANs based on the integrated voice and data PABX can also provide interstation communication at lower (<64Kbps) data rates than those obtained on the most popular LAN architectures. It may also be interesting to observe that LANs using fiber optical buses can provide interstation communications at very high (<= 1000Mbps) data rates. For such environments, the relative merit of a token ring may not be important. In that case, the cost of interface devices may be the criterion of choice. The interface cards for a token ring/bus are generally more expensive than those for CSMA/CD. The cost of an interface card for the IVD PABX is generally hard to quantify for identical LAN operational environs.

4.3.2 Network Response Time Analysis

On occasions, a designer must compare several data networks in terms of their end-to-end response times. There are four types of WANs of interest for performing end-to-end response time analyses:

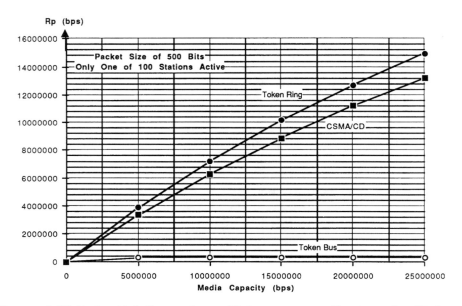

Figure 4.12 Potential throughput (R_p) versus media capacity (R_o) and some useful LAN design parameters: 500 bits per packet and only 1 of 100 stations active.

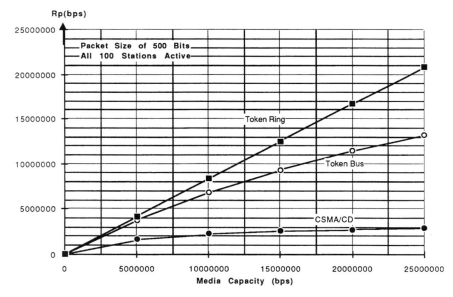

Figure 4.13 Potential throughput (R_p) versus media capacity (R_o) and some useful LAN design parameters: 500 bits per packet and all 100 stations active.

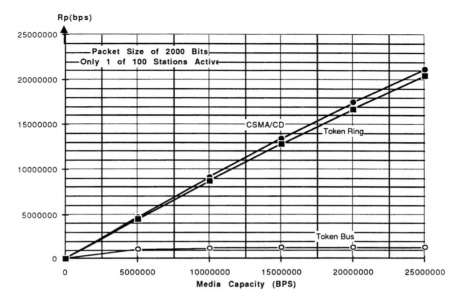

Figure 4.14 Potential throughput (R_p) versus media capacity (R_o) and some useful LAN design parameters: 2000 bits per packet and only 1 of 100 stations active.

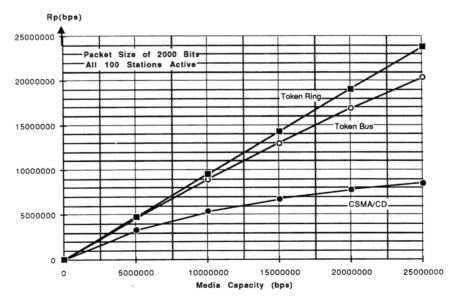

Figure 4.15 Potential throughput (R_p) versus media capacity (R_o) and some useful LAN design parameters: 2000 bits per packet and all 100 stations active.

1. Private networks based on IBM's SNA technology using the traditional multidrop topology

2. Public networks employing PS nodes and PADs (cf. X.25)

3. Private networks using AT&T's ISN/Datakit nodes

4. Public data networks based on fast packet switching (e.g., ESR, Frame Relay, and ATM)

4.3.2.1 End-to-End Response Time Analysis for a Classical Multipoint Network

According to Figure 4.16, contention for the use of a multidrop link is resolved through the use of a roll-call type polling of N cluster controllers. The cluster controllers are polled either sequentially or according to a preset order. Several types of messages flow between the host and a cluster controller before the host sends the desired message. See Figure 4.1 for an illustration of the communication protocol. One can study a useful quantity defined as the elapsed time between the moment a packet arrives at the controller from a local port and the moment that packet begins transmission toward the host. One can also study the elapsed time between the moment one presses the Send button for the inquiry message and the moment one receives the first packet of the host message. It is equivalent to the expected delay. Schwartz (Schwartz, 1987) has shown that the average access delay is a function of multidrop line loading (ρ), the number of drops (N), the average message arrival rate, λ, the second moment of message duration, $Avg(m^2)$, and the system walk time (W) as shown here:

$$E\ (D) = (W/2)\ [1\text{-}\rho/N]/(1\text{-}\rho) + [N\lambda\ Avg(m^2)]\ /\ 2(1\text{-}\rho) \qquad (4.21)$$

where

$$W = 4N*T_{np} + 2N\ *T_{pm} + 2N*\ T_m + \tau\ (N+1)/2 \qquad (4.22)$$

where T_{np} is the nodal processing time for each poll message, T_{pm} is a fixed poll message transmit time, T_m is modem time (which should be twice the half-modem time used for analyzing ESR netlinks), and τ is the total two-way, round-

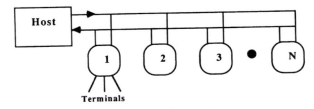

Figure 4.16 Model of multipoint polling for a network with *N* controllers.

trip propagation delay on the netlink. The factor of 2 implies the two directions of data flow as shown in Figure 4.1. The factor of 4 implies four nodal processing functions occurring in the two nodes related to each idle poll. The propagation delay assumes that the N controllers are distributed equally on the netlink.

For the case of hub polling where each controller passes the poll to the next controller after being serviced, the value of W can be expressed as follows:

$$W_h = 4N*T_{np} + 2N*T_{pm} + 2N*T_m + \tau \tag{4.23}$$

By substituting the value of W_h in Equation 4.23, one can solve for the expected value of access delay for the case of hub polling.

4.3.2.2 End-to-End Response Time Analysis for a Public PS Network (PSN)

A typical service on a PSN involves the establishment of a virtual circuit prior to the transmission of data packets associated with the protocol messages. For example, the ASYNC protocol will require an inquiry message and a reply from a host. Another example deals with the transmission of a data file from one host to another. For most applications, one can define an end-to-end response time as the elapsed time between the sending of the first character of the message or packet and the receiving of the first packet of that message by the destination. The end-to-end communication path is shown in Figure 4.17. Figure 4.18 shows the various processes/queues in every network PS node.

The end-to-end response time is the sum of end-to-end queuing delays and fixed system delays. The end-to-end queuing delay (T_q) can be expressed as follows:

$$\text{Avg}(\,T_q\,) = \text{Avg} \sum_{i=1}^{N} (\,T_{in}\,) + \text{Avg} \sum_{j=1}^{L} (\,T_{jl}\,) \tag{4.24}$$

$$\text{Var}(\,T_q\,) = \sum_{i=1}^{N} \text{Var}(\,T_{in}\,) + \sum_{j=1}^{L} \text{Var}(\,T_{jl}\,) \tag{4.25}$$

where
T_q = end-to-end queuing delay
Avg(T_q) and Var(T_q) are the average and variance of T_q
T_{in} = delay in the ith node (input process)
T_{jl} = delay on the jth internodal link (output process)
N = number of nodes in the transmission path
L = number of internodal links in the communication path

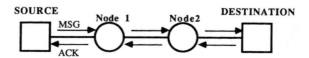

Figure 4.17 Model of a communication path in a PS network.

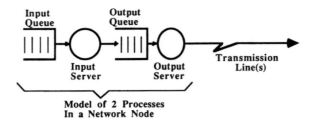

Input Queue

Output Queue

Input Server

Output Server

Transmission Line(s)

Model of 2 Processes
In a Network Node

Figure 4.18 Queuing models within a PS node.

The fixed system delay, T_{fs}, consists of modem delays (if modems are used), propagation delay, and input/output port delays. End-to-end fixed system delays can be expressed as follows:

$$T_{fs} = T_{pg} + NT_{pt} + NT_m \qquad (4.26)$$

where

T_{pg} = propagation delay for the entire path length
T_{pt} = processing delays (buffering for ACK/NAKs/packetization/switching)
T_m = modem delay if modems are employed
N = number of PS nodes in the path of communication

For a PSN, user ports are generally served by a PAD. For that case, the PAD node should be added to both ends of the communication path. Equation 4.26 assumes identical PS nodes and user ports. If not, the processing delays must be computed for the input, the output, and en route nodes independently. The end-to-end response time can be represented as the sum of $Avg(T_q)$ and T_{fs}. Since there is no such thing as a typical communication path and there are too many variables, it is quite difficult to present useful curves. The reader is advised to develop a meaningful model for each communication path and solve for end-to-end response times. In a succeeding section dealing with actual case histories, comparisons between different systems will be made. The expression for end-to-end response time can be utilized to compute another useful response time defined as the elapsed time between the moment the user sends an inquiry to a host and the moment it receives the first character or packet of the response message. We can define it as the system response time expressed as follows:

$$T_{sr} = 2(T_q + T_{fs}) + HTT = \text{system response time} \qquad (4.27)$$

where
HTT = host think time spent in analyzing the user request and receiving the response from its secondary storage

The quantity T_{sr} is sometimes called the bid-to-start (BST) time. Furthermore, the quantity $2(T_q + T_{fs})$ is also expressed as the block/packet turn-around time, T_{ta}.

4.3.2.3 End-to-End Response Time Analysis for AT&T's ISN PS Network

The methodology for computing the end-to-end and system response times for a private network based on AT&T's ISN is identical to that presented in the previous subsection. The only difficulty lies in computing the nodal delays. As mentioned earlier, no generalized equation for computing the delays in an ISN node can be employed to compute the delay through a node serving an arbitrary number of subscriber lines with varying capacities and loads. Consult Section 4.2.2.2 for the recommended nodal delays for an ISN environment. These recommended values can then be added together in a manner as shown in Equations 4.24, 4.25, 4.26, and 4.27 to compute the desired end-to-end response times.

4.3.2.4 End-to-End Response Time Analysis for Fast Packet Switched Networks

It is quite an involved process to derive rigorous expressions for end-to-end response times for all types of FPS networks since many of the standards and the extent of applications are still undefined. Fortunately, significant work has been done on the ESR fast packet switching technology. We will consider only two popular operations. The Asynchronous point-to-point operation has many similarities to those for Frame Relay and ATM, whereby the nodal delays obey the queuing model of Figure 4.10. The Virtual MultiPoint (VMPT) operation may also have similarities to ATM and other architectures whereby several data streams can be superimposed on a single physical media while obeying the nodal delay model of Figure 4.11.

Case 1. Asynchronous point-to-point operation (ASYNC).

Doelz (Doelz, 1985) has derived expressions for the average of turnaround queuing time, T_{aq}, and the bid-to-start time, BST, for the ASYNC operation as follows:

$$T_{aq1} = [\rho_0 / 2(1 - \rho_0) + [NUM_2 / DEN_2]*T_s \tag{4.28}$$

$$BST = T_{ta} + T_{it} + T_{ir} + T_{rt} + T_{fc} + T_{rb} + HTT + T_{sd} \tag{4.29}$$

where
$NUM_2 = [\rho_i / N][1 + (N-k) \rho_0/N + (k-1) \rho_i/N]$
ρ_i, ρ_0 = the actual input/output loads at the node
$DEN_2 = 2[1 - (N-k) \rho_0/N - k(\rho_i/N)]$
N = number of drop nodes (not including the node with host port)
k = kth node from which the turnaround queuing delay is measured
T_{ta} = the turnaround time on the ESR as defined in Equation 4.36
T_{it} = inquiry transmit time on the user port
T_{ir} = inquiry receipt time on the host port
T_{rt} = the effective time to transmit a response block on the host port
T_{fc} = the time to receive first response character on the user port
HTT = host think time
T_{sd} = the shipping delay

See Figure 4.19 for a timing chart showing the events occurring on the transmit (T) and receive (R) sides of the user (DTE or Destination) and host ports (or Source), and the two associated ESR (or PS) nodes for the ASYNC operation in a network path as illustrated in Figure 4.17. The durations of the two block transmission times are determined by the capacities of user ports (or SL) and the shared netlinks (or AL). Figure 4.19 assumes a 1:4 ratio between the two capacities. It should be emphasized that timing charts of Figure 4.19 are extremely useful in modeling the turnaround times T_{ta} in each and every complex data communication network.

Case 2. VMPT operation.

This application allows the superimposition of several individual virtual multipoint circuits on the same shared physical media. Although the following model has been developed for ESR applications only, it should be useful for Virtual LANs and future ATM applications. Each virtual MPT circuit can be designed to handle N_{cu} logical control units. The total number of physical control units (N) is generally higher than N_{cu}. Doelz (Doelz, 1985) has derived the values of turnaround queuing delay (T_{aq2}) and the bid-to-start (BST) delay for a VMPT operation as follows:

$$T_{aq2} = \{[\rho_0 / 2(1-\rho_0)] + [\ NUM_3 /DEN_3]\}* \ T_s \qquad (4.30)$$

$$BST = EPL + PMST \qquad (4.31)$$

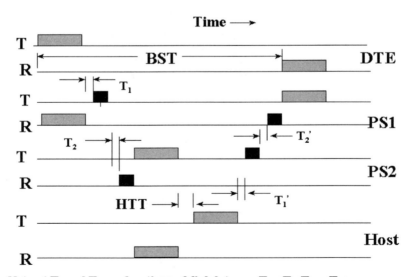

Notes: 1. T_1 and T_2 are functions of link latency, T_{np}, T_m, T_{mm}, T_q or τ,
   2. T and R represent the nodal   Transmit and Receive functions
   3. The SL capacity is about 1/4th of the AL capacity

Figure 4.19 A chart showing the major events occurring at the nodes served by an ESR for ASYNC operations.

Here ρ_i, ρ_0 are as illustrated in Figure 4.11, N = number of drop nodes on the VMPT netlink, $NUM_3 = [\rho_i/N] [1 + \rho_0 + (k-1)\rho_i/N]$, $DEN_3 = 2[1 - \rho_0 - k\rho_i/N]$, k = kth node ID from which the turnaround queuing delay is measured, EPL is the expected value of the polling latency, and PMST is equal to the poll-to-message-start time. The value of EPL can be determined in terms of the idle-poll-time (IPT) and PMST as follows:

$$EPL = 0.5(N_{cu})(IPT)[1 + \lambda(PMST + ERT)] \qquad (4.32)$$

$$IPT \approx T_{pt} + T_{pr} + T_{ta} \qquad (4.33)$$

$$PMST = HTT + 2 T_{ta} + 147 + 168000(1/UPR + 1/HPR) \text{ milliseconds} \qquad (4.34)$$

$$ERT = \max\{\text{response message receipt time on ESR or user port}\} \qquad (4.35)$$

See Section 2.4.2 for computing the average value of ERT (as equal to the average message transmission time multiplied by AMEF):

$$T_{ta} = 4T_{np} + 2T_{sl} + T_{aq} + 4(N-1)T_{hm} + 2(K_{pg})L_{esr} + T_{nbp}(N-2) + 2T_{eb} \qquad (4.36)$$

For these equations we employ the equivalences: N_{cu} = actual number of control units on the VMPT circuit, λ= actual message arrival rate on the particular VMPT circuit, T_{pt} and T_{pr} are the poll message transmit/receive times on the ports with data rates of HPR and UPR respectively, T_{np} = nodal processing time, T_{sl} =slot latency time, T_{hm} = half modem time, K_{pg} = propagation constant, L_{esr} = ESR length, T_{nbp} = nodal bypass time, N = total number of drop nodes on the ESR, T_{eb} = effective ESR block transmit time, and T_{aq} is applicable end-to-end queuing delay.

The quantity $(2*EPL=N_{cu}*IPT)$ valid for no activity, is sometimes called the ESR system walk time. It is equivalent to the minimum time required to poll all of the stations on a virtual circuit.

It should be instructive to compare the equations for E(D) or BST for the ESR and the conventional multidrop network. Using Equations 4.32, 4.33, and 4.36, the value of W for ESR can be defined as follows:

$$W_{esr} = N_{cu}*IPT$$

$$= N_{cu}*(T_{pt} + T_{pr} + T_{ta})$$

$$= N_{cu}*(T_{pt}+T_{pr}+4T_{np}+4(N-1)T_{hm}+2(K_{pg})L_{esr}+2T_{eb})$$

$$= N_{cu}*[T_{pt}+T_{pr}+4T_{np}+4(N-1)T_{hm} + \tau + 2T_{eb}] \qquad (4.37)$$

where the value of τ has been substituted for $[2(K_{pg})L_{esr}]$.

A comparison between Equations 4.22 and 4.37 shows much similarity. However, it does appear that the value of W_{esr} should be much higher than W for the conventional MPT networks, for identical values of N and N_{cu}. The additional terms of T_{pt} and T_{pr} representing the poll message transmission times on the user/host ports were completely ignored in Equation 4.22. These values are negligible for most cluster controllers, which employ generally high-speed ports. The value of T_{eb} should be identical to T_{pm} for identical capacity of shared media. The major part of the difference is influenced by the term $[4N_{cu} *(N-1)* T_{hm}]$. The difference grows rapidly with increasing values of N and N_{cu}. Using the fact that ESR allows the superimposition of several virtual multipoint circuits on the same physical media, the value N_{cu} is rarely greater than 4 for most applications of ESR applications. Since the value of N for a conventional multipoint circuit will generally be quite high, when compared to N for ESR applications, and since N is not allowed to exceed 6 for most ESR applications, W_{esr} should not be much higher than W for most practical cases. Furthermore, the ESR should always outperform the conventional MPT network at high traffic loads due to its unique queuing mechanism. It was quite interesting to observe that Equation 4.31 provides identical results to those obtained from Equation 4.21 if one employs identical values of walk times and the appropriate values of queuing delays for both types of systems. See Chapter 8 for some response time analyses of the two architectures for meaningful design parameters and assumptions.

We have now studied the nodal and end-to-end response times for WANs. We are now left with the end-to-end delay performance of LANs. Deriving simple expressions for the end-to-end response times for different LANs for identical operating environs is generally an impossible task. Since each WAN can be visualized as an interconnection of several LANs, the task becomes even more difficult. An attempt to derive a simple, qualitative comparison between the end-to-end response times for LANs and WANs is discussed in the next section.

4.4 End-to-End Response Time Comparisons for LANs and WANs

Since the emphasis of this book is on network topologies and their synthesis, we will only present a qualitative discussion of the end-to-end response times for LANs and some WANs. The reader should consult the works of Schwartz (Schwartz, 1987), Stuck (Stuck, 1983), and Stallings (Stallings, 1984) for an insight into the LANs' operations and their performance. The curves of Figure 4.20 were derived from their works to illustrate normalized response times to media (shared or otherwise) loading.

The end-to-end response time is defined here as the elapsed time between the moment a packet begins transmission at the source node and the moment the packet starts to arrive at the destination node. Only Token Ring and CSMA/CD LANs are shown along with the WANs based on the Aloha, Slotted Aloha, X.25 PS, and ESR access techniques. The netlinks based on the X.25 and ESR architectures were assumed to serve 10 nodes in tandem for each path of communication. The number of nodes served by a LAN indirectly affects certain design parameters such as α and β for LANs based on the concepts of CSMA/CD

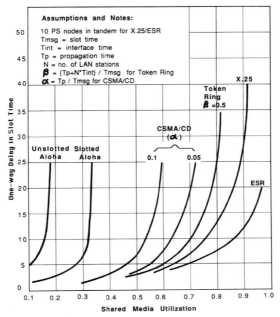

Figure 4.20 One-way delay versus shared media utilization in LANs and WANs.

and token ring. It must be emphasized that the curves are idealized to illustrate the limits of their performance. Each one of the systems illustrated has an upper bound for media utilization. According to the results plotted, the ESR-based system is the best and the Aloha-based WAN is the worst of all considered in terms of the media utilization. Other performance issues such as the cost of transmission media and hardware/software interface modules should also be considered.

Detailed expressions for the response times for each type of network can only be derived for the exact environments under which a network operates, and there are an infinite number of valid possibilities. Furthermore, such design possibilities are never the same for any two practical systems. It is for that reason that most references cited earlier shy away from discussing such expressions in detail. Even those papers that derive complex equations for a single system do so for a very limited set of idealized operational situations. The actual set of design variables will generally fall out of the limits assumed. A successful methodology employed by the author is as follows:

1. Measure the desired response times for several useful values of media loading.

2. Develop an approximate analytical model for end-to-end response time.

3. Produce some useful curve relating response times to various design parameters such as media utilization.

4. Develop spreadsheets that will give answers as a function of design parameters on a "what-if" basis.

In order to get meaningful results, the preceding methodology should be repeated for each network architecture and set of operational environments.

Exercises

4.1 Create a chart similar to that in Figure 4.4.

4.2 Create a chart similar to that in Figure 4.5.

4.3 Characterize the M/M/N and M/D/N queue types and define the total time spent in each type of queue in terms of its components.

4.4 Compute the average queuing delay experienced by a packet after passing through the input queue and the output queue of packet switching (PS) node as shown in Figure 4.18. The two queues are of the M/D/1 and M/D/2 types, respectively, and characterized by λ_1= 50 PPS, T_{s1}=10 milliseconds, λ_2=50 PPS, EML_2=1000 bits, each server capacity (C_{s2}) =50,000bps. (*Hint:* Compute P_0 for each case separately.)

Bibliography

Beizer, B. 1978. *Micro-Analysis of Computer System Performance.* New York: Van Nostrand Reinhold.

Doelz, M., and R. Sharma. 1984. "Extended Slotted Ring Architecture for a Fully Shared and Integrated Communication Network," Proc. MIDCON 1984 Conference held in Dallas, TX.

Doelz, M. 1985. Personal communications.

Feller, W. 1957. *An Introduction to Probability Theory and its Applications.* New York: John Wiley.

Fraser, A. 1983. "Towards a Universal Datatransport System." *IEEE J. on Selected Areas in Communications,* vol. SAC-1, no. 5, November 1983.

Martin, J. 1972. *System Analysis for Data Transmission.* New York: Prentice-Hall.

Schwartz, M. 1987. *Telecommunications Networks.* New York: Addison-Wesley.

Stallings, W. 1984. "Local Networks." *ACM Computing Surveys,* vol. 16, no. 1, pp. 3–41.

Stuck, B. 1983. "Calculating the Maximum Data Rate in Local Area Networks." *Computer,* May 1983, pp. 72–76.

Syski, R. 1960. *Introduction to Congestion Theory in Telephone Systems.* Edinburgh and London: Oliver and Boyd.

5

Algorithms for Synthesis of Network Topologies

5.1 Introductory Remarks

There is no such thing as a closed form algorithm for optimizing a network topology for any voice or data applications. The designer must employ a combination of heuristic algorithms and "what if" type of analyses in an iterative manner until the best solution is obtained. The network designer must be deeply involved in the network design process at every stage, and it is the network designer who dictates the various iterations. The network design software is only a tool. A PC-based tool can be helpful in performing fast computations economically. See Chapter 2 for a discussion of such a methodology. Basically the solution consists of the following steps:

1. Computing the lower bound on the number of switches at each level of network hierarchy as dictated by their maximum capacity to handle calls/messages/terminations.

2. Selecting the optimal locations of a required number of switches at each level of hierarchy.

3. Assigning of customer premise equipment (CPE) locations to lower-level switches and the assignment of lower-level switches to next-level switches.

4. Connecting the CPE locations to lower-level switches, connecting the lower-level switches to the next-level switches with a correct number of

links according to the desired topology, and connecting the tandem switches together via trunk bundles to form a backbone network.

5. Computing the total cost of the network for the allowable sets of design parameters.

6. Selecting the optimum solution with the correct set of design parameters.

To illustrate the preceding process at a very high level, the total network costs were plotted as a function of the number of tandem switches in Figures 5.1(a) and 5.1(b) for two typical voice/data networks, and Figure 5.1(c) for a typical ACD network. By studying similar plots of the total monthly costs for several mixes of link types in the access/backbone subnets, the designer should be able to choose the most optimum network topology. These charts show that the hardware and NMC costs do not generally influence the selection of an optimum topology. To illustrate, the monthly costs due to hardware in Figure 5.1(b) have almost tripled when compared to similar costs as shown in Figure 5.1(a), but the optimum topology hasn't changed. Similarly, the labor costs in an ACD network are very large compared to other network components and increase with the number of switches as shown in Figure 5.1(c). Consequently, the best network topology will consist of a very few ACD switches. Hardware costs have no influence on network topology. One may justify two or three ACD switches on the ground of network survivability alone. Figure 5.1(d) is obtained for a voice/data MAN with high costs of hardware and owned wiring system. Similar charts apply for voice/data LANs with lower transmission costs. The number of switches/hubs is generally determined by reliability.

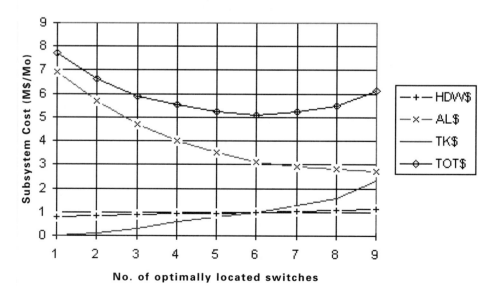

Figure 5.1(a) Distribution of costs in a typical voice/data network with low hardware costs.

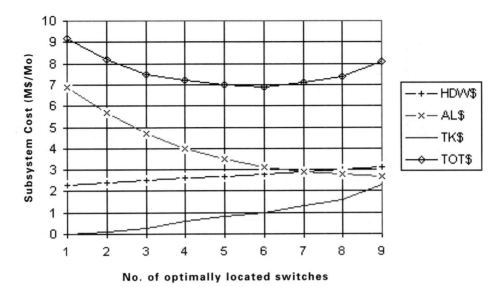

Figure 5.1(b) Distribution of costs in a typical voice/data network with high hardware costs.

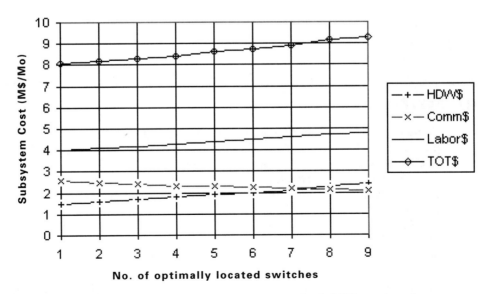

Figure 5.1(c) Distribution of costs in a typical ACD network.

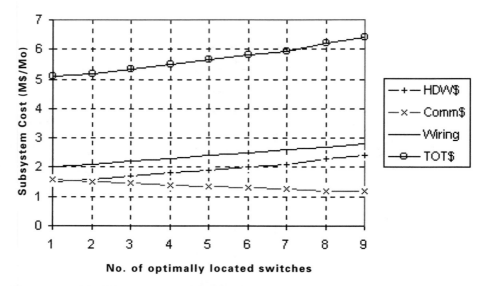

Figure 5.1(d) Distribution of costs in a typical voice/data MAN.

It should be stressed that the word *optimum* used in this book refers only to the best solution found among those considered (and not among all the possible solutions). For those who consider themselves as purists, this revelation may be a shocking one. For those who consider themselves as pragmatists, such a revelation only adds beauty to the art and science of networking. We will now discuss each of the steps mentioned earlier that can be applied to voice, ACD, data, and integrated voice/data applications.

5.2 Computing the Lower Bound on Number of Switches

Using the traffic engineering techniques of Chapter 2, it is an easy matter to compute the total number of calls/messages/bits handled during the busy hour (BHR). By studying the vendor data dealing with the capacity of the switches to handle maximum traffic and/or number of lines and/or trunks, the designer can quickly determine the minimum number of switches required to design the network at any level.

For voice applications, the CPE location is generally served by either a PABX or a CO-Centrex. An unusual requirement may involve knowing only the number of telephones at each customer location. In that case, the first task may be to compute the optimum number of PABXs and/or CO-Centrexes at that site. One can now buy a PABX or a tandem switch (and even a combined PABX/Tandem switch) with terminations ranging from a small number to sometimes over 20,000 telephones. Such a range of terminations therefore does not prevent a location to be served by a single PABX. For some small offices with only a few telephones, economics may dictate to serve such offices as an off-net location. The design of an optimum network is then guided by computing the number and locations of tandem switches according to a least-cost criterion.

For data applications, the number and type of data LANs at various locations is generally given. In some cases, only the number and type of workstations (WSs) may be given. That may require the choice of a data LAN or a lower-level PS as a CPE. These days, most enterprises don't even consider the use of dumb concentrators anymore. They generally select LANs as CPEs and packet switches (PSs) at all other levels of network hierarchy. As mentioned earlier, each switch also acts as a traffic concentrator. Therefore, in the following pages, a concentrator will always be equivalent to a lower-level switch. The first major task is to compute the minimum number of lower-level PSs by studying their capacities to switch packets or messages or ports. The next task is to vary the number of tandem PSs (starting from the minimum number of PSs required) for serving a given number of lower-level PSs until the least-cost criterion is met. This process may then be repeated for a different number of lower-level PSs. This will result in several sets of plots as illustrated in Figure 5.1 before the designer obtains an optimum network topology. Since many vendors sell lower-level PSs that can be connected according to DL or MD topologies, the designer may be required to consider many mixed topologies to find an optimal data network topology. This is where the mind of the designer plays a key role. Availability of a truly autonomous and automatic tool can only be dreamed of at this time.

Cost may not be the only criterion for choosing the correct number of switches as suggested by the curves of Figure 5.1. An additional criterion should be based on system availability or reliability. If the choice is between two or three switches and if the cost is not much different, one should choose a network topology with three switches. The third switch will enable the network system to handle the future growth better. Furthermore, with three switches, the network will be able to handle a critical portion of trunk traffic in case a switch or some transmission facilities fail. With two tandem switches, the major CPEs will fail to communicate with one another in case of a switch or a trunk bundle failure. In general, the selection of an additional switch will always enhance the reliability of the system. The reliability criterion becomes even more significant when working with an ACD system (see Chapter 8 for a detailed design methodology) that employs virtual transmission facilities. Even the addition of a second switch will enhance the reliability of the 1-node ACD system by rerouting the incoming traffic to the functioning ACD location. Such an option is available from most providers of virtual services for a small fee.

5.3 Algorithm for Optimally Locating Switches

For any given number of switches at any hierarchical level, the designer must locate the switches at an appropriate subset of existing customer locations (instead of cornfields) to achieve the least cost. For voice and data applications, the problem is to select an optimum subset of CPEs as sites for locating switches.

The writer has developed a simple algorithm for locating a given number of switches at a subset of the CPE locations as an initial solution. This algorithm employs the concept of a center of gravity (COG) and successive decompositions and it can be executed as follows:

1. Define the V- and H-coordinates for each of the N CPE locations.

2. Find the V-and H-coordinates of the mathematical COG for all N locations using following expression:

$$V(COG) = [\sum_{i=1}^{N} V_i * W_i] \, / \, [\sum_{i=1}^{N} W_i] \qquad (5.1)$$

and

$$H(COG) = [\sum_{i=1}^{N} H_i * W_i] \, / \, [\sum_{i=1}^{N} W_i] \qquad (5.2)$$

where

V_i and H_i are the V and H coordinates of the ith location and W_i is time consistent average (TCA) of traffic intensity (TI) or simply the traffic weight of the ith location

3. Find the CPE location closest to the preceding mathematical COG. This location is suggested as the optimum location for a single switch.

4. To find the locations of two COGs, one must first draw a vertical line passing through the mathematical COG. Then enumerate the locations on the left side and the right side of the vertical line and find the mathematical COG for each of the two sets of locations using the procedure of Step 2. Again, find two locations closest to these new COGs.

5. To find the locations of four additional COGs, one must first draw two horizontal lines through the two COGs and obtain four sets of CPE locations belonging to the newly decomposed regions. Again, find the four mathematical COGs and the closest CPE locations to these four COGs.

6. Repeat Steps 4 and 5 by decomposing additional regions and computing the COG locations and the corresponding closest CPE locations. The alternation of vertical and horizontal decompositions is necessary to eliminate any influence of geographical boundaries. The earlier scheme of starting with a horizontal decomposition (e.g., Step 4) is recommended for a country such as Canada or the United States. In some cases, starting with a vertical decomposition is preferred. In other cases, both types of initial decomposition may be tried to find the best solution for questionable boundaries.

The process of finding these COGs and the associated closest CPE locations in an iterative manner can be illustrated in Figure 5.2. Once the solution space

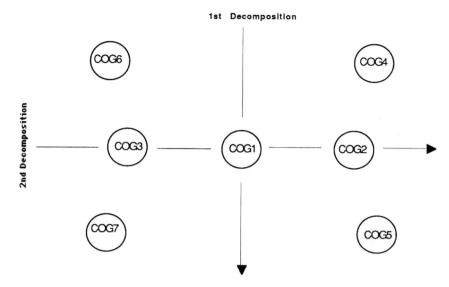

SWITCH LOCATION OPTIONS:
One switch at COG1
Two Switches at COG2 and COG3
Three Switches at COG1, COG2 and COG3
Four Switches at COG4, COG5, COG6 and COG7
Five Switches at COG4, COG5, COG6, COG7 and COG1
And so on....

Figure 5.2 Decomposition algorithm for locating concentrator/ switching nodes.

is known, it is a simple matter to select 1, 2, 3,....M locations for switches. As illustrated in Figure 5.2, the choice of 1, 2, 3, 4, 5, 6, and 7 switch locations is obvious when 1, 2, and 4 COGs and the closest locations are computed successively using the methodology presented earlier.

The previous method can be summarized by stating that one can get a set of M candidate locations for switches such that M=2*d+1 and where d=2^x-1 (x=0, 1, 2, 3,.....integer). As an example, the allowable values of d are 0, 1, 3, 7, 15 for x=0, 1, 2, 3, 4, respectively. The EcoNets software package allows one to get an output file defining M COGs and candidate locations.

The network designer will find that the previous algorithm provides a rather good solution for most cases. However, there may be some pathological cases where the initial solution can be improved. One quick approach is to find other nodes close to the mathematical COGs with heavy traffic and try these iteratively to find any savings. An exhaustive method is to study each of the M clusters and relocate a corresponding network center/switch at all neighboring nodes such that the sum of the costs for all access lines of that cluster is minimized as follows:

$$\text{Cost} = \min_i \left[\sum_{j=1} d_{ij}\, w_j \right] \dots\dots\dots \tag{5.3}$$

where

d_{ij} = distance or cost between the ith center and jth location/node

w_j = weight of jth location/node in terms of erlangs or bps

An algorithm based on an exhaustive search can yield an optimal solution but it will be extremely time consuming. Many other approximate algorithms for locating switches have been described by Sharma, De Sousa, and Ingle (Sharma, De Sousa, and Ingle, 1982). Since none of these algorithms can yield an absolute optimal solution, it is difficult to compare their efficiency with that for the afore-mentioned simplified algorithm. Some day it may be possible to compare all known heuristic algorithms in a scientific manner. At this time, no single algo-rithm can be proven as the best for all applications.

5.4 Algorithms for Connecting CPE Locations to Switches

The designer must employ the bottom-up approach to connect all the nodes to one another. One must always start with the lowest level (e.g., CPE nodes) of unconnected nodes and connect these to the switch nodes belonging to the next hierarchical level according to an applicable topology. These switch nodes now take up the role of unconnected nodes that must be connected to the switch nodes of the next level. This process stops when the tandem switch level is reached. Then these tandem switches are connected to one another by trunk bundles. There are several housekeeping chores that must be accomplished as the network evolves. These can be listed as the following numbered tasks:

1. Accumulate the traffic intensity at each switch level.

2. Accumulate the cost of network facilities (nodal hardware and transmis-sion line) at each switch of network hierarchy.

3. Eliminate the lower-level nodes from the nodes yet to be connected.

There are several applicable topologies for synthesizing voice and data net-works at any stage of the aforementioned process. The star topology is useful for designing almost all voice networks. The star topology is also useful for designing data networks that generally provide the fastest response times. Some packet switched networks based on AT&T's ISN and CCITT's X.25 architecture also deploy the star topology for very high traffic loads. Star topology is also quite effi-cient for LAN or MAN applications for reliability and lower response times. The star topology is also becoming the correct choice for multimedia applications char-acterized by heavy traffic. The multidrop (MD) topology was originally employed to reduce transmission cost through the sharing of low-speed lines. Consequently, it was mainly used for designing WAN data networks based on the SNA architec-ture. Some data networks based on X.25 also employ a mixture of the multidrop and star topologies. The directed link (DL) topology is useful for designing LAN, MAN, or WAN data networks based on some fast packet switching technologies (e.g., Doelz's ESR) and many modern frame relay systems. The network design process employing these topologies is illustrated in Chapters 7 through 14.

The bus and the ring topologies for LANs don't require any design effort since the connectivity is already predefined. The cost of the entire LAN is generally influenced by the media interface modules and not by the network topology. The LAN links are generally owned and not leased from the local exchange carrier (LEC). One may reduce the one-time network cost through an optimal placement of telephone closets, mainframe distribution (MFD) units, or the switching equipment in a large building or a multilevel ship or a campus. Another technique for reducing the network cost is through replacing the expensive vendor-recommended transmission media with the existing twisted copper wires through the use of Baluns (see Chapter 9 for a proven methodology) or other special devices. The design of a premise wiring system (PWS) generally requires a lower-level design of voice/data network based on the star topology.

5.4.1 A Generalized Algorithm for Designing All Network Topologies

For most voice and data applications, the algorithm must perform the following numbered tasks:

1. Select the number (=C) of lower-level switches.

2. Find the optimum locations for C lower-level switches.

3. Assign the unconnected nodes at the lower level to the various switch locations based on the closest distance criterion. This creates C clusters. In some cases, it may be advised to employ a more complex assignment criterion based on the from-to traffic intensity and the cost of all the paths. In the discussions that follow, only the simplified assignment scheme will be considered.

4. Execute the steps of an algorithm required for designing any topology (see the subsequent paragraphs for the steps required for each of the Star, DL, or MD topologies). Sizing of each AL and trunk bundle is computed in Steps 5 to 11.

5. Compute the total traffic (erlangs/bps) handled by each switch level by adding the traffic between each nodal pair (or traffic on each access line, AL).

6. Compute the number of circuits required to handle the required traffic on each AL bundle while obeying the design constraint of either allowed blocking or maximum allowed line traffic weight of W_m.

7. Compute the cost of each AL bundle as a function of the distance, number of circuits in the bundle, and the applicable tariff. Chapter 2 describes a methodology for modeling tariffs of leased and fully owned

facilities. Chapter 6 describes a very flexible and modular approach of entering several tariffs into a tariff file and choosing the applicable one for computing the cost of any AL or a trunk bundle.

8. Compute the cost of each region and the entire system.

9. Repeat Steps 1 through 8 for each level of switch until only the backbone network remains to be designed.

10. Synthesize a fully connected backbone network topology with S switches using the techniques presented in Chapters 2, 3, and 4.

11. Compute the size of each trunk bundle as a function of the traffic handled while obeying the design constraint of either allowed blocking or maximum allowed trunk utilization of MTKU.

12. Compute the cost of each trunk bundle as a function of the distance, number of trunks in each bundle, and the applicable tariff. Also compute the cost of the entire trunk network.

13. Repeat Tasks 1 through 12 for each combination of lower-level switches and tandem switches, and for various combinations of transmission and nodal hardware technologies.

14. Make a chart similar to that in Figure 5.1, plotting network-related costs (e.g., cost for total AL, trunks, hardware, and NMC/TC). Select the most optimum network topology consisting of the correct number of lower-level switches and tandem switches for a correct combination of nodal and transmission technologies.

15. Display all of the resulting network topologies in the form of graphics with a summary of network costs.

The preceding process will generally require several human interventions, especially when judgments are to be made regarding the chosen sites for lower-level switches and tandem switches and the link types with applicable tariffs (one for each transmission technology). But in any case, most of the other steps must be executed on a computer to save human labor. Chapter 6 presents a desktop computer technology to handle most of the steps. Several attempts are being made by researchers to develop a technique based on the so-called expert system (ES) or artificial intelligence (AI). However, ES and AI are useful only when cost-effective algorithms are unavailable and the entire intelligence or experience is available only in the form of a large relational database (e.g., in the area of human illness diagnosis). The approach taken in this book mainly relies on a rich source of cost-effective algorithms. Only a few human interventions are needed at critical points. Actually, the entire process as illustrated earlier can be accomplished within a few hours after the required input files (as defined in Chapter 6) are available to run the network design program. Since most networks have only two or three levels of nodes (i.e., CPE locations, lower-level switches, and tandem switches), only two passes are required to complete the first nine tasks of the described design process. In the first pass, the CPE locations are connected to the selected number of lower-level switches, and in the second pass, the lower-level switches are connected to tandem switches before designing the backbone network.

5.4.2 Algorithm for Designing a Star Topology

Only Task 4 in the generalized design process of 5.4.1 needs to be altered for connecting the nodes to associated switches locations according to the star topology. This step is defined as follows:

1. Connect the locations in each cluster to the associated lower-level switch or tandem switch location by an access line (AL) bundle. This step yields a multicenter, multistar topology network as illustrated in Figure 1.8.

5.4.3 Algorithm for Designing a Directed Link (DL) Topology

Only Task 4 in the generalized design process described in Section 5.4.1 needs to be altered for connecting the given nodes to lower-level switches or tandem switch locations according to the DL topology. This algorithm was patented (U.S. Patent 3 703 006 awarded to Sharma) in 1972. A paper by Sharma and El-Bardai (Sharma and El-Bardai, 1970) defines the same algorithm (as a part of the design of a multidrop topology) in a rigorous manner. A DL topology can be synthesized if the following numbered steps are executed in that order:

1. Divide the ith nodal traffic weight W_i by the allowable maximum link weight W_m and compute the integer quotient Q_i for all i. Connect node N_i to the switch by Q_i special circuits to handle excess traffic for each node. Replace the original W_i by the value $(W_i - Q_i*W_m)$.

2. Find the node (N_i) among the unconnected node that is farthest from the given switch node S_j.

3. Find an unconnected node (N_j) closest to N_i.

4. Add the traffic weights of these two nodes to get a running total as $W_t = W_i + W_j$.

5. If $W_t <=$ the allowable maximum, W_m, then connect N_j to N_i and remove these from the unconnected list of nodes. Call N_j as N_i. Repeat Steps 3 and 4 until $W_t > W_m$.

6. If $W_t > W_m$ then connect N_i to the switch node.

7. Repeat Steps 2 through 6 until all nodes have been considered and connected.

The preceding algorithm is illustrated in Figure 5.3. If one ignores the special circuits created in Subtask 4.1, the network now consists of several netlinks, each of which may be shared by several nodes. The concept of a multicircuit access line bundle for the star topology does not apply to a shared netlink based on the DL or a multidrop topology. The remaining steps of the network design process of Section 5.4.1 remain unchanged. If W_m becomes too small, one may get too many special links compared to the number of netlinks. The resulting network can effectively reduce to a star topology in the limit.

5.4.4 Algorithms for Designing a Multidrop (MD) Topology

Kruskal (Kruskal, 1956) and Prim (Prim, 1957) are credited with the development of algorithms for connecting a set of nodes to one another according to a minimal spanning tree (MST) topology. Such a topology guarantees a minimal sum of length associated with all links or connections employed to connect the nodes. An MST topology represents a true multidrop topology. The original MST topologies were synthesized with no regard to capacity constraints; therefore, the algorithms of Kruskal and Prim require modifications for designing practical networks.

The first practical algorithm for designing an MST with capacity constraints was developed by Esau and Williams (Esau and Williams, 1966). See Esau and Williams for a description of their algorithm. Several practical algorithms were also developed by the writer. Consult Sharma and El-Bardai (Sharma and El-Bardai, 1972) for a description of three algorithms. Two of these algorithms always produced more expensive CMST networks when compared to the third algorithm that later became the basis of U.S. Patent 3 703 006. Unfortunately, most of the published papers refer only to one of the first two algorithms discarded by the writer. Since 1970, several other authors have developed additional algorithms for synthesizing MD topologies with capacity constraints. See the works of Frank et al. (Frank, Frisch, Chu, and Van Slykes, 1971), Frank and Chu

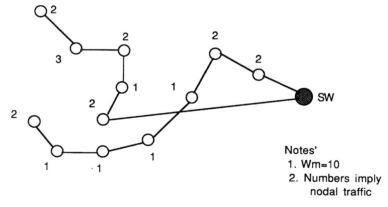

Figure 5.3 Sharma's algorithm for optimizing a DL topology.

(Frank and Chu, 1972), and Kershenbaum and Chu (Kershenbaum and Chu, 1974) for a summary and a comparison of several algorithms including the one discarded by the writer.

Kershenbaum and Chu (Kershenbaum and Chu, 1974) also present a unified, parametrized algorithm for designing an MD topology with capacity constraints. Since no single algorithm based on heuristics can be guaranteed to always yield the best results, the designer must try several parametrized approaches to obtain an optimum solution. As a result, their algorithm becomes quite cumbersome. Furthermore, their paper does not document the cost reductions for some practical networks. Ideally, it should be possible to compare all known good algorithms against a fixed set of input data and design criteria and then choose the best. But that has been impossible until now. For first time, the author will attempt to compare some of the well-known algorithms in Chapter 8 for identical inputs and constraints imposed by practical networks only. The reader is encouraged to compare these algorithms with others by using the input data presented in Chapters 7 through 10.

5.4.4.1 Algorithm for Designing Minimal Spanning Tree (MST) Network with No Capacity Constraints

We will describe the basic Prim's algorithm (Prim, 1957) considering no capacity constraints. Kruskal's algorithm (Kruskal, 1956) will yield identical results for no capacity constraints. The major steps of designing an MST network with no capacity constraints using Prim's Algorithm can be described as follows:

1. Start out with all unconnected (or isolated) nodes and no clusters.

2. Consider the closest node to an unconnected node. If the closest node is also unconnected, make the connection between the two nodes. This step when repeated for all isolated nodes will yield either isolated nodes or clusters of connected nodes.

3. Connect every isolated node to the closest cluster by the shortest connection.

4. Connect two closest clusters by the shortest connection.

5. Repeat the preceding steps until only a single cluster is formed.

The previously mentioned steps are illustrated in Figures 5.4 (a), (b), (c), and (d) for a case starting with nine unconnected nodes. This algorithm will be applied to several cases of interest in later chapters. In the discussions that follow, Prim's algorithm will be consistently used to synthesize each netlink with an MD topology when the nodes have already been assigned to that netlink by one of the two algorithms described next. In other words, the selected algorithm will always be used first to assign a set of nodes to a netlink, and then Prim's algorithm will be used to derive the actual MD or MST topology. It should be stressed that the computer time required by Prim's algorithm becomes excessive for very large (> 40) numbers of nodes. Fortunately, the number of nodes per netlink allowed by practical implementations is not generally excessive.

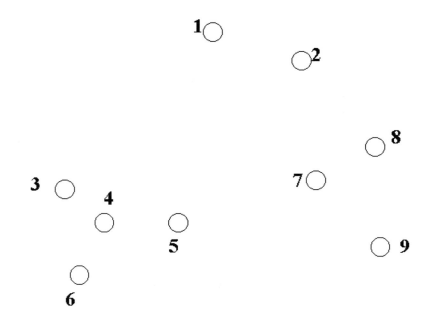

Figure 5.4(a) Prim's algorithm for synthesizing an MST; starting with unconnected nodes.

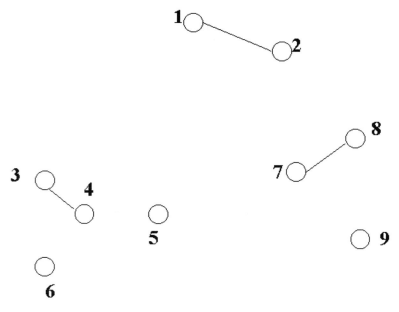

Figure 5.4(b) Prim's algorithm for synthesizing an MST; connecting all pairs of unconnected nearest nodes.

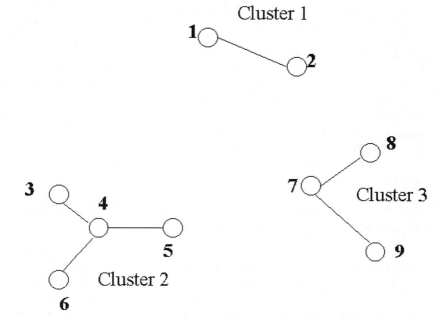

Figure 5.4(c) Prim's algorithm for synthesizing an MST; connecting isolated nodes to closest cluster by shortest connection.

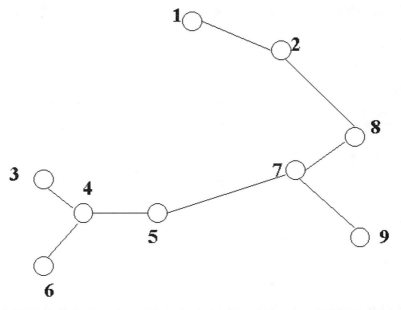

Figure 5.4(d) Prim's algorithm for synthesizing an MST; connecting closest clusters by shortest connections for final MST.

Kershenbaum and Chu (Kershenbaum and Chu, 1974) have employed algorithms of Prim and Kruskal for designing capacitated minimal spanning trees (CMSTs). However, the network costs always exceeded those synthesized by using the Esau-Williams or Sharma algorithm.

The algorithm developed by Prim should not be confused with another closed form algorithm for minimizing the total length of connectorized wiring that scientists have been trying to derive for several centuries. Such an algorithm allows the addition of junction nodes for providing connection to regular network nodes. Unfortunately, such an algorithm is not useful for computing the link costs for most of the modern telecommunication networks. It should be useful in problems dealing with the design of ultrafast, very large-scale, integrated (VLSI) circuits.

5.4.4.2 Esau-Williams (EW) Algorithm for Designing MD Topologies with Capacity Constraints

While the original algorithm was explicitly described by Esau and Williams (Esau and Williams, 1966), the writer is aware of many variations of their algorithm. Each worker who did develop a unique variation of this famous algorithm claims to have done so for improving the original algorithm. Since the Esau-Williams (EW) algorithm is based on heuristics, it is impossible to always guarantee better results than those achieved from the original algorithm. We will attempt to describe the original algorithm in the language we have developed thus far. Before describing the various steps of this algorithm, one must first define a function T(X,Z) that is the basis of the Esau-Williams algorithm. One can define the T(X,Z) function in terms of two nodes X and Z of the starting network topology of Figure 5.5 as follows:

$$T(X,Z) = [(DX + DZ)] - [\text{Min }(DX, DZ) + XZ] \tag{5.4}$$

$$= DX - XZ \quad ....\text{applicable for Figure 5.5}$$

where
    DX = distance or actual cost of line DX
    DZ = distance or actual cost of line DZ
    Min(DX, DZ) = minimum of the two quantities DX, DZ

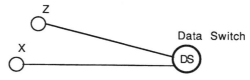

Figure 5.5 A starting topology for computing the T (X:Z) function involving three nodes, using the Esau-Williams algorithm.

If a single tariff applies to the network and the costs are proportional to distance, the distance should be adequate for computing the T function for the network. The objective of this algorithm is to synthesize netlinks consisting of more than two nodes sharing a link and satisfying the maximum link capacity, W_m. The algorithm can be described in the form of numbered steps or tasks:

1. Compute the maximum T function [equal to Max (T)] for all but the switch node. This means the computation of the best secondary node Z for each primary X node.

2. Form (N-1) components (or subsets or netlinks) with only two nodes in each component or netlink, one primary node, and the data switch. See Figure 5.5 for a starting network with only three nodes.

3. Search for the unselected primary node X with the Max (T) function and consider the associated component (or subset or netlink).

4. Consider the secondary node Z and the associated component.

5. If the Max (T) is =< 0, then remove the primary node X from the table. Go to Step 3.

6. If the sum of component traffic loads exceeds the maximum capacity W_m due to prior changes, then recompute Max (T) function and compute a new Z. Repeat Steps 4 through 6. If the sum of component traffic loads is <= W_m, then merge the two components or netlinks with only one connection to the data switch. Delete this X from the table.

7. Repeat Steps 3 through 6 for all primary nodes until all primary nodes have been considered and deleted from the table.

8. Apply Prim's MST algorithm to each set of components to obtain M netlinks based on the MD or CMST topology. See Figures 5.6 and 5.7 for illustration of these two steps for three and four nodes to be connected to the data switch in the network.

Since this process may be difficult to fathom in the written form, a flow chart of Figure 5.8 has been prepared to illustrate the EW algorithm. This flow chart replaces Task 4 of the generalized network design process of Section 5.4.1. In the succeeding chapters, several networks based on three variations of the EW algorithm will be synthesized and their costs compared to other MD solutions based on different algorithms and topologies.

Figure 5.6 A three-node MD network topology using the Esau-Williams algorithm.

Figure 5.7 A four-node MD network topology using the Esau-Williams algorithm.

The author believes that a large number of variations of the original EW algorithm exist in the world. It is not very difficult to develop a variation of the original algorithm that reduces the network cost slightly for a particular application and declare it as the ultimate solution. But it is another matter to create a new variation of the EW algorithm and prove that it is always better than the original version. Such a goal may be an impossible task considering the heuristic nature of the algorithm. A better alternative may be to use the original algorithm and its variation in rapid succession for a large number of applications and see

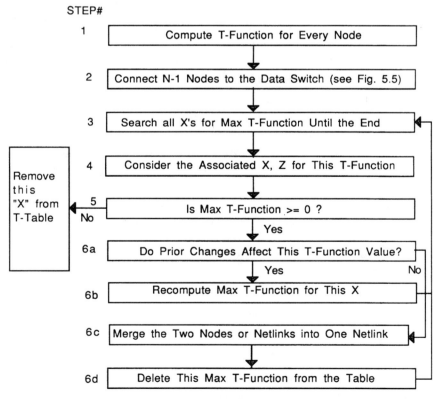

Figure 5.8 A flow chart defining the Esau-Williams algorithm for optimizing an MD topology.

the overall trend. Only this way can one choose the better of the two. One can also apply the nodal Exchange algorithm as described in the next section on any EW network and achieve additional sizable savings in most cases. This approach is analyzed, along with two popular variations of the original EW algorithm in the next section for cost savings, in Chapter 9.

5.4.4.3 Sharma's Algorithm for Designing MD Topologies with Capacity Constraints

This is a two-pass algorithm as described by Sharma and El-Bardai (1970). During the first pass, a feasible network is realized by applying the MST algorithm to each netlink that was obtained through the use of the DL algorithm as described in Section 5.4.2. The second pass employs a nodal exchange algorithm for reducing the cost of the first feasible network. Experience shows that cost savings of up to 8 percent can be realized for only an ordinate amount of additional computer time. Each run for the second pass requires about the same time as that required for a normal EW algorithm. This process is stopped when no additional cost reductions are achieved. Kershenbaum and Chu (1974) had wrongly reported that cost savings of only 1 percent could be realized at the expense of enormous additional computation time. The numbered steps of this algorithm should replace Task 4 of Section 5.4.1. These are defined as follows:

Pass One (Feasible Network Design Algorithm)

1. Using the algorithm of Section 5.4.2, define M unique sets of nodes for each netlink.

2. Using the MST algorithm of Section 5.4.3.1, obtain the first feasible network consisting of M netlinks, each obeying the multidrop or the minimal spanning topology.

3. Compute the cost of each netlink C_i (i=1 to M), and the entire network, C_f.

Pass Two (Exchange Algorithm)

1. Find the farthest node, N_i, from the switch (SW) among the unconsidered nodes belonging to a link L_i.

2. Find a node N_j belonging to a link L_j (i ≠j) and closest to N_i. Find the cost of two links, $C_o = C_i + C_j$.

3. Add node N_i to link L_j. If the weight of new link W_j' exceeds the maximum allowable W_m, add one of more nodes (closest to the SW) of L_j to L_i until the two link weights are $\leq W_m$. Now compute the cost of the newly created links, $C_1 = C_i' + C_j'$.

4. Repeat Step 3 by adding N_j to L_i. Consider the cost of the two newly created links, $C_2 = C_i'' + C_j''$.

5. Replace the original two links with the set of two newly created links that yield the lowest cost, Min (C_1,C_2).

6. Repeat Steps 1 through 5 for each of the other nodes of L_i.

7. Repeat Steps 1 through 6 for all other links of the network. Note the new network cost, C_n.

8. If $C_n < C_f$, then repeat Steps 1 through 7; otherwise, stop the exchange process.

It should be emphasized that the costs of each newly created link is computed only after applying the MST algorithm to the set of nodes belonging to the links. For some cases, additional cost reduction was observed if after each successful nodal exchange, the nodes of each netlink were ordered according to their rank (or their distance from the SW). This algorithm has been described in a precise language by Sharma and El-Bardai (Sharma and El-Bardai, 1970). Since the Exchange algorithm may be hard to fathom in the written form, a transformation of a 2-link feasible network is shown graphically in Figure 5.9 through nodal exchange.

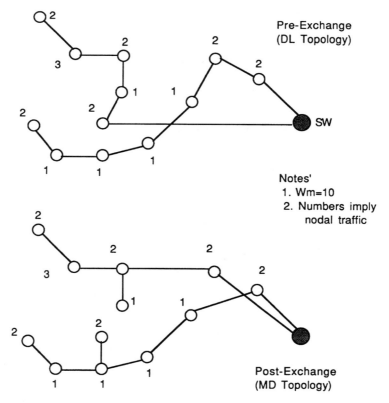

Figure 5.9 Sharma's exchange algorithm for optimizing a network with two netlinks.

It is interesting to observe that the Exchange algorithm can also be applied to any network synthesized through the use of the EW algorithm. The writer has experienced a sizable (1 to 4%) cost reduction with only a nominal increase in computational time. Such a cost reduction is significantly higher than what has been achieved through the use of other so-called greedy algorithms. Consult a paper by Kershenbaum and Chu (Kershenbaum and Chu, 1974) for further details. The writer has also successfully used the Exchange algorithm to reduce the cost of networks based on the DL topology. It was observed that only large but sparsely populated networks produced significant cost saving.

It must be stressed again that the steps required to apply each useful algorithm for an applicable topology (e.g., Star, DL, or MD) should replace Task 4 to obtain a generalized, multilevel network design process in Section 5.4.1. Prim's MST algorithm is only employed to synthesize each netlink of the EW or Sharma's network.

Several meaningful network design related databases will be developed and many networks will be designed for the Star, DL, and MD topologies using the preceding algorithms and studied in the chapters that follow. Only a careful study of those results can help the reader appreciate the usefulness of each algorithm when considering the applications for voice, data, or integrated broadband.

But before studying the resulting topologies, one should consider a user-friendly software package that executes the aforementioned steps for each applicable network topology. To execute these steps by hand will take forever. The next chapter describes a software package that enables the designer to optimize the network topology quickly and compare the associated network costs for all applicable algorithms described in this chapter. Our literature does not provide examples of such a comparison for many of the well-known algorithms. Most of the published articles usually consider only one heuristic algorithm favored by the writer.

Exercises

5.1   Characterize an MST topology. What are the main uses of uncapacitated MST topologies?

5.2   Characterize a DL topology. What are the main uses of a DL topology?

5.3   Characterize an MD topology. What are the main uses of an MD topology?

5.4   Describe the steps required for synthesizing a DL topology.

5.5   Define the V & H coordinates of a COG.

5.6   Relate the number M of COGs to the total number of decompositions d by an equation.

5.7   Is the COG solution always optimum for locating the switch? How can one improve the solution?

Bibliography

Frank, H., I. Frisch, W. Chu, and R. Van Slykes. 1971. "Optimal Design of Centralized Computer Networks." *Networks*, vol. 1, no.1, pp. 43–57.

Frank, H., and W. Chu. 1972. "Topological Optimization of Computer Network." *Proc. IEEE*, vol. 60, November 1972, pp. 1385–1397.

Karnaugh, M. 1972. "Multipoint Network Layout Program." IBM Internal Report No. RC3723.

Kershenbaum, A., and W. Chu. 1974. "A Unified Algorithm for Designing Multidrop Teleprocessing Networks." *IEEE Trans. Communications*, vol. COM-22, No. 11, pp. 1762–1772.

Kruskal, J. 1956. "On the Shortest Spanning Subtree of a Graph and the Traveling Salesman Problem." *Proc. American Math. Society*, vol. 7.

Prim, R. 1957. "Shortest Connection Networks and Some Generalizations." *BSTJ*, vol. 36, pp. 1389–1401.

Sharma, R., and M. El-Bardai. 1972. "Sub-Optimal Communications Network Synthesis." *Proc.* 1970 Intl. Conference on Communications, at San Francisco. *Also see* U.S. Patent No. 3, 703, 006 "Algorithm for Synthesizing Economical Data Communications Network." Issued to R. Sharma on November 14, 1972.

Sharma, R. L., P. T. De Sousa, and A. D. Ingle. 1982. *Network Systems*. New York: Van Nostrand Reinhold.

Sharma, R. 1983. "Design of an Economical Multidrop Network Topology with Capacity Constraints," *IEEE Trans. on Comm.*, vol. COM-31, no.4, pp. 590–591.

6

A Description of EcoNets Software for Network Design

6.1 Introductory Remarks

Previous chapters describe the (1) basic telecommunications concepts, (2) meaningful analysis/design tools, and (3) appropriate algorithms required for the synthesis of voice, data, and integrated voice/data networks. By now, it should be clear that the job of designing even a small network is not a trivial one. In the past, the large number of inputs and computations required for designing a network have always discouraged most experienced analysts. The complexities of statistical traffic theory made the task of network design appear even more insurmountable. Consequently, the network design process had always required the use of powerful mainframes hidden in secure buildings, away from the reach of network specialists. And for those simple reasons, the design process has been a mysterious one even for the majority of DP and telecommunications managers.

But thanks to the recent advances in computer hardware and software, the situation has changed in favor of the humble network designer. Now everyone can afford a powerful personal computer, the desktop or the laptop variety. The only missing link was a brand-new user-friendly network design approach. The network designer needed the simplicity of a scientific calculator. The EcoNets software is one of many packages that can provide a user-friendly interface. This package enables the designer to enter a large number of design parameters and change some key design parameters effortlessly, for the sole purpose of optimizing a network topology through the process of iterative design as described in Chapter 2 and illustrated in Figure 2.2. This chapter will describe a design process that has taken several years to develop and has already proven to be extremely flexible for optimizing and displaying network topologies as these are being synthesized. Although the contents of this chapter may seem out of place when compared with those of earlier chapters, the philosophy behind and the operation of such a software package must be clearly understood before the network design process is mastered. These concepts will become clear when EcoNets is actually employed to synthesize optimum multilevel voice, data, and integrated broadband network topologies and derive deep insights as described in the following eight chapters. For discussion purposes, this software package has been named as EcoNets, which is short for *economical networks*.

6.2 The Philosophy Behind the EcoNets Package

In the past only large mainframes were used for network design. The reasons behind this fact were many. Powerful and affordable desktop workstations were not available. Most network design programs required the use of a large database related to the tariffs. Consequently, the time for computing the cost of a network topology was too high even on a mainframe. Furthermore, the MIS department prevented the distribution of computing power to workers and thus helped perpetuate the mystery of centralized data processing (DP) and network design. Some vendors began allowing their customers to access their mainframe-based network design program through desktop PCs. But the cost of telecommunications far exceeded that for licensing the network design program. Consequently, such a solution was not only costly but also user unfriendly.

The availability of reasonably priced fast desktop computers for network design is finally a reality. One no longer needs to fight the MIS department for the scheduling of RJE terminals and batch processing times on the expensive mainframes. The availability of friendly man-machine interfaces on some of the desktop computers is reducing the training period required to master their operations and use. The availability of bitmapped graphics on desktop computers has already eliminated the dependency on mainframes for topology/map printouts during the "what-if-analysis" mode. The easy availability of a high-quality graphics printer is also enhancing the quality of textual and graphical presentations. The entire network design cycle can now be reduced to a few hours or days instead of months. Several other developments have also been instrumental in bringing the network design capability to the desktop.

Since the divestiture of AT&T in 1984, the forces of deregulation have been steadily destroying the stability of tariffs. Nowadays, one can lease DDS and T1 circuits from a bulk carrier for a song compared to the FCC regulated tariffs. Large customers can also lease private lines from AT&T at a discount (e.g., Tariff 12) or special rates (e.g., Tariff 15). Then there are competing tariffs from OCCs such as MCI and Sprint. Furthermore, the use of bypass of BOC (and in some cases, OCC) facilities creates additional opportunities to ignore the published tariffs. Thus the stability of tariffs as we have known for years has been eroded permanently. There are indications that the FCC may completely eliminate the tariffs, as we know them today. Users are going to be forced to negotiate each and every tariff. This will challenge the capability of large network design packages available now. Van Norman (Van Norman, 1990) and Fike and Jacobsen (Fike and Jacobsen, 1991) have published papers describing the network design tools available in the market. Most of the larger ones will be forced to create simpler models of tariffs. This author predicts a new two-stage network planning process in the offing. The first stage will deal with the initial network topology optimization process using only simplified models of tariffs. The second stage will employ the available optimum network topology to compute the network transmission costs to a high degree of accuracy. This goal can be achieved by either using commercially available PC-based line pricer packages or by asking one or more common carriers to compute the monthly costs for the optimum topology as an incentive to get your business. This two-stage approach is especially attractive if one considers that the network topology is influenced primarily by the network architectural considerations and network design parameters/algorithms, and never by the small perturbations in a selected tariff. This fact will be amplified in the chapters that follow. There will be many cases when the optimum network topology obtained through the use of EcoNets is more than adequate. Some large networks may also require a large-scale computer simulation of many of its subsystems for gaining insights into their detailed performance.

The rapidly changing tariffs and networking technologies create the need for ongoing planning and management of your own network as if it were your only strategic resource. One can reduce the cost of transmission facilities significantly through the process of negotiations with OCCs, bypass, and in-house topological optimizations. The network design process must be a continuous one to keep up with the changing tariffs, evolving switching hardware, bypass technologies, and corporate goals/strategies. Mainframes by their very nature and complexity prevented this real-time planning process. The availability of a powerful desktop workstation, on the other hand, provides the designer a totally flexible and productive tool for network topology optimizations on an ongoing basis.

The aforementioned considerations were employed to create a very user-friendly software package for modeling, design, and analysis of multilevel networks based on several topologies to serve a mix of applications such as voice, data, and video. The name EcoNets was selected to represent its capability to design economical networks to a high degree of optimization. It consists of many subroutines that have been assembled to produce a holistic package for modeling, design, and analysis of either a simple 2-level network or a multilevel, integrated network for both voice and data. The software package for IBM-compatible PCs was created using the GFA Basic Compiler for the Basic source code language. Portions of these codes are copyrighted (1985, 1986, 1987, 1988, 1989, 1990, 1991, and 1992) by GFA Software Technologies, Inc. The source codes for EcoNets are copyrighted (1990 and 1995) by Roshan L. Sharma. This package runs on the IBM-compatible desktop computer with Windows operating system, equipped with about 16MB of RAM. We will describe first the organization of the software package and then its detailed operations dealing with the preparation of input data files, viewing/updating of input data files, viewing of output and help files, and using a host of proven utilities and analytical tools.

6.3 *Organization of the EcoNets Software Package*

The EcoNets software package consists of many compiled, structured subroutines written in the GFA-Basic source language. It employs the principle of keeping the input files separate from the solution kernel of algorithms and analytical tools used to synthesize a network topology. The designer has the option of using any set of applicable disk-resident input files. The software package automatically stores the output files (text or and graphics) under predefined names on the hard disk. The names of these output files must be changed to desired names if the user wants to prevent overwrite. Operating systems for IBM compatible workstations also allow this renaming operation. These output files (or their renamed versions) can be opened later for viewing and final printouts. An overview of the organization of EcoNets and its capabilities are shown in Figure 6.1. According to this illustration there are three major subsystems:

1. Input Files that define all the inputs to the software system

2. EcoNets Package with all of its operational capabilities

3. Output Files that describe the output results of modeling, design, and analysis

The input data are read from the set of input files, and these determine the type and quality of outputs. If one is not careful, one can easily get meaningless output data. Alert messages have been designed to prevent most major illegal combinations of input files and useless output data, but it is impossible to guard against all combinations. The designer must have the ability to test the validity of the output data. The famous GIGO (garbage-in-garbage-out) adage still applies here in all its glory. On the other hand, a basic knowledge of the key

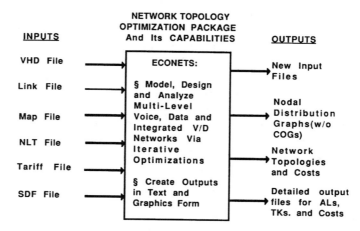

Figure 6.1 An overview of the organization and capabilities of a software package used for network topology optimization.

telecommunications concepts as elaborated in the first five chapters of this book will help the designer to create a meaningful network topology and interpret all of the output data correctly. Just as a hammer does not help a carpenter to create a beautiful house every time, the availability of a good network design program can never always guarantee a fully optimized network. The specialist will always remain as the ultimate key to a good network design. And the designer must always remain vigilant and in full control of the network design process.

The EcoNets package provides all the capabilities in terms of creating new input files and creating output files containing detailed results obtained from modeling, design, and analysis of telecommunications networks. The package also provides additional capabilities related to viewing and printing input/output files in addition to other utility/analysis tools. The package requires a minimum hardware system consisting of a 486-type desktop computer operating on Windows operating system and equipped with at least 16MB of RAM and a hard-disk space of at least 10MB for EcoNets' use, a high-resolution monitor, a Windows-compatible printer, a keyboard, and a mouse.

The outputs are automatically stored on the hard disk under special titles such as TOPF (for network topology), ALFGN, ALFSDi, ALFDLi, ALFMDi, ALMST, ALFVN, ACDF, and TKF for access line output data and trunks, and specialized files such as COGF and DBF for databases for possible switch locations and input file data. Each output text file contains a detailed description of the network topology and costs in terms of each link or connection and other aspects of the analysis. The output text TOPF file describes the overall network topology. This network topology also contains a summary of ALs, trunks, and total transmission costs. In most cases, one does not require to store each automatically produced output text file under a different name. In most cases, as experience has shown, a quick study of the summary costs appearing on the screen is more than sufficient to maintain a useful direction of attack. To save these automatically produced output files under different names, one may use either the Rename capability of the File Manager, or Save As... capability of the File menu for an already opened file, or Duplicate capability of the File menu.

We will now discuss each of the three EcoNets subsystems in some detail. This discussion should familiarize the designer with the structure/meaning of input files, the capabilities of EcoNets, and the structure/meaning of output files. This knowledge, coupled with the understanding obtained from reading the earlier chapters, should equip the designer to prepare all the inputs required to model, design, and analyze any meaningful network.

6.4 A Description of the Input Files

Each input file is of the sequential type. A negative number terminates each input file. The initialization input file called FILES names 17 input files. The software generally requires only a subset of these input files to model or design a network. Each of these input files has unique logical/virtual structure in terms of the number of columns and rows; columns are equivalent to the number of elements in each vector and rows define the number of vectors. The software realizes their virtual structures through the first two letters of each input file name. Therefore, each input file must be named with the following 2-letter prefixes: VH, LI, MA, NL, TA, SD, NA, FT, LA, FI, CS, UT, WU, MU, RS, DT, SW. These input files are listed as follows in the same order in which these are named in FILES (the order cannot be violated):

1. The VHD file defines the V&H coordinates and peak period traffic intensities for each site or location.

2. The LINK file defines the link type for connecting the location to the network nodes for most networks.

3. The MAP file defines the map boundaries for illustrating the resulting network topology.

4. The NLT file defines up to 10 link types in terms of capacity, maximum rate allowed (Wm), the number of voice conversations carried, related tariff number, and private facility design parameter.

5. The TARIFF file defines up to 10 tariffs that can be used for computing network costs.

6. The SDF file that defines up to 60 system design parameters required for influencing system performance in terms of response times, cost, and presentations.

7. The NAME file defines the names of locations according to a 3-symbol city code and a 2-symbol state code represented as (CCCST). The NAME file is used only to create a database (DB) with nodal names.

8. The FTD file may be required to define the peak period internodal traffic flows or internodal connections for modeling a given net.

9. The LATA file is generally required to define a 3-digit code for the Local Access and Transport Area associated with each CPE. Its values are employed to compute the costs of either interoffice channels (IOCs) or local channels (LCs), if such channels are leased separately from a Regional Bell Holding Company (RBHC) on an intraLATA or interLATA basis. A LATA file is also employed to represent the geographical region numbers (1,2,..,60) that are used to derive the customer service area bands (CSABDS) for computing the cost of virtual services.

10. The FILES file is always required to name all of the 17 input files that may be required to design a network. It is a unique start-up file whose entire name cannot be changed.

11. The CSABDS file is a 60X62-sized matrix that provides a Customer Service Area Band (CSABDS) for each combination of two region numbers associated with the source and destination.

12. The UTBL file defines the usage rates for the virtual 800-Service.

13. The WUTBL file defines the usage rates for the virtual Wide Area Telecommunication Service (WATS).

14. The MUTBL file defines the usage rates for the virtual MegaCom 800-Service.

15. The RSTBL file defines the Rate Steps for known CSABDS and Region numbers. It is only used by 800-Service and WATS.

16. The DTP file defines the daily traffic profile used for computing the monthly costs and average costs of a call minute or cost for carrying one megabit of traffic. It defines the fraction of daily traffic handled during each hour of the day.

17. The SWF file defines the number of switches (S) and nodal ID of each switch location.

We will now describe each file in terms of its structure and its meaning.

The VHD file is a sequential file with non-zero 4-element vectors chained together. Each 4-element vector defines the assigned site number, the associated vertical (V) coordinate, the associated horizontal (H) coordinate, and the peak period traffic intensity, respectively. The number of elements in each vector is also equated to the number of columns. The number of vectors (also equated to the number of rows) defines the number of sites/nodes in the network. The V&H coordinates are obtained from the database tables maintained by the telecommunications manager or DP manager. These tables (as illustrated in Figure 6.3) define DB elements define parameters such as site name, area code-CO number (NPA-NXX), part of the telephone number associated with each site, associated V&H coordinates, and so on. (An enterprise that doesn't maintain such tables isn't geared to perform network planning.) The traffic intensity is represented as milliErlangs for a corporate voice and ACD networks or bits per second (bps) for data applications. The units used for peak period traffic intensities must be remembered when interpreting the contents of output files.

The LINK file is a sequential file consisting of non-zero 1-element vectors where the number of vectors equals the number of sites in the VHD file. Each value defines the link type (i.e., 1,2,...) that will connect the associated site to the network nodes, such as a lower-level switch or a tandem switch. The attributes of each link type are defined in the NLT file as described later. A provision exists to ignore the LINK file. In that case, the SDF parameter ALT is assigned to all sites. To illustrate, a LINK file may define the Link Type of "1" for 90 nodes/sites and define a Link Type of "2" for 10 nodes. This will enable the EcoNets package to connect the 100 CPEs to the closest switch among the 10 low-level switches/concentrators by a link type of "1". Each lower-level switch/concentrator is then connected to the closest tandem switch among those defined in the SWF by a link type of "2". The older EcoNets version allowed the use of "0" links types, but not anymore. All link types in a LINK file must now be non-zero values. Furthermore, one doesn't have to select only consecutive values of link types as was the case for the older version of EcoNets.

The MAP file is a sequential file with non-zero 3-element (or 3-column) vectors chained together. There may be as many vectors needed as to represent the map and state boundaries. A coarse map boundary for the United States requires about 80 vectors. The first element is either a 1 or 2 representing whether the map point is a starting point of a new boundary or a continuing point of the previously defined boundary, and the last two values of each vector represent the V&H coordinates associated with the map point.

The NLT file is a sequential file with non-zero 6-element (or 6-column) vectors chained together for up to 10 link types. The first element defines the link type number (=1,2,3,....10). The second element defines the raw link capacity C in bits per second. The third element defines the maximum link rate (Wm) allowed on the link type. The fourth element defines the link-multiplexing factor (MF). It is equivalent to the number of voice conversations that can be multiplexed on this type of link. The fifth element defines the corresponding tariff type (one of up to 10 tariffs as defined in the selected TARIFF file). The sixth element defines the multiplying factor (F_{pf}) for each link type. The factor F_{pf} is=1 for leased lines and a unique real number for each fully owned facility. The older version of EcoNets allowed the use of a "0" link type when a LINK file was not used for a data network. Furthermore, the MF factors for ALs and TKs had to be defined in the SDF file. The new NLT file structure defines the MF for any link type (1,2,3,..) used for synthesizing AL and TK bundles.

The TARIFF file is a sequential file with 1-column vectors chained together for each tariff ($\leq$10 tariffs). Each tariff is defined by 16 values: the first value defines the average monthly cost of the two local channels on both ends of an Inter-Office Channel (IOC); the next five values define the distance limits employed to compute the monthly lease costs of a private line; the next five values define the fixed monthly costs (FMC) associated with a private line whose length is between two distance limits; and the last five values define the cost per mile (CPM) of that particular private line. Experience shows that five tariffs are more than sufficient to analyze any meaningful multilevel private network for any country. One may construct several tariff files, each with a different combination of individual tariffs, including those for owned facilities employing digital coaxial cables, microwave, and fiber optical cables. A special design parameter, F_{pf}, is available in the NLT file to account for a dependency on the bandwidth or data rate associated with the owned facility.

The SDF file is a sequential file with 1-column vectors chained together to define up to 60 design parameters. These system design parameters are used for synthesizing network topologies and computing system response times and network costs. These design parameters/elements can be described as follows:

- The 1st element defines the analysis type (ATP) used for computing the response time in a data communication network. ATP=1 signifies an asynchronous (ASYNC) protocol used in FPS, Frame-Relay, and ATM systems; ATP=2 signifies a virtual multipoint (VMPT) protocol as used by some Doelz networks; ATP=3 represents a BSC or SDLC protocol employed in a polled, multidrop network; and ATP=4 is reserved for X.25 PS networks.

- The 2nd element defines the user port rate (UPR) in bps for data networks.

- The 3rd element defines the host port rate (HPR) in bps for data networks.

- The 4th element defines the input message length (IML) in bytes for data networks.

- The 5th element defines the output response message length (RML) in bytes for data networks.

- The 6th element defines the number of control units (Ncu) per virtual multipoint (VMPT) circuit of a data network based on the Doelz data network architecture.

- The 7th element defines the average message arrival rate (Rmph) per hour on each VMPT circuit in a data network based on the Doelz data network architecture.

- The 8th element defines the host think time (HTT) in seconds for a data network.

- The 9th element defines the optimization factor (Fopt) that directs the program to reduce the cost of the first feasible data network by using Sharma's Exchange Algorithm.

- The 10th element defines the average nodal processing time (T_{np}) per packet in milliseconds (ms).

- The 11th element defines the half-modem time (T_{hm}) in ms for data networks.

- The 12th element defines the propagation constant (K_{pg}) in ms/mile for data networks.

- The 13th element defines the block (or packet) length (BLKL) in bytes for data networks.

- The 14th element defines the information characters (ICPB) per packet/block of a data network.

- The 15th element defines the traffic growth factor (TGF) to study the network costs or performance as a function of traffic growth.

- The 16th element defines the factor (F_{lk}) that directs the program to read the LINK file when set to 1.

- The 17th element defines the factor (F_{nn}) that directs the program to read the NAME file when set to 1.

- The 18th element defines the factor (F_{lt}) that directs the program to read the LATA file when set to 1.

- The 19th element defines the FTD-use factor (F_{ftd}) (=1 for using an FTD file).

- The 20th element is not assigned (NA).

- The 21st element defines the access link type (ALT) for a voice/data network for the case when a LINK file is not read.

- The 22nd element is not assigned (NA).

- The 23rd element defines the allowed blocking factor for voice network access lines (B$_{al}$).

- The 24th element defines the economical call cost (ECC) for five minutes of a virtual/switched network service in dollars.

- The 25th element defines the expected call duration (ECD) of a voice/ACD call in seconds.

- The 26th element defines the delay required (DREQ) in seconds for ACD design/analysis.

- The 27th element defines the probability of delay exceeding DREQ (PEXD). It must be less than 1.

- The 28th element defines the cost of labor (C$_{lbr}$) in dollars per hour.

- The 29th element defines the fixed rate-setting factor (F$_{rst}$) for virtual tariffs used in costing an ACD. To illustrate, Frst=0.85 implies a 15% discount.

- The 30th element defines the ACD type to be designed (ACDTs of 1, 2, 3, and 4 represent an 800-Svc, MegaCom800-Svc, WATS, and the new distance-related tariffs, respectively).

- The 31st element defines the link type associated with a trunk (TKLT) for both voice and data nets.

- The 32nd element is not assigned (NA).

- The 33rd element defines the allowed blocking factor for voice trunks (B$_{tk}$).

- The 34th element defines the full-duplex factor (Ffdx) for trunks (1 for an FDX, 2 for an HDX trunk).

- The 35th element defines the maximum data trunk utilization factor (MTKU) ($\leq$1).

- The 36th element defines the backbone trunking (traffic) factor (BBTF) for influencing the amount of interswitch traffic. It is a non-zero, positive real number equal to or larger than 1.0.

- The 37th element defines the minimum value of the V-Coordinate (Vmin) for mapping.

- The 38th element defines the maximum value of the V-coordinate (Vmax) for mapping.

- The 39th element defines the minimum value of the H-Coordinate (Hmin) for mapping.

- The 40th element defines the maximum value of the H-Coordinate (Hmax) for mapping.

- The 41st element, Fvc0, is a design parameter used for a special purpose. Its value can be either 0 or 1. It is generally 0. If it is 1 and $F_{lk}=1$, then sites for a StarDataNet are connected to the tandem switches by the AL types as defined by the LINK file.

- The 41st to 48th elements define the eight virtual control factors for many special purposes. Fvc0 is used for using the LINK file for specifying link types for a StarData network. Non-Zero Factors Fvc1, Fvc2, and Fvc3 can be used to represent the hourly labor costs for the three shifts; Fvc4 is used to represent the average monthly cost of a local business line (if employed), and Fvc6 is used to represent the exact number of days (default value is always 30) modeled in the ACD network design, respectively. The Fvc5 element when set to 1 prevents the creation of an FTD file each time a backbone trunk network is modeled. Its normal value is set to 0. Fvc7 can be used to represent the cost per minute surcharge for FX line usage for ACD networks.

- The 49th element defines a self-healing factor (Fsh) for designing a Doelz network if it is fixed at 1. This will also prevent the computation of labor costs in an ACD design.

- The 50th element (F_{np}) directs the program to print nodal number on the network map.

- The 51st element (DPM) defines the equivalent typical days in a month. It is used to compute the network costs and performance summaries on the network topology maps. Its value is generally maintained as 30, whereas Fvc6 could vary depending upon the particular task.

- The 52nd element defines the distance factor (F_{dis}) for computing distances on arbitrary V&H maps. It is equal to 1 for the United States and real numbers for other countries.

- The 53rd element is not allocated (NA).

- The 54th element defines a factor (FXCM) for a fixed cost multiplier for FX lines.

- The 55th element defines the number of decompositions (NDEC) for COG modeling.

- The 56th element defines the type of first decomposition (0,1) for COG modeling.

- The SDF parameters 57–60 are available for future expansion of SDF.

The NAME file is a sequential text file (1-column vectors) consisting of N site names where N is the number of sites in the VHD file. Each name may observe the structure CCCST where the 3 bytes CCC denote an abbreviated city/location/site name and ST is the standard 2-letter state designation. For example, Los Angeles, California, is designated as LAXCA and Dallas, Texas, is designated as DALTX. The NAME file can be used to complete the database before it is created and printed. Other structures with lengths not exceeding 10 symbols may also be used.

The FTD file is a sequential file consisting of M, 3-column vectors chained together. The first element defines the From-node ID number, the second element defines the To-node ID number, and the third element defines the TCA traffic intensity (in milliErlangs or bps). If an FTD file is employed to model a Given Network Topology, the third element represents the number of circuits in the bundle connecting the From- and To-Nodes. (See Chapter 12 for complete instructions for using an FTD file for modeling a Given Net.) If the 18th SDF parameter (F_{ftd}) is a "1", then the program computes the load on all access line (AL) and trunk (TK) bundles and also computes the actual traffic loads handled by each node. The contents of an FTD file can be viewed and/or updated using the View-UpdateInputFile item of the File menu. In many cases, this data is unavailable for new systems. For such cases, TCA traffic intensities are computed for each site for both incoming and outgoing directions (voice) and the higher of the two (data). The methodology for computing the traffic intensities for each AL and TK bundle is fully described in Chapter 2.

The LATA file is a sequential file (with 1-column vectors chained together) and consists of the Local Access and Transport Area (LATA) numbers for each of the N sites as listed in the VHD file. The numbers can be obtained from commercially available tariff data handbooks. Such books have very good maps of each state, along with the boundaries of associated LATAs. The LATAs are essential for computing the costs of leased circuits for a intra-LATA network. For an ACD application, the LATA file lists the state/region number for each location as defined in the CSABDS file.

The FILES file is a sequential file with 1-column vectors chained together. Each element defines the name of the input file in a strict order defined as follows: VHD, LINK, MAP, NLT, TARIFF, SDF, NAME, FTD, LATA, FILES, CSABDS, UTBL, WUTBL, RSTBL, DTP, and SWF. The FILES file can be viewed and updated by using the ViewUpdateInputFile item of the File menu. Furthermore, the name FILES cannot be changed, in part or as a whole. Each time a Networking menu item is invoked, the program first reads the FILES input file. If any named input file is not in the directory, EcoNets directs the designer to correctly name the ith file before proceeding to the next step.

The DTP file is used to define the fraction of daily traffic occurring during each of the 24 hours. This input file is used to compute the total traffic handled during a typical day, in a defined month, and the average cost of a voice call minute or of moving one megabit.

The SWF is used to specify the number of switches, S, and the nodal ID of each switch. To illustrate, an SWF for three switches can be named as SWF3 and its contents may be as follows: 3, 15, 45, 89.

The ACD network design requires some specialized input files such as CSABDS, UTBL, WUTBL, MUTBL, and RSTBL. The structures of these specialized files are too complex to be detailed here. See Chapter 8 for their individual structures. The contents of these files can also be viewed and updated using the ViewUpdateInputFile item of the File menu. The software package derives the actual structure of each specialized input file by looking at the 2-letter prefix of each input file name.

See Figure 6.2 for a view of some typical (e.g., VHD17, NLT, TARIFF, and SDF) input files in their natural sequential form. These input files were viewed as output files in order to show their sequential nature. One can view each of these input files in their structured form (with proper number of columns and rows) by using the ViewUpdateInputFile item of the File menu as described later on. To print most of the important input files in their structured form, one must first create a database (DBF) file and then print it as shown later.

By now, the reader should get the idea of how the input files are structured. In the next section we discuss the ways in which the contents of these input files can be created, viewed, and updated. The view/update capability is particularly important if one has to perform topological optimizations in quick succession.

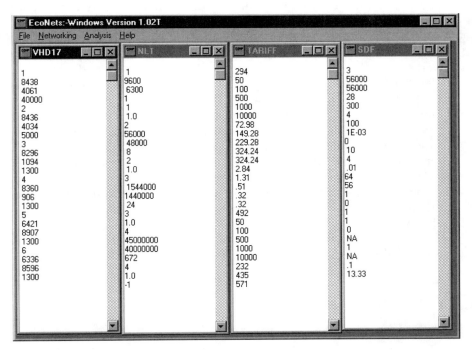

Figure 6.2 An unstructured view of four common input file data (using View Output File item of FILE menu).

6.5 Operational Capabilities of the EcoNets Software Package

The EcoNets application performs like any other Windows application. A double-click on the EcoNets icon with the mouse will open the application. In the latest version, the program goes directly to the main menubar. It is recommended that the user employ the View/UpdateInputFile item of the File pull-down menu to list the names of the 17 input files from the initialization (or start-up) file named FILES. The designer is given the chance of naming all the input files required for modeling, design, and analysis of a desired network. If one finds the contents of the input file being viewed as correct, one should click on the Close Window button. To update the contents of any cell of FILES, one must first select the cell by clicking the mouse into it. One can now edit its contents using the familiar editing process. Once done, one should then press the Return key. The editing cursor will disappear to allow the user to go to the menu . If some of these input files are not present in the folder/directory, the designer must either create each input file using the CreateInputFile item of the File menu or update an existing input file using the SaveFileAs.. item of the File menu.

It must be stressed that all of the input files named in FILES must be in the same directory in which the EcoNets application resides. The EcoNets package always begins with four pull-down menus titled File, Networking, Analysis, and Help.

The File menu is used for all operations dealing with input and output files. The Networking menu is used for modeling and designing several types of networks. The Analysis menu is used for studying or simulating the analytical models described in the book or employing these to test a given hypotheses regarding the performance of any subnetwork. The Help menu is used for referencing the contents of most important input and output files and steps for designing various types of networks without having to look at the book. Each menu when selected shows a vertical column consisting of a unique set of menu items that can be chosen to accomplish a desired task. When a menu item is selected, several windows may appear in sequence, prompting the user to either input some additional data or make some design choices. After the final task is accomplished, the main menubar appears again for the user to select another task or quit the program altogether.

We will now discuss each of the four pull-down menus that encompass the total operational capability of the EcoNets package.

6.5.1 The File Menu

The File menu has 10 items described here:

1. *CreateInputFile* for creating a new input file

2. *ViewUpdateInputFile* for viewing or updating an existing input file

3. *SaveFileAs..* for saving an existing text file under a desired name for backup

4. *Duplicate File* for duplicating any existing text file for backup

5. *VHDFilesMerge* for merging two VHD files

6. *VHDTrafficMultiplier* for multiplying the BHR traffic loads by a constant

7. *MAP/FTFilesMerge* for merging two MAP or FTD files

8. *ViewOutputFile* for viewing any text file in its natural structure

9. *PrintOutputFile* for printing any already opened output file

10. *Quit* for leaving the program altogether to go to the Desktop

6.5.1.1 Create Input File

One should use the menu item CreateInputFile to create any of the required input files. It is recommended that the designer has a source data sheet with all the values already printed on a sheet as suggested in Figure 6.3. Although such a table may take a long time to prepare, the effort is well worth the time spent on it. Most DP/telecommunications managers already require such tabulated data as part of the corporate inventory or a database (DB). An enterprise that doesn't maintain such a DB is not generally geared to perform network planning. Many users of EcoNets have shown a desire to replace the V&H coordinates with the NPA-NNX (equal to the first six digits of a site's telephone number) and let EcoNets compute the V&H coordinates internally. The writer has resisted this approach for two reasons. First, the philosophy behind EcoNets, as described in earlier paragraphs, will be violated on the grounds of excessive number crunching. Second, since the database changes each time a new NPA or a CO is created, the EcoNets package will need constant updating. This also violates the need for a software that is totally separated from any moving targets such as tariffs. Chapter 2 described a new approach of using postal Zip Codes that are less volatile. One can use the enclosed application for converting a Zip Code's Longitude and Latitude into associated V&H coordinates. The new global positioning system (GPS) technology should be harnessed to directly provide the V&H coordinates for any site in any country.

The DB inventory table, as just described, will always be needed to check the accuracy of the databases produced by EcoNets. Reading from such a table will also minimize errors during the input file creation mode. If an error is made during the file creation mode, one should not get too concerned. The designer can either scroll back and forth into the spreadsheet or use the View-Update item later to quickly update the erroneous parameters of any input file.

When the CreateInputFile item is selected, a dialog window of Figure 6.4 appears with three edit fields and two dialog buttons. One may click on the CANCEL button (or the Window Close box) to exit from the menu item altogether. The user should use the Tab key to go to the desired edit field. Enter the number of columns associated with input file type. The number of columns

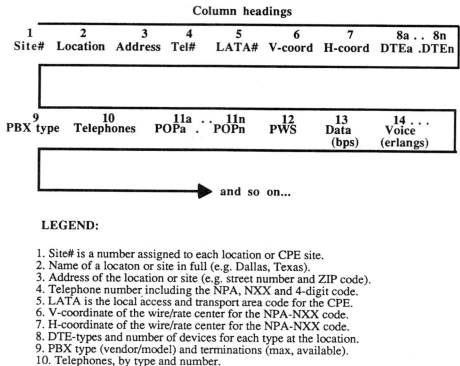

LEGEND:

1. Site# is a number assigned to each location or CPE site.
2. Name of a locaton or site in full (e.g. Dallas, Texas).
3. Address of the location or site (e.g. street number and ZIP code).
4. Telephone number including the NPA, NXX and 4-digit code.
5. LATA is the local access and transport area code for the CPE.
6. V-coordinate of the wire/rate center for the NPA-NXX code.
7. H-coordinate of the wire/rate center for the NPA-NXX code.
8. DTE-types and number of devices for each type at the location.
9. PBX type (vendor/model) and terminations (max, available).
10. Telephones, by type and number.
11. Closest POPs by OCC and their V&H coordinates.
12. Premise Wiring System (PWS) with type and size.
13. Data traffic in bits per second during a time consistent busy hour.
14. Voice traffic in erlangs during a time consistent busy hour.

Figure 6.3 A suggested source data sheet used for preparing input files.

for each input file type is listed at the bottom of the CreateInputFile window. The second edit field may then be selected by using the TAB key. Enter the number of rows associated with the input file. The number of rows is generally equal to the number of sites. In some cases, it may equate to the number of hours in the day (e.g., DTP) or number of switches (e.g., SWF), or number of nodal pairs (e.g., FTD). The program provides 10 additional rows just in case one makes a mistake. In any case, the insertion of a negative number determines the actual number of rows. After these two entries are made, the user may click on the OK button.

After one clicks on the OK button, a fully scrollable spreadsheet (SS) appears with edit fields ordered according to the structure of the input file to be created. See Figures 6.5 to 6.14 for studying the data of commonly used input files in a structured form. One should enter the appropriate values in the edit fields after reading these from Figure 6.3 described earlier. After entering a negative number in the edit field following the last correct entry, one should click on the Save

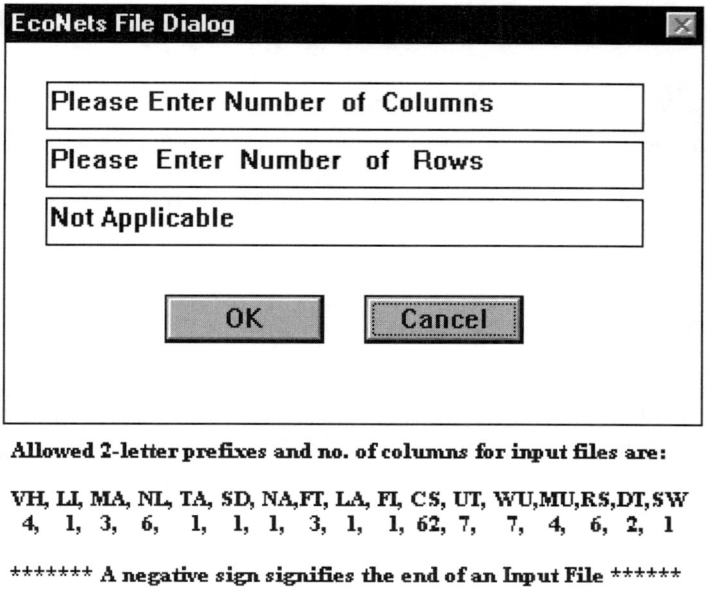

EcoNets File Dialog

| Please Enter Number of Columns |
| Please Enter Number of Rows |
| Not Applicable |

OK   Cancel

Allowed 2-letter prefixes and no. of columns for input files are:

VH, LI, MA, NL, TA, SD, NA,FT, LA, FI, CS, UT, WU,MU,RS,DT,SW
 4, 1, 3, 6,   1,   1,   1,   3,   1,   1, 62, 7,   7, 4,   6, 2, 1

******* A negative sign signifies the end of an Input File ******

Figure 6.4 EcoNets window employed for creating an Input file.

button located in the top-left corner of the spreadsheet. The user is then provided a familiar Fileselect Dialog window of Figure 6.15 that enables one to select the appropriate name for saving. The name must have the appropriate 2-letter prefix as described earlier. The program also warns the user if there already exists

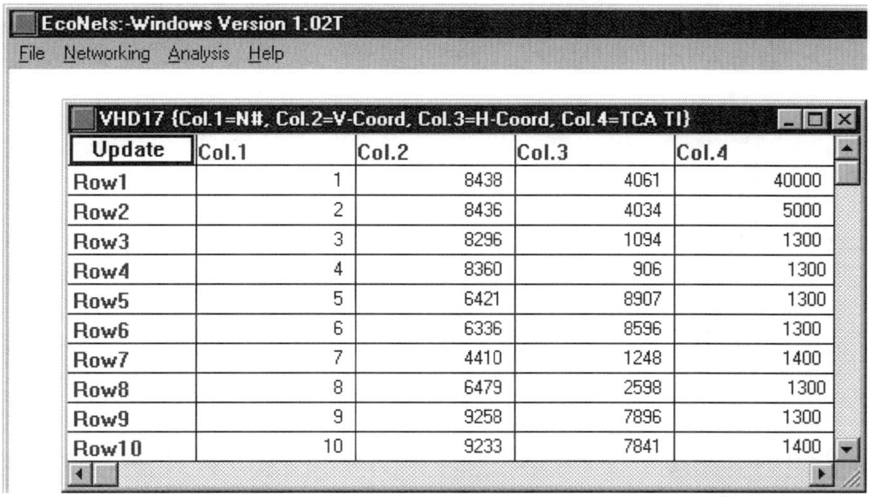

EcoNets:-Windows Version 1.02T

File   Networking   Analysis   Help

VHD17 {Col.1=N#, Col.2=V-Coord, Col.3=H-Coord, Col.4=TCA TI}

Update	Col.1	Col.2	Col.3	Col.4
Row1	1	8438	4061	40000
Row2	2	8436	4034	5000
Row3	3	8296	1094	1300
Row4	4	8360	906	1300
Row5	5	6421	8907	1300
Row6	6	6336	8596	1300
Row7	7	4410	1248	1400
Row8	8	6479	2598	1300
Row9	9	9258	7896	1300
Row10	10	9233	7841	1400

Figure 6.5 VHD file data in a structured form, as obtained from using the ViewUpdateInputFile item.

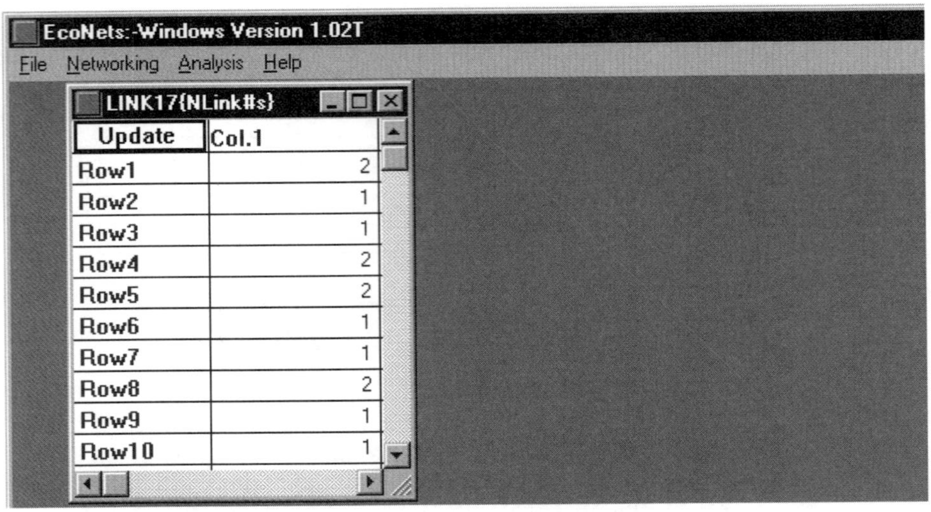

Update	Col.1
Row1	2
Row2	1
Row3	1
Row4	2
Row5	2
Row6	1
Row7	1
Row8	2
Row9	1
Row10	1

Figure 6.6 LINK file data in a structured form, as obtained from using the ViewUpdateInputFile item.

a file with that name. In this manner, one can create any one of the 17 input file types. The only constraints are (1) the structure, and (2) the proper 2-letter prefix, and the terminating negative number in the cell after the last meaningful row of the input file.

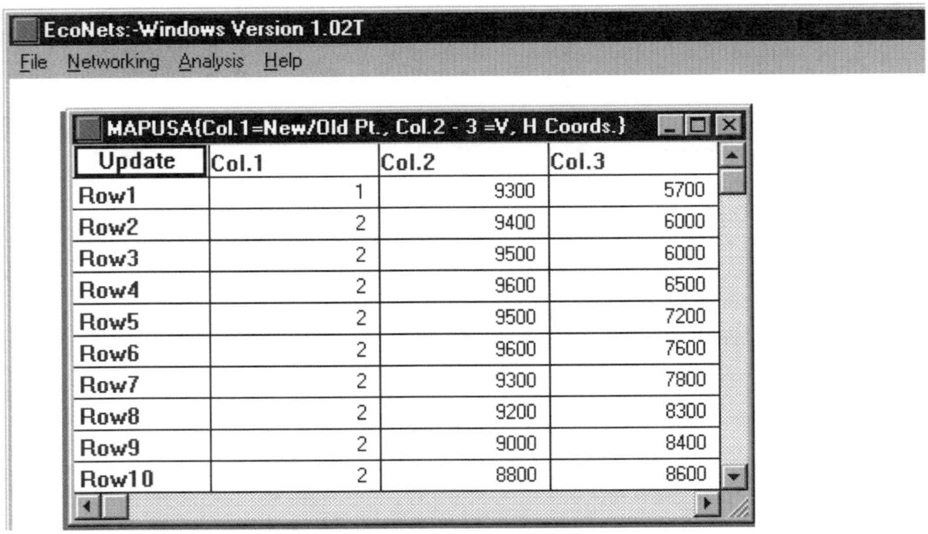

Update	Col.1	Col.2	Col.3
Row1	1	9300	5700
Row2	2	9400	6000
Row3	2	9500	6000
Row4	2	9600	6500
Row5	2	9500	7200
Row6	2	9600	7600
Row7	2	9300	7800
Row8	2	9200	8300
Row9	2	9000	8400
Row10	2	8800	8600

Figure 6.7 MAP file data in a structured form, as obtained from using the ViewUpdateInputFile item.

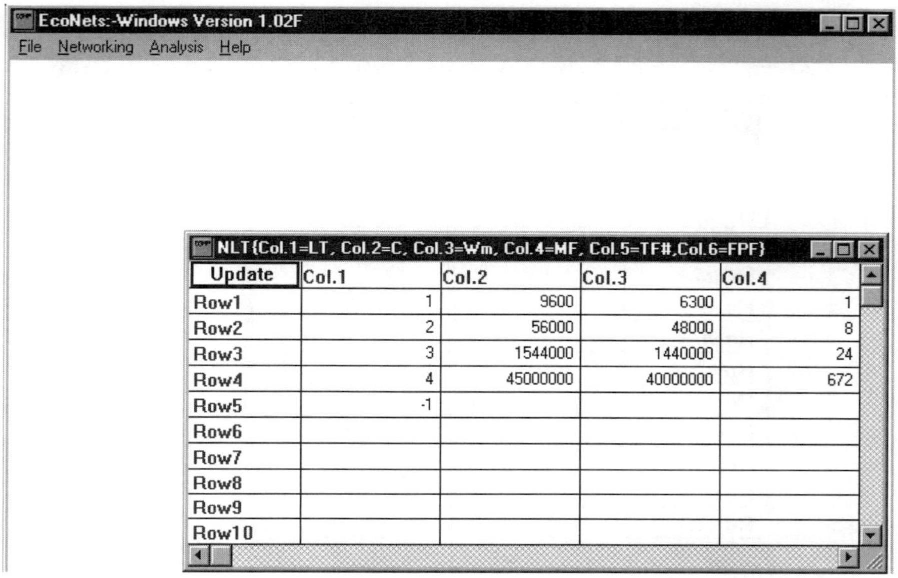

Figure 6.8 NLT file data in a structured form, as obtained from using the ViewUpdateInputFile item.

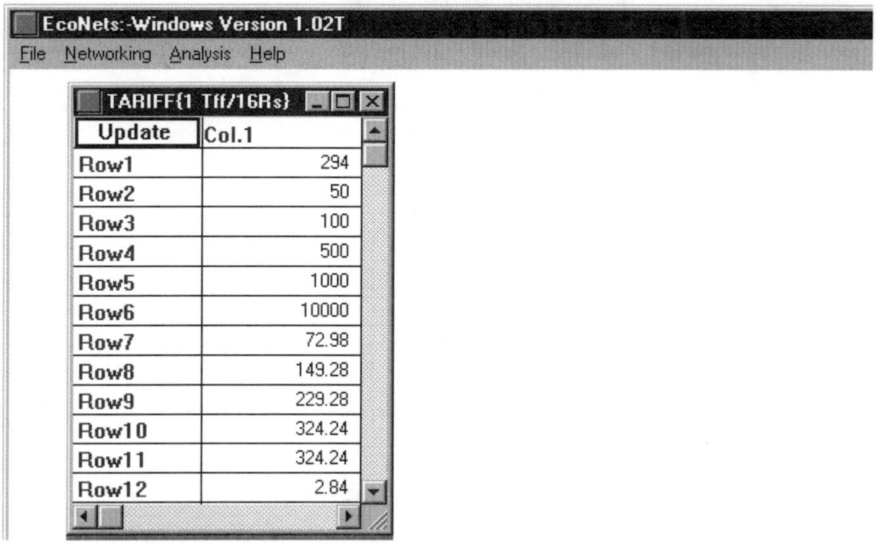

Figure 6.9 Tariff file data in a structured form, as obtained from using the ViewUpdateInputFile item.

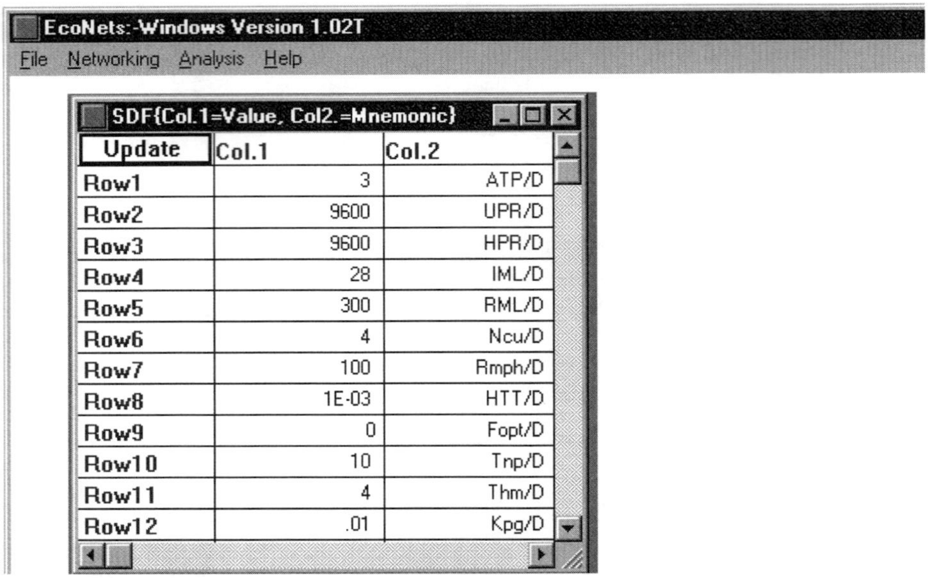

Update	Col.1	Col.2
Row1	3	ATP/D
Row2	9600	UPR/D
Row3	9600	HPR/D
Row4	28	IML/D
Row5	300	RML/D
Row6	4	Ncu/D
Row7	100	Rmph/D
Row8	1E-03	HTT/D
Row9	0	Fopt/D
Row10	10	Tnp/D
Row11	4	Thm/D
Row12	.01	Kpg/D

SDF{Col.1=Value, Col2.=Mnemonic}

Figure 6.10 SDF file data in a structured form, as obtained from using the ViewUpdateInputFile item.

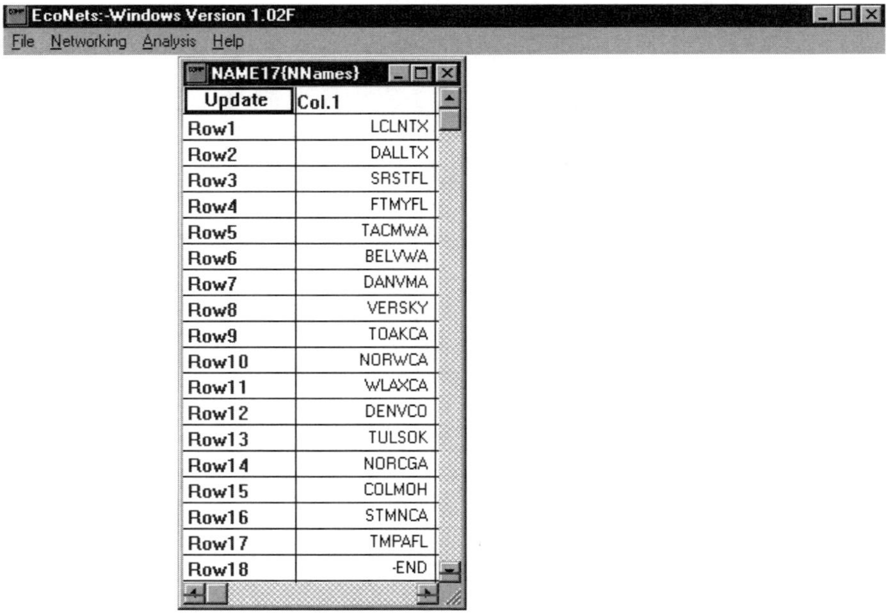

Update	Col.1
Row1	LCLNTX
Row2	DALLTX
Row3	SRSTFL
Row4	FTMYFL
Row5	TACMWA
Row6	BELVWA
Row7	DANVMA
Row8	VERSKY
Row9	TOAKCA
Row10	NORWCA
Row11	WLAXCA
Row12	DENVCO
Row13	TULSOK
Row14	NORCGA
Row15	COLMOH
Row16	STMNCA
Row17	TMPAFL
Row18	-END

NAME17{NNames}

Figure 6.11 NAME file data in a structured form, as obtained from using the ViewUpdateInputFile item.

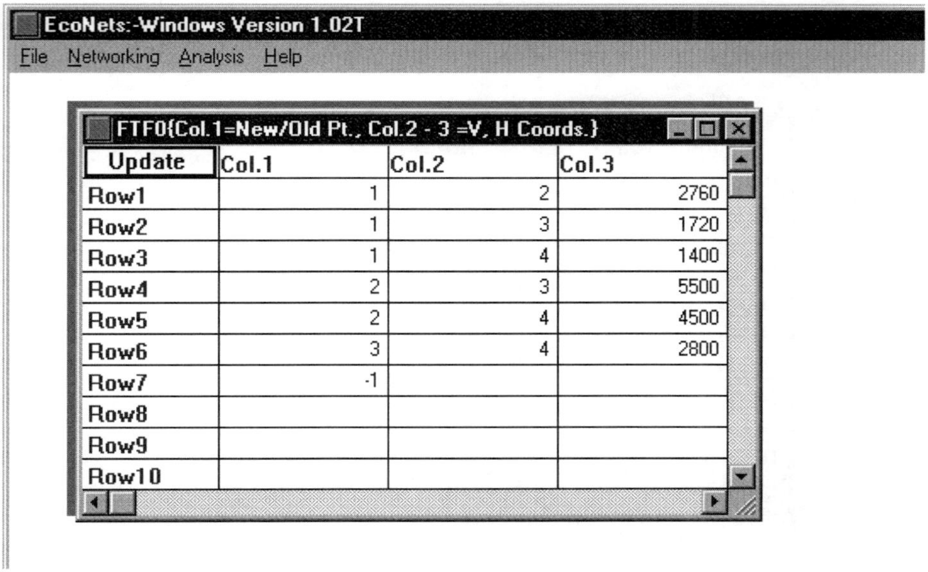

FTF0{Col.1=New/Old Pt., Col.2 - 3 =V, H Coords.}			
Update	Col.1	Col.2	Col.3
Row1	1	2	2760
Row2	1	3	1720
Row3	1	4	1400
Row4	2	3	5500
Row5	2	4	4500
Row6	3	4	2800
Row7	-1		
Row8			
Row9			
Row10			

Figure 6.12 FTD file data in a structured form, as obtained from using the ViewUpdateInputFile item.

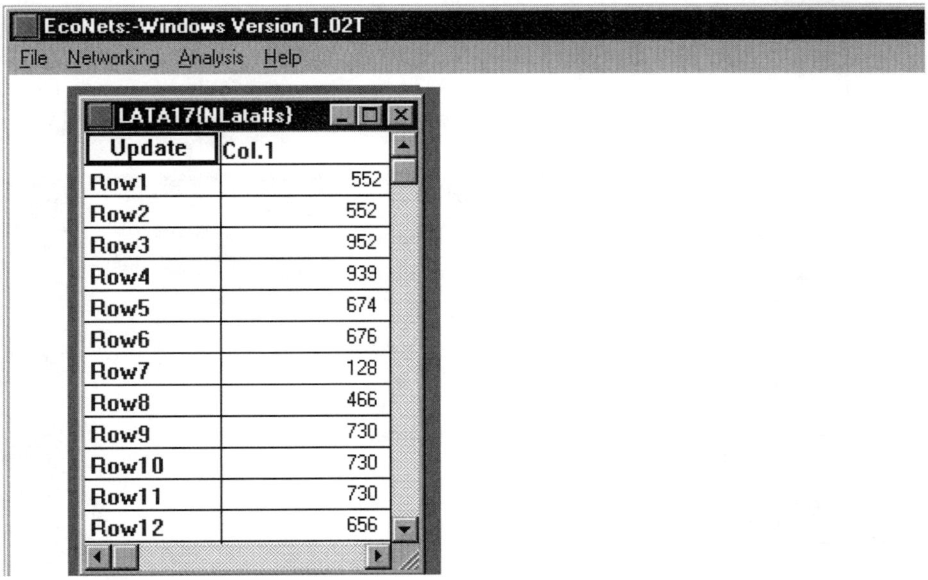

LATA17{NLata#s}	
Update	Col.1
Row1	552
Row2	552
Row3	952
Row4	939
Row5	674
Row6	676
Row7	128
Row8	466
Row9	730
Row10	730
Row11	730
Row12	656

Figure 6.13 LATA file data in a structured form, as obtained from using the ViewUpdateInputFile item.

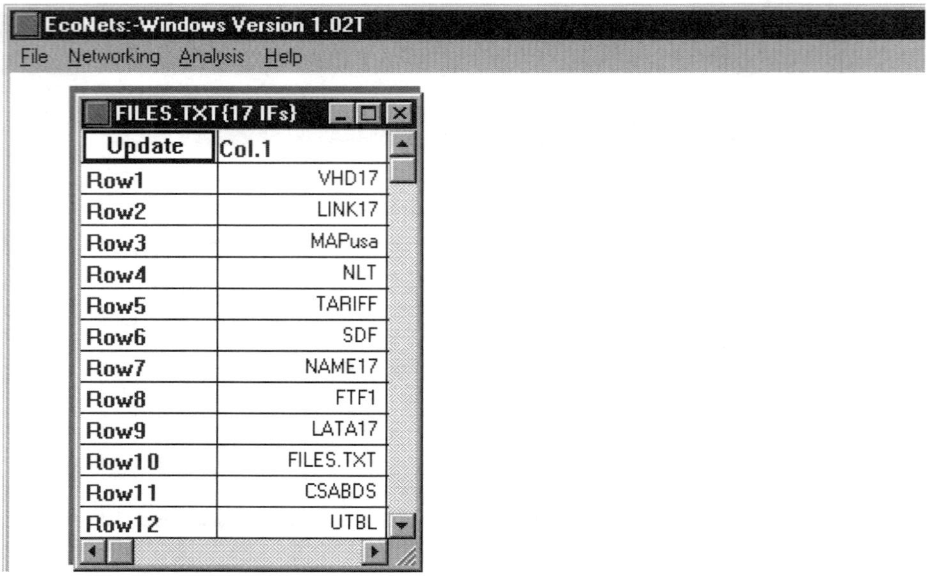

Figure 6.14 FILES file data in a structured form, as obtained from using the ViewUpdateInputFile item.

6.5.1.2 *ViewUpdateInputFile*

The user interface for this item employs a familiar Fileselect Dialog window (see Figure 6.13) that enables the user to select the desired input file and then employs a fully scrollable SS to display the contents of the existing input file. The title of the display also shows the name of the input file and meaning of each column. The icon of this input file must be in the same folder (or directory) where the application is residing. The program deduces the structure of each input file

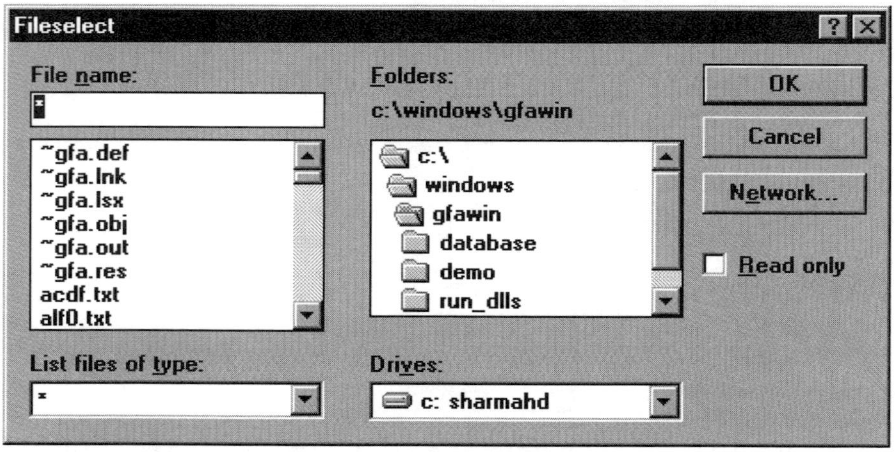

Figure 6.15 The Fileselect window for choosing or saving a file.

by studying its 2-letter prefix and provides the proper SS. After viewing the input file contents, the user may go to the menu by simply clicking on the Close Window icon. The user may also elect to alter the contents of one or more cells and then click on the Update button located in the top-left corner of the spreadsheet. The user has also the option of saving the altered input file under another name by using the SaveFileAs.. item.

6.5.1.3 SaveFile As..

When the user selects this item, the user is supplied with the familiar Fileselect Dialog window that enables the user to save the already opened file (input or output) under any desired name. This item can also be used to make some changes in the opened input file and then saving it under a different name. The original input file will remain unchanged in the directory. For some designers, it may be an ideal way to create a new file with the same structure as that of an opened input file. Since an output text file can't be edited, one can only save an original output file under a different name.

6.5.1.4 Duplicate File

When this item is selected, a dialog window of Figure 6.4 appears with three edit fields and two dialog buttons. This version requires the use of the TAB key to get to the first edit field. After entering the name of the input/output file in the first edit field, one may then select the second edit field by using the TAB key and enter the desired name of the backup file. Once this is done, one may click on either the OK button to accomplish this task or the Cancel button to return to the main menu immediately.

6.5.1.5 VHD Files Merge

When this item is selected, a dialog window of Figure 6.4 appears with three edit fields and two dialog buttons. This version requires the use of the TAB key to get to the first edit field. Enter the name of the first VHD file. After this one may select the second edit field by using the TAB key and enter the name of the second VHD file. After this one may select the third edit field and enter the name of the new merged VHD file. Once this is done, one may click on either the OK button to accomplish this task or the Cancel button to return to the main menu immediately. This approach can be used to create a very large VHD file from smaller VHD files created by different workers.

6.5.1.6. VHD Traffic Multiplier

When this item is selected, a dialog window of Figure 6.4 appears with three edit fields and two dialog buttons. This version requires the use of the TAB key to get to the first edit field. Enter the name of the first VHD file. After this, one may select the second edit field by using the TAB key and enter the name of the new VHD file. After this, one may select the third edit field by using the TAB key and enter the value of the fixed multiplying factor. Once this is done, one may click on either the OK button to accomplish this task or the Cancel button to return to the main menu immediately. This approach can be used to create a new VHD file with altered BHR traffic intensities.

6.5.1.7 MAP/FTFiles Merge

The user interface for this item is identical to that for the VHDFilesMerge item except that one is now dealing with 3-columned MAP or FTD input files.

6.5.1.8 ViewOutputFile

The user can open any text or network topology map file. In this version, output filenames are constrained to start with the prefixes AL, TK and TOP. When this menu item is selected, a familiar file-select window with a vertical scrollbar appears with many dialog buttons (e. g., Cancel, Open, Drive, Eject). By scrolling up and down, one can find the name of the desired output file (text or graphics) and double-click on it to open it. The contents of the output file appear on a window equipped with a vertical scrollbar. The scrollbar can enable the user to view any portion of the output text file. Up to four output files can be viewed concurrently. One can close any of the open windows by clicking on the Close button. One can select any one of the opened windows for viewing by simply clicking on the desired window.

6.5.1.9 PrintOutputFile

The Print item of the File menu can be used to print an already open output file. The Page Setup window will appear after the Print item is detected and request the user to select either the portrait or the landscape mode of printing. The text file generally requires a portrait mode, while the network map will generally require a landscape mode. After the printing mode is selected, a Print window appears on the screen and prompts the user to select the range of printing (e.g., one copy of pages 1 to 4), printer type, and so forth. After the output file is printed, the control is reverted back to the open output file.

6.5.1.10 Quit

This allows the designer to quit the EcoNets application altogether to go to the Finder or Desktop.

6.5.2 The Networking Menu

The Networking menu allows the designer to model, analyze, and design data, voice, and integrated broadband networks. In order to model, analyze, and design a network, one must first define all the applicable input files in FILES. The format of each input file has been defined in Section 6.3. The input file named FILES defines 17 input files. The order in which the input files are named is very important. Any error message related to an input file number always mentions the file number in FILES. For example, if the VHD file is not available according to the name specified in FILES, an alert message will inform the designer that the file # 1 is not available. The following capabilities are provided for modeling, analysis, and design:

1. *FindCOGs* for locations as candidates for switches at all levels of network hierarchy

2. *StarDataNet* for modeling, analysis, and design of data networks with a star topology

3. *DLNet* for modeling, analysis, and design of data networks with a directed link (DL) topology

4. *MDNet* for modeling, analysis, and design of data networks with a multi-drop topology

5. *MSTNet* for modeling, analysis, and design of an unconstrained minimal spanning tree topology

6. *VoiceNet* for modeling, analysis, and design of 2-level voice networks with star topology

7. *ACDNet* for modeling, analysis, and design of a voice network for incoming call management (also known as the Automatic Call Distribution)

8. *GivenNet* for modeling, analysis, and design of a given network topology using any arbitrary link types and numbers of links connecting a set of nodes

9. *CreateDB* for creating a database of most important design parameters in a structured manner for debugging anomalies in the output files

6.5.2.1 FindCOGs

It requires only three input files, VHD, MAP, and SDF as named in FILES. The program helps the designer find M=2*d+1COGs for locating 1,2,3,...M switches among the available N CPE sites for a specified number of decompositions, d, and the starting type of decomposition (horizontal or vertical). The allowed values of d are 0,1,3,7,15... since d is equal to 2^x -1 where x is any integer. The COG design parameters are defined by the SDF file elements 55, 56. For a country shaped as a flattened rectangle (e.g., the United States or Canada), it is recommended to start with the horizontal type. Of course, the designer can experiment with the other type and analyze the penalty, if any. The recommended site for switch location is generally optimal. But in some rare cases, another site that is close to the COG and carries heavy traffic may be a better choice. The program saves an output file named COGF containing a complete set of COG-related data and a network topology map named TOPF that illustrates the COG locations. These output file can be saved under a desired name by using the File menu item "SaveFileAs.." to avoid overwriting.

6.5.2.2 StarDataNet

StarDataNet models, analyzes, and designs multilevel data networks based on the connected star topology. It requires at least eight input files: VHD, MAP, NLT, TARIFF, SDF, LATA, DTP, and SWF. Additional files such as LINK, NAME, FTF, and LATA may be needed for some applications. This program employs bits per second as the unit for representing time consistent average (TCA) of peak traffic intensity. This program first analyzes the F_{lk} parameter to see if a LINK file is required for a multilevel net. Depending on that fact, the program connects the CPE nodes to the closest next-level switch or a tandem switch.

Some preprocessing is necessary. Traffic, Wi at the each ith node is noted. It is converted to {Wi - [INT(Wi / Wm)]}bps. [INT(Wi / Wm) + 1] links are used to connect that node with the next-level switch. The value Wm is defined in the NLT file for each LINK type as the allowable rate in bps. If no LINK file is used, the STARNET has only two levels: ALs (their type defined by SDF parameter ALT), and trunks in a BBNet. By using the LINK file, one can employ several switch levels. To illustrate, some nodes may have LINK numbers as "1"s and some have "2"s. The nodes with LINK numbers as "2"s will be considered as next-level switches. At the first pass, the nodes with LINK numbers as "1"s will be connected to the closest switch according to a star topology.

This process is repeated until the last level of switches (or tandem switches) as specified by the SWF is encountered. The NLT file defines the link type and the associated tariffs numbers associated with the nodal LINK number. As the last step, a fully connected backbone trunk network will be designed.

By making Fvc0=1 and F_{lk}=1, one can allow each CPE to be connected to the nearest tandem switch by a link type defined by the LINK file. This capability is especially useful for broadband and multimedia networks employing a star topology with only a single AL hierarchy.

If the FTF is not specified, the backbone trunk is designed assuming a symmetrical traffic flow proportional to the SWITCH node traffic loads. The methodology is fully described in the last section of Chapter 2 of the book. By using the FTD file the program will ignore the traffic loads as specified in the VHD file and recompute the new nodal traffic loads and the correct interswitch traffic flows before designing the fully connected backbone network. EcoNets can enable the designer to further improve the backbone network in an iterative manner by using the 13th item of the Analysis menu.

The program produces the final network topology with a summary list of AL and trunk costs and average response times. The program also prints the average cost of transporting a megabit of data on the network based on the daily traffic profile (DTP) and average number of days per month (DPM). The program automatically saves the network map and detailed AL and trunk output files for later analysis and printout. These files are temporarily named as TOPF.TXT, ALFSDi.TXT (ith level), and TKF.TXT, respectively. By varying the design parameters such as Traffic Growth Factor (TGF) and BackBone Trunking Factor (BBTF), the designer can study their effects on system costs. These output files should be saved under desired names if one must prevent overwrites.

The analysis also provides the average end-to-end response times in the network. The design parameter ATP determines the type of analysis: 1 for an ASYNC point-to-point for FPS(or Frame Relay), 2 for Doelz's virtual multipoint (VMPT with FPS), 3 for multipoint BSC/SDLC protocols, and 4 for X.25-based PSN.

6.5.2.3 DLNet

DLNet models, analyzes, and designs a data network with netlinks based on the directed link (DL) topology. The modeling analysis and design process is identical to that described for a StarDataNet. The outputs also provide the average value, maximum value per subnet, and maximum value per entire network of end-to-end queuing delay, turnaround delay on a netlink, and bid-to start delay (also called the access delay) in milliseconds. The design parameter ATP determines four types of analysis: 1 for an asynchronous FPS, 2 for Doelz's virtual multipoint (VMPT), 3 for multipoint BSC/SDLC protocols, and 4 for X.25-based PSN. The result is stored in temporary output files. If the traffic intensity from a site exceeds Wm, one or more special links are created, each handling Wm bps. The site now carries the remainder of the traffic (= Wi Mod Wm). These special links are also considered for computing the response times. One can get into a similar situation when one uses a high value of traffic growth factor (TGF) as defined in SDF. If the value of TGF is too excessive, the value of response times may be slightly lower since the software may ignore a certain number of special links, each with Wm weight. The network costs will not be affected. The program automatically saves the AL output files under the temporary names of ALFDLi (ith level). These output files should be saved under desired names to prevent overwrites. Output files for the network map and trunking analysis are also saved as described earlier. Since a netlink based on the DL topology employs only a single link type, the LINK file will not be used to connect CPEs with varying types of link types. It will be used only to model one or more levels of switches.

6.5.2.4 MDNet

MDNet models, analyzes, and designs a data network with netlinks based on a true multidrop topology. The modeling, analysis, and design process is identical to that described for a DLNET. See Chapter 8 for a comparison of several data network topologies. The program automatically saves the AL output files under the temporary names of ALFMDi.TXT (ith level) just as we did for the DLNet topology. These output files should be saved under desired names to prevent any overwrites.

6.5.2.5 MSTNet

MSTNet connects all the locations according to a minimal spanning tree topology assuming no capacity constraints. The StarNet, DLNet, and MDNet were characterized by the link capacity constraint of Wm. The program requires the six input files as mentioned earlier. The MSTNet module provides the AL costs based on a specified tariff (per ALT parameter of SDF) in addition to the topology map. This module provides a valuable insight to the designer as to the manner in which various nodes may get interconnected to minimize cost. The program automatically saves the AL output files under the temporary names of ALFMST.TXT. These output files should be saved under desired names if one must prevent overwrites.

6.5.2.6 VoiceNet

VoiceNet models, analyzes, and designs a voice network with ALs connected to the nearest switches using a star topology. MilliErlangs are employed to represent traffic load. To model, analyze, and design a voice net one also needs at least five input files, namely VHD, MAP, NLT, TARIFF, and SDF. Additional files such as LINK, NAME, FTF, and LATA may be required for some applications in a manner described for the StarNet. A LINK file may be used to specify ALTs for each node. The designer needs only to specify the number of switches and their respective ID designations. The SDF parameters ALT (for $F_{lk}=0$) and TKLT are also important. The parameters F_{pf} and the number of voice channels multiplexed (=MF) for each link type are fully defined by the NLT file. Also important are ECC for a five-minute call, the global Traffic Growth Factor, a daily traffic profile, and Backbone Trunking Factor (BBTF). Since ECC normally represents the cost of a five-minute call on the virtual private network (e.g., software defined network), one can control the number of private lines leased for ALs and trunks in a hybrid network. If ECC is made artificially too high the entire voice net will use dedicated lines. If ECC is too low the entire network will be synthesized using the virtual facilities. The cost of a backbone trunk network will also depend on the value of BBTF. The maximum trunking cost will correspond to BBTF=1. Values of BBTF larger than 1.0 will imply higher intranodal voice traffic and lower internodal traffic. The value of TGF can be varied to see its influence on the network cost. Values of AL blocking (Bal) and trunk blocking (Btk) can also be varied to study the relationship between network cost and grade-of-service. The output consists of a network topology map with summary AL and trunk costs, and the average cost of a call minute based on the supplied daily traffic profile (DTP) and average model days per month (DPM). The program automatically saves a detailed AL output file under the temporary names of ALVN.TXT. These output files should be saved under desired names if one must prevent any overwrites. Output files for the network map and trunking analysis are also saved as described earlier. One can also design an integrated voice and data network using the methodology as described later.

6.5.2.7 ACDNet

ACDNet provides a cost analysis of a corporate network designed to manage incoming voice calls using virtual facilities such as the 800- and Megacom-Services, and to handle outgoing calls on the Out -WATS facilities. At present, four types of ACDs can be modeled: (1) 800 service for handling incoming calls, (2) MegaCOM-800 service, (3) WATS for handling outgoing calls, and (4) newer virtual circuit tariffs based on mileage bands. To be precise, the selection of the ACD type must be based on the structure of the tariff as determined by the number of columns and what is designated by the first column. To illustrate, some new 800 services employ a tariff that is similar to that used by the older MegaCOM-800 service. Therefore, the ACDT=2 must be used. The older 800 service employed a tariff structure using 7 columns. One can select the correct virtual service by varying the SDF parameter ACDT as 1, 2 , 3, and 4. The program needs several additional input files in addition to the five input files mentioned earlier. The ACDNet program should not employ the FTD file due to the use of virtual circuits. A LINK file can be used to specify the facility type that serves a node: A "0" specifies only a virtual facility, a "1" specifies a correct choice of either a leased FX/LL line or a virtual facility, and a "2" specifies that only a leased FX or a Local Business Line is used to serve the node. Although this capability allows this module to synthesize a hybrid voice network with leased lines and virtual facilities, one must bear in mind that one must always view and update the SDF parameters ALT and TKLT in a manner similar to the modeling and design for a VoiceNet. The program employs a LATA file that assigns to each location a state/region number. It uses a large table to compute the customer service area based on the originating and destination state/region numbers of the two locations involved. Its size is 60X62 and is called the CSABDS. The rate-step table (RSTBL) is used for computing a rate step for each service band. It is a 62X7 matrix. The ACD module also employs a hourly usage-rate table UTBL (which is a 26X7 matrix for computing the costs of 800 or WATS) or a MUTBL (which is a MX4 matrix; M=6 for MegaCOM or an arbitrary number of mileage bands that are generally specified for newer tariffs for all virtual circuit types such as 800-, WATS-, and MegaCOM- type services). Since these tariffs generally list rates on a "call-by-call" basis by specifying first-call-seconds and the subsequent call-second rates for each mileage band, one must first convert these MX7 matrices into hourly-rate MX4 ones for each known value of ECD using the formula $(3600/ECD)*[C1+((ECD-TL1)/ TL2)*C2]$ where $(TL1, C1)$ & $(TL2, C2)$ are the two time-limit and cost pairs. One should study the older and newer MegaCOM tariff tables before attempting the conversions. The program computes the cost of the virtual facilities and also the total number of agents (if applicable) required to handle the incoming calls during each hour. It is accomplished through the use of a daily traffic profile (DTP). This determines the number of erlangs handled during each hour of the day including the busy hour (BHR). The DTP also computes the average number of equivalent peak hours per day. The designer can also select any number of switches or ACD centers and GOS parameters such as expected call duration (ECD), delay required (DREQ), probability of delay greater than DREQ, and trunk blocking factor.

One can define the cost of a labor hour to compute the total cost of labor. This is defined by the SDF parameter, Clbr, if a single value is required. Otherwise, Fvc1, Fvc2, and Fvc3 of SDF must be specified as the cost of an agent hour in the midnight, regular, and evening shifts, respectively. One must also specify Fvc4 to define the monthly cost of a local business line, Fvc6 to specify the exact number of model days in a month (only the values 1, 7, and 30 are allowed; if the value of 1 is selected, the value of Fvc0 can be represented as 0 or "non-zero" to represent a work day or a weekend-day), and Fvc7 to specify the cost (in U.S. dollars or applicable currency) for each FX call minute. One can make the SDF parameter Fsh (element#49) = 1 to eliminate the detailed agent manning analysis. This will provide only the transmission subsystem analysis and make the modeling of the ACD network quite fast since the computation of agents required for each hour is time consuming. All of the specialized input files required by the ACD module can be created, viewed, and updated by using the CreateInputFile or ViewUpdate items of the File menu, respectively. The output results are automatically saved in ACDF.TXT and TOPF.TXT. The TK output data are included in the ACDF.TXT.

6.5.2.8 GivenNet

GivenNet is quite useful in modeling any given network topology as defined by the FTF file as named in the FILES. There is a distinct difference between the specialized FTD (or FTF) file used by this menu item and the FTD(or FTF) file used for representing the actual traffic flows between all network nodes. The listing of "From" and "To" nodes is identical but the number of circuits connecting the two nodes replaces the listing of actual traffic intensity (milliErlangs or bps). The type of link (as required by the network topology) is defined either by SDF's ALT (if $F_{lk}=0$) for all links, or link type (if $F_{lk}=1$) of the "From-Node" as defined by the named LINK file. This menu item always reads the named FTF file in FILES while ignoring the SDF parameters F_{ftd}. The user is automatically provided with two output files named TOPF.TXT and ALFGN.TXT, respectively. The output file can help the user model each link bundle (as defined by the associated nodes, number of circuits, their length, and cost).

6.5.2.9 CreateDB

CreateDB allows the designer to create a database of VHD, LINK, NLT, TARIFF, SDF, NAME, FTD, LATA, FILES, DTP, and SWF files for each modeling analysis and design attempt. This is the only way to see all the important files in a structured form. It is a very useful exercise especially when one is varying a large number of design parameters to see their combined effect on the network cost/performance. The results are temporarily saved in an output file named as DBF.TXT. Since this file is so important for debugging anomalies in outputs, DBF is now automatically created each time a networking menu item is exercised. This file should be saved under a different name to avoid an overwrite.

6.5.2.10 Designing Integrated Voice/Data (IV/D) Networks

The networking capabilities described earlier appear to handle separate, stand-alone applications. That can be very deceptive. The user can design a network for integrating many applications to achieve significant economies of scale. We can illustrate such a design process for an integrated voice/data (IV/D) network for two situations: (1) where each network node has only one type of traffic, and

(2) each network node has sources of both voice and data traffic. The second situation will be the order of the day as time progresses. To model a T1or T3 broadband network in which voice and data traffic is integrated only in the backbone network (BBN) facilities, one can first model the appropriate data network designed with a desired number (greater than 1 to realize a BBN) of switches and then model a voice network with the same number of switches. Of course, each network design will require the creation of a separate VHD file. By varying the type of BBN facilities (T1 or T3) and studying the two trunk-related output files, the user can compute the cost of the combined BBN and hence the IV/D network. To model a truly IV/D network where each location has many types of traffic source, one must first compute the total number of bits generated per second by both voice and data (considering data encryption and compression) traffic at each location creating an integrated VHD file and then modeling the appropriate data network topology. Such a process can be generalized for any mix of traffic types, such as voice, data, image, and video. As traffic increases, one will find that a data network based on the star topology will tend to be optimum from both cost and response time viewpoints.

6.5.3 The Analysis Menu

The EcoNets package provides several proven tools that can enable the designer to analyze several networking situations in terms of sizing, performance and reliability, delay distributions, response times, and monthly payments for owned hardware/transmission facilities without leaving the application. It can also enable one to optimize a backbone network of an already modeled voice/data network and find routes in an already modeled given network. All of these tools are based on the material contained in Chapters 2, 3, and 4. These tools are very user friendly since each of these employs a fully scrollable and annotated spreadsheet. The designer can change any set of variables several times to see the effect on the answers. When the user has entered all the desired unknown parameters in the appropriate cells of the SS, one can then click on the Compute button located on the top-left corner of the spreadsheet to obtain the answers. The analysis menu items can be employed to prepare an unlimited number of useful charts for illustrating the behavior of telecommunications subsystems as a function of one or more meaningful design parameters.

The Analysis menu has 14 items described as follows:

1. FindB-Erl B/Poisson

2. FindN-Erl B

3. FindB-Finite M

4. MMN Queue

5. MDN Queue

6. Composite ACD

7. MultiPriority Queue

8. AvgMsgExpansion

9. LANPerformance

10. PSS Delay

11. Reliability

12. Return-On-Investment

13. BBNetOptimization

14. Find Routes

6.5.3.1 Find B-Erl B/Poisson

This analysis tool will enable the designer to compute the blocking factor, B, for a loss system characterized by A offered erlangs of traffic intensity and N servers using the Erlang-B and Poisson formulae. The tool also provides the average server utilization computed for the Erlang-B formula.

6.5.3.2 Find N-Erl B

This analysis tool will enable the designer to compute the number of servers for a loss system characterized by A offered erlangs of traffic intensity and a desired blocking factor of B using the Erlang-B formula. The tool also provides the average server utilization computed for the Erlang-B formula.

6.5.3.3 Find B-FiniteM

This analysis tool will enable the designer to compute the blocking factor for a loss system characterized by A offered erlangs of traffic intensity, N servers, and the number of sources, M. The offered erlangs should not be too large to prevent overflows. The correct procedure is to select the values of N and M and then vary the real variable, A, until a desired B (e.g., 0.01) is achieved. Using this procedure, one can draw several curves relating A to N for any desired values of M and B. The tool also provides the average server utilization computed for the Erlang-B formula.

6.5.3.4 MMN Queue Analysis

This analysis tool enables one to analyze the M/M/N queue (also known as Erlang-C formula and the worst queue) as characterized by random arrivals, negative exponential service times, Ts, and N servers as defined in Chapter 4. After the values of A and N are entered and the Compute button is clicked on, the program computes the probabilities of delay exceeding 0, Ts/4, Ts/2, 0.75*Ts, and Ts. It also provides the average time spent in the queue or Avg (T_q), and standard deviation of T_q.

6.5.3.5 MDN Queue-Analysis

This analysis tool enables one to analyze the M/D/N queue (also known as the best queue) as characterized by random arrivals, constant service time (normalized here to equal 1), and N servers as defined in Chapter 4. After the values of A and N are entered and the Compute button is clicked on, the program computes probabilities of delay exceeding 0, Ts/4, Ts/2, 0.75*Ts, and Ts. It also computes the average time spent in the queue or Avg (T_q), and standard deviation of T_q.

6.5.3.6 CompositACD

This analysis tool provides the probability of observed delay exceeding a required delay (DREQ) and the probability of blocking on the associated trunk bundle for a composite ACD system (that includes the ACD delay system and the trunk-related loss system) when the values of ECD and the number of busy hour calls are known. This is a very useful tool for computing the grade-of-service (GOS) of a composite ACD system.

6.5.3.7 MultiPriority Queue

This analysis tool enables the designer to compute the average queuing delay in a multipriority system when the number of priorities and the respective loads are known. The outputs are meaningful only when the sum of queuing loads is less than 1. This tool is valid for one or more priorities and a M/M/1 type queue characterized by random arrivals, negative exponentially distributed execution times, and a single server.

6.5.3.8 MSG Expansion

This analysis tool enables the designer to compute the factor by which the raw message transmission time must be multiplied in order to compute the actual message receipt time at a receiving channel which is also being contended for by additional sources of data. Assuming each message consists of N packets, S sources of traffic all vying for the receiving channel with a loading ρ, and first-in-first-out service discipline pertaining to a M/M/1 queue, one can compute the value of message expansion (AMEF). This approach is essential when the bid-to-end-of-message is required.

6.5.3.9 LAN Performance

This analysis tool provides a performance comparison among the three most popular LAN architectures: token ring, token bus, and CSMA/CD. The designer must provide the message length in bits, propagation delay on the LAN media, processing delay on the LAN interface unit, number of stations served by LAN, and transmission capacity of the LAN media in bps in order to obtain the maximum transmission rate on the LAN media. The answers are provided for two operational situations: (1) only one station is busy, and (2) all stations are busy. These answers can then be used to estimate the average access time on the LAN.

6.5.3.10 PSS Delay

This analysis tool provides an end-to-end delay distribution for a Packet Switching System (PSS) characterized by several queues (type M/M/N or M/D/N) in series. Several types of transactions with unique execution times are taken into account. Whereas the M/D/1 type queue is generally valid for the input queue in a PS node, the M/M/N type queue is generally valid for an output queue of PS node served by N transmission lines. The output distribution is expressed as 90, 99, and 99.9 percentile values of the end-to-end delay. Up to five subsystems in series can be handled at one time. Such results can then be combined to model a path of any length.

6.5.3.11 Reliability

This analysis tool provides the reliability analysis of each subsystem and the entire system for the given values of mean-time-between-failures (MTBF) and mean-time-to-repair (MTTR) of each component, and its configuration topology (Serial, Parallel, or Required K-out-of-N) is known. Several subsystems, each with a unique configuration topology, can be connected in series. The reliability analysis is in the form of subsystem/system availability, MTBF and MTTR. Up to 20 subsystems arranged in series can be handled.

6.5.3.12 Return-On-Investment (ROI)

This analysis tool provides the equivalent monthly costs for hardware purchased at a fixed price using the cost of money, the assumed life cycle of the system, and annual interest rate. This amount must be added to the monthly transmission costs computed using the EcoNets package. The ROI analysis generally deals with the payoff period, which can be computed as the ratio of [cost of money computed using this tool], to [monthly savings achieved by using EcoNets outputs]. This tool can also be used to model the equivalent monthly tariff for a privately owned transmission facility (e.g., fiber optical or a digital microwave facility or a VSAT hub).

6.5.3.13 Backbone Net Optimization

After an entire network has been modeled with both AL and TK facilities costs and an FTF0 file has been created automatically (by enforcing Fvc5=0) for a fully connected BBNet, one should first create a new VHD file for the switch nodes only and update the FILES with this new file, along with FTF0, NLT, and Tariff files required for modeling the BBNet. In order to model/optimize the network, one must also make the design parameter $F_{ftd}=1$ and Fvc5=1. Run the original application with switches located at all the original places (named as 1, 2, 3,...S in SWF). One should now see the BBNet topology on the screen with the cost of the fully connected BBNet as a summary. If the backbone network cost is not the same as defined in the original trunk output file, one should correct the error before proceeding to the next step. The stage is now set to optimize the BBNet using the 13th item of the Analysis menu. This will open a special but a familiar analysis spreadsheet, asking the designer to enter the node IDs of the BBNet link to be removed and the node ID of the reroute path. By clicking on the Compute button one will see the previous cost for comparison and the new cost of the BBNet. This process can be continued until the desired BBNet topology or appropriate cost reduction is obtained. Each step creates an FTD file named as FTF0i, each of which can be named in FILES and the corresponding BBNet topology modeled, one at a time, if desired, to study the costs and topologies. In this iterative process, one can design either a pure ring or a meshed-rings for use in Sonet type networks for enhanced reliability.

6.5.3.14 Find Routes

After modeling a Given Network, one can employ this Analysis menu item to find routes between any two network nodes of the network. The spreadsheet (SS) interface invites the designer to enter the two nodes of the already modeled Given Network and then click on the Compute button to define the major desirable routes between the From-node I and the To-node J. For each ith click of the Compute button and ith combination of I and J, an output file named RTBLi is created and stored in the working directory of the hard disk. These output files can be employed to prepare a set of routing tables for the design of a switching node. A Given Network topology is derived from the FTD (or FTF) file and such a table is literally interpreted to find most of the directed (no loops allowed) paths between all pairs (I, J).

6.5.4 The Help Menu

The Help menu provides the designer instant reference to the structure of input files (as required by the networking tasks), the organization of data in output files (as created by the networking menu items), description of networking menu items and network design procedures, and enumeration of techniques for using the Analysis menu items. Only one of the menu items can be invoked at a time. The Help menu has five items listed as follows:

1. About EcoNets

2. About Inputs

3. About Outputs

4. About Networking

5. About Analysis

6.6 A Description of Output Files

EcoNets creates output files containing five categories of design data defined as follows:

1. Access Line (AL) design data for networks with star topology. This covers both voice and data networks falling into the category of StarDataNet, VoiceNet, and GivenNet.

2. Access Line (AL) design data for multidrop data networks (e.g., MSTNet, DLNet, and MDNet)

3. Backbone trunking design data for all networks

4. A Specialized output file for an ACD network

5. A Database (DB) output file for each network modeled, analyzed, or designed

6. COG data for locating the switches optimally

7. Output text file for creating a network topology

Each of these output files can be viewed by using the ViewOutputFile item of the File menu. EcoNets doesn't allow the editing of output text files. Their large sizes prevent EcoNets from creating editable windows. However, some well-known applications can also be used to view, edit, and print the output text files. The output files present a voluminous amount of useful data for each network topology. The data is presented in a very structured fashion. The experienced designer can judge quickly as to the usefulness of the data. Experience will also help the designer to find the source of a problem causing a generation of random numbers in summaries. We will not discuss the actual structures of output files in this chapter, but we will illustrate their detailed structures in Chapters 7, 8, and 9. We will also discuss each portion of the output file in those chapters. Here we will attempt to define the scope of the data presented in each type of output file.

The output text file for creating a network topology is very proprietary for EcoNets. It is easy to create a graphics file in bitmap after the topology is drawn on the screen by pressing the Alt and PrtScr keys. This copies a bitmap graphics file to the clipboard. The Windows applications such as Paintbrush or Paint can be used to paste this file on the screen and store it under a desired name. One can also store it as either a 16-color or a 256-color bitmap graphics file. The first type requires about 150 Kbytes and the second type requires about 300 Kbytes. We will now describe the structure of each output table.

6.6.1 Output Files for Star Topology Network Access Lines (ALs)

As soon as the network topology is completely shown on the screen, the program automatically saves the output files for AL data. Each output file is saved under the fixed names ALFSDi for StarData networks, ALFVN for voice networks, and ALFGN for a given network topology. The user must rename these files to prevent overwrites by using the "SaveFile As..." item of the File menu. Each output file starts with a delineation of some important design parameters such as AL blocking factor (B_{al}), economical call cost (ECC), AL link type (ALT), expected call duration (ECD), and AL multiplexing factor (ALMF) for a voice network, maximum data rate allowed (W_m) per data AL and traffic growth factor (TGF).

Each output file then presents the AL design data with 6 columns entitled as node#, switch node# that serves the node, number of links (voice or data) in the AL bundle, the mileage of the AL, the cost of each AL, and the total cost of each AL bundle. There are as many entries as there are AL bundles or sites. The cost of each AL bundle is determined by the LINK type of the AL bundle, which in turn determines the tariff type through the NLT file. For voice nets, the number of ALs is determined by the AL blocking factor and the AL multiplexing factor (MF). In some cases, the number of ALs may be zero while the AL Bundle cost may be non-zero. This suggests a virtual Private Network (or Software Defined Network) option chosen by the program to save money. For a data network with star topology, the number of ALs required is computed as [$INT(W_i/W_m)$ +1]. A StarData network will also allow the designer to design an AL network for each Link type, starting with the "1" type. The designer should design a LINK input file in such a way that the nodes with a "1" Link type will be connected to the nearest lower-level switch located at node with Link type "2", and so on. The final switches can be located anywhere and must be defined in SWF. The output file also provides a summary of the results for each switch/concentrator at each level. The summary presents useful information in four columns entitled as Switch#, Switch Size (in terms of terminations), total traffic (bps) handled by the switch, and the total cost of ALs terminated on that switch. The summary also contains the total AL cost and TK cost. This summary is identical to that appearing on the network topology map. The summary also prints out the grand average of all response times on all netlinks of a StarDataNet.

6.6.2 Output Files for DLNet, MDNet, and MSTNet Data Network Access Lines (ALs) Data

The program always saves output files for AL data under the names ALFDLi, ALFMDi, and ALFMST, reflecting the topology and the applicable ith level hierarchy of the network. The user must rename these files to prevent overwrites. A typical output file first presents a summary of some design parameters such as number of switches and their locations, W_m, ATP, ALT/F_{lk}, and TGF. The output files then present data about each network link (or netlink) in terms of the Switch#, the Link (or netlink) number, actual traffic load handled in bps by the netlink, average queuing time (AQT) in MS, the average total turnaround time (ATAT) in MS on the netlink, expected value of response time $E(T_r)$ on the netlink, and the list of nodes that are served on the particular netlink. The summary also includes the maximum and the average values of T_r observed on each subsystem (served by a switch) and a system average value for T_r. The output file also summarizes the cost of all netlinks, special links, the total mileage for netlinks and special links, and the cost of trunks.

The output file then presents detailed data for each connection in the form of six columns entitled as the N1#, N2#, the link mileage, the link cost, the running total cost, and the running total for link mileage for each netlink. There are as many lines of data as there are multidrop connections. The output file then lists the data about any special links created to connect certain nodes with the switch to handle traffic loads in excess of Wm (maximum allowable data rate for the netlink). Each special link is defined in a similar manner as a connection with attributes such as mileage and cost. It should be mentioned that the special links also influence the response time computations on a statistical basis.

6.6.3 Output Files for Backbone Trunking Design Data

The program always saves an output file for the backbone trunking data under the name TKF.TXT. This follows soon after the network topology is fully designed and plotted on the active window. The user must rename (or back up) this file to prevent an overwrite.

This output file first lists some of the most important design parameters such as the Trunk-Link Type (TKLT), allowed trunk blocking (B_{tk}), economical call cost (ECC) for a virtual private network, trunk multiplexing factor (MF) for voice application, backbone traffic factor (BBTF), traffic growth factor (TGF), maximum allowed trunk utilization(MTKU), grand total (GTT) of internodal and intranodal traffic (erlangs for voice and bps for data, and the number of switching centers (S or N_c) and their location IDs.

The output file then shows the original (N_cXN_c) trunk traffic matrix. Each T_{ij} element defines the traffic intensity flowing from the ith switch center to the jth switch center. The units for matrix elements are always erlangs for a voice net and bps for a data net. If an FTD file was not used, the internodal traffic values are derived using a simplified traffic model with symmetric but proportional data flows. According to this model, the traffic intensity on the i-j trunk can be expressed as $(TT_i*TT_j)/GTT$ where TT_i and TT_j are the total traffic handled by the ith and jth switches, respectively, and GTT is the grand total of all traffic (internodal and intranodal) handled by the backbone network. The computed values of internodal traffic must then be divided by BBTF and the difference added to the intranodal traffic. The larger the value of BBTF, the lower the value of internodal traffic. It should be clear that BBTF is a simple device to influence the amount of internodal trunk traffic.

If an FTD (or FTF) file was employed for network design, only the exact values of traffic handled by each node and trunk are computed. In that case, the value T_{ij} may not equal T_{ji}. The value of BBTF (SDF parameter #36) is also ignored when computing the actual values of traffic flows.

The next (N_cXN_c) matrix presents the modified values of traffic. For a voicenet, the traffic loads Tij and Tji are simply added. For data networks, the values are added only if the trunks connecting the nodes i and j are of the half duplex (HDX) variety. One can define the trunks to be HDX or FDX through the 34th SDF parameter. The intranodal traffic values along with one side of the internodal value are zeroed since these do not contribute to trunk requirements. The next (N_cXN_c) matrix presents the modified values of the trunks required between each pair of nodes. Only the top-right part of the matrix is populated with values. The other side does not add to the information. Since the intranodal traffic does not require trunks, the diagonal values are also zeroed. For voice, the Erlang-B formula is employed to compute the number of trunks required for a given value of trunk blocking (or B_{tk} per 33rd SDF element). For data networks, the number of trunks is computed using the allowed maximum trunk utilization factor (or MTKU per 35th SDF element). The last matrix defines each trunk bundle by specifying the two associated switch nodes, the circuit miles, cost per circuit, and the total cost of each trunk bundle.

The last lines of the output file for trunking data summarizes the total access line cost, the total trunking costs, and the total system costs. The cost of a single five-minute voice call or moving 1 megabit of data on the network is also listed. The contents of the output file for trunking data can be viewed by the ViewOutputFile item of the File menu. Up to four output files can be viewed concurrently.

It should be mentioned that an FTD (or an FTF) file is automatically created if Fvc5=0 at the time a networking menu item is used. The name of this file is FTF0. This can then be employed to model and optimize the backbone network at a later time while maintaining Fvc5 equal to 1.

6.6.4 Output Files for a Database

The output file representing an integrated database for any set of input files listed in FILES can be created by using the Networking menu item CreateDB. A DBF is also created automatically each time a Networking task is completed. One should make the 16th, 17th, and 18th parameters of the SDF file equal to "1" to read the listed LINK, NAME, and LATA files so that their data can also be incorporated into the database. A complete database presents the entire network design data in a very structured form that is easy to read. Its data can be referred to at a later date when analyzing all of the output files data or debugging for any anomaly in the output results.

The Output file for a database first lists the entire VHD file in a structured fashion with nodal number, the associated V&H coordinates, nodal traffic (actual or computed from an FTD file), LATA number, LINK type, and NAME. At the end of this nodal data, the DB also summarizes the total nodal traffic handled by the system during a peak period.

The DB then lists the entire NLT file in a manner that is easy to read. Each line lists the link type number, the link capacity in bps, maximum traffic weight or allowable data rate (W_m) on the AL, the multiplexing factor (MF) or number of voice conversations that can be multiplexed in the AL, the associated tariff number, and the private facility factor, F_{pf}. There are as many rows as there are node/link types (<10).

The DB then lists each of the tariffs contained in the TARIFF file used for network design. For each tariff, the monthly local-loop costs for both ends of the leased circuit, five mileage band limits, five fixed monthly costs for each mileage band, and five costs per mile for each mileage band are listed in a manner that is easy to read. The DBF then lists the 60 elements of the SDF input file along with the parameter mnemonic. Each parameter symbol is also characterized by a suffix (/D,/V, /A, or /C) to designate the associated applications to data or voice or ACD or common to all networks.

The DBF also lists the contents of DTP and SWF input files. The DBF also lists the Networking Item employed for any networking task. If the SDF parameter $F_{ftd}=1$ then the contents of an FTD file is also listed. The contents of a DB can be viewed by the File menu Item "ViewOutputFile".

6.6.5 Output Files for COG Data

An output file can also be created while working with the FIND COGs Networking menu item. It first lists the name of the VHD and MAP input file employed, the number of decompositions (=d), and the type of first decomposition (i.e., Horizontal/Vertical) employed as defined in SDF. It then lists the V&H coordinate of each mathematical COG, the closest network node/site, and its associated V&H coordinate. Such an output file should be created and consulted for selecting switch/concentrator locations. This output file can also be viewed by using the File menu item "ViewOutputFile". Again this output file must be saved under another name to prevent overwrites by using the "SaveFileAs.." item of the File menu.

6.7 Computing the V&H Coordinates for Countries Other Than the United States

It was in 1960 when AT&T began using the V&H coordinate system to determine the cost of interstate DDD calls and leased private lines for North American countries, Canada, Mexico, and the United States. V&H coordinates introduced a great deal of simplicity in computing line-of-sight mileage. Previously, one used the latitudes and the longitudes and a complicated formula to compute the surface mileage. Databases are now available that relate the NPA-NXX portion of each telephone number to the V&H coordinates of an associated wire (or rate) center. Maps based on Donald's projection are also available. Using these, it is an easy matter to compute the V&H coordinates of U.S. and Canadian locations for the VHD input file and map boundaries of the three countries and their associated states or provinces for the MAP input files. Although the need for designing private networks for other countries and continents is increasing, no unified databases are available for associating telephone numbers to V&H coordinates for other countries. However, one can employ an approximate method for creating simple maps employing equivalent V&H coordinates valid for other countries in order to avoid the use of latitudes and longitudes.

To accomplish this, obtain a good atlas with maps of countries drawn on any appropriate projection system (e.g., Bonne, Conic, etc.) with linear scales of miles and kilometers. Overlay a transparent grid with each inch divided into 10 intervals. Number the major scales (i.e., inches) in such a way that the vertical scale values range between 3,000 and 10,000, and the horizontal scale values range from 0 to 10,000, similar to the V&H maps for North America. The boundaries of the country and its provinces or states can thus be read to create MAP files. One can also read the V&H coordinates for all the locations to create any VHD file. This method was employed to create MAP and VHD files for Australia and India. It was found that one required a multiplication factor of 1.557 to compute the distances in kilometers for Australia, and a multiplying factor of 0.86 to compute the distances in miles for India if one used the maps appearing on pages 40 and 34 of the 1980 edition of the *Hammond World Atlas*. It was a coincidence that the two projections employed for Australia and India provided distances that are closer to miles than kilometers using the approximate method. The multiplier for distance computation and the ranges for drawing map boundaries can be specified by the SDF parameter #52 (Fdis) and design parameters #37,38,39, and 40 (Vmin, Vmax, Hmin, and Hmax), respectively. See Figures 6.16 and 6.17 for maps obtained for Australia and India using the technique just

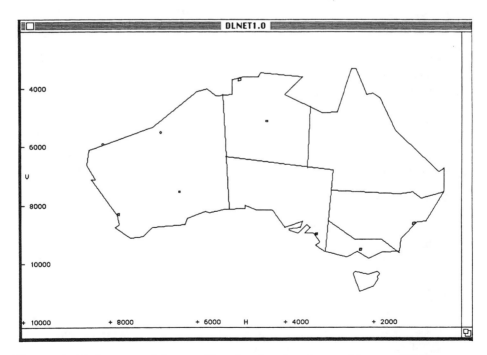

Figure 6.16 A map of Australia obtained using arbitrary V&H coordinates.

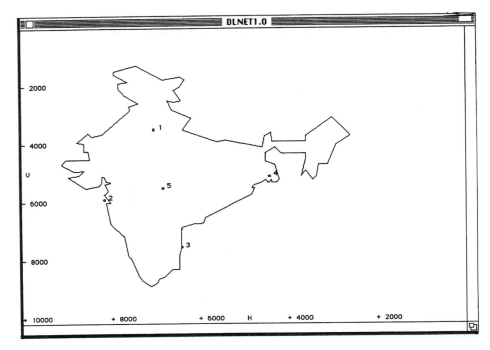

Figure 6.17 A map of India obtained using arbitrary V&H coordinates.

described. One can also create maps for U.S. states using a map based on Donald's projection. In order to use such state boundary maps, one must adjust the map-related SDF parameters, Vmin, Vmax, Hmin, and Hmax, as described earlier. The same requirements apply for using arbitrary maps of provinces of other countries.

6.8 Creating Models of Tariffs for All Countries

It should now be clear that EcoNets is primarily an interactive tool for optimizing network topologies very rapidly in an iterative manner as dictated by the mind of a network designer. Such a goal was partly achieved through (1) avoiding the use of a large tariff database during the topological optimization process, (2) using a site database with already computed V&H coordinates, and (3) using meaningful but simplified tariffs for computing circuit costs on a monthly basis.

In order to avoid the use of large databases, we employ the computed V&H coordinates for each site (or node) directly in the VHD file using the corporate database of Figure 6.3. This minimizes the computation of the site's V&H coordinates each time a circuit is priced by analyzing the NPA-NNX part of the site's telephone number. Since a given network topology employs only a few link types (or tariffs) during a network design process, one doesn't need a database of all tariffs (as offered by AT&T, US Sprint, or MCI). Instead, one can create a set of unique NLT and TARIFF files defining most link types and corresponding tariffs such as AT&T's Tariffs 9 for VG, 56Kbps/DDS, ACCUNET T1.5 and T3 circuit-types, and use these to study a large number of network topologies. Only when a final network topology is achieved, one can use a private line pricing (PLP) package to compare all the available IEC tariffs to select a particular set of common carriers. Most IECs will also be glad to compute the monthly costs of all the leased lines associated with an EcoNets-derived optimum network topology.

Situations also arise when an enterprise must consider owning its own transmission facilities. In particular, many state and federal agencies already own rights on laying the digital fiber optical (DFO) cables or digital microwave (DMW) routes on certain paths. By creating simplified models of onetime costs and using the Return-on-Investment model already described, one can compute the equivalent monthly costs of such privately owned facilities. Each model generally consists of two fixed constants, K and FC_i, and two multiplying variables, CPM_i and F_{pf}, to handle cable/path length as shown in the following expression:

$$\text{Monthly cost(\$)} = [\ K + FC_i + CPM_i*(\text{circuit/path length in miles})\ ]*F_{pf} \tag{6.1a}$$

or simply

$$\text{Monthly Cost (\$)} = MC_1 * F_{pf} \tag{6.1b}$$

where

K = average cost of two LCs (as defined in the TARIFF file)

FC_i = fixed cost associated with the ith mileage band (as defined in the TARIFF file)

CPM_i = cost per mile cost associated with the ith mileage band (as defined in the TARIFF file)

F_{pf} = factor associated with the private facility (as defined in the NLT file)

MC_1 = model of the monthly cost of an equivalent leased line as defined by Equation (6.1a) or fully defined by the 17 elements in a TARIFF input file.

Figure 6.18 plots monthly costs of four MCI leased lines (VG, 56Kbps/DDS, ACCUNET T1.5, and ACCUNET T45 as defined in Table 2.6 of Chapter 2) and a fully owned, private digital fiber optical (P-DFO) facility providing a bandwidth (BW) of three times that of a T3. It is interesting to observe that the P-DFO facility compares favorably with a T3 facility for short distances; its monthly costs are much lower than those for a leased T3 line for longer distances. Similar models can be created for other foreign tariffs for either leased lines or fully owned facilities. The existing structure of a tariff (i.e., a constant [=K], five mileage bands, five fixed costs, and five costs-per-mile values for each interval) and NLT file's parameter related to corresponding link type(=Fpf) should be adequate to compute the monthly costs for a fully owned facility. If a country employs more than five (5) mileage bands, one can easily approximate the monthly lease costs by

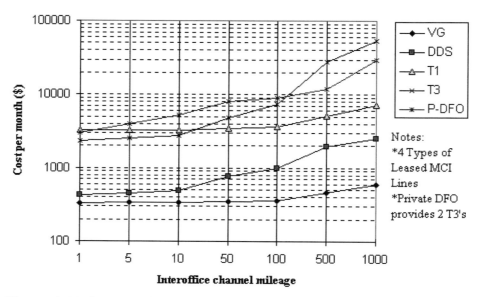

Figure 6.18 Cost models of some leased and owned transmission facilities.

using only five mileage bands. The parameter, Fpf, is always equal to 1.0 for all leased facilities. It should be pointed out that a single NLT file can now define a small mix of both leased and privately owned facilities.

6.9 Performance Metrics for EcoNets

A designer is generally interested in the execution times and the random access memory (RAM) requirements for any given software package. Excessive execution times will prevent the software package from being called an interactive one. Excessive memory requirements may force the package to be either impossible to implement on a desktop computer or too expensive to afford, despite the recent drops in RAM prices. The following paragraphs define two useful performance metrics for EcoNets that are valid for a 66MHz, 80486 DX2 IBM-compatible PC desktop computer.

Several typical network configurations were run to compute the execution times for various network topologies and architectures related to the Networking menu. The results are plotted in Figure 6.19 as a function of the number of locations (N) and for one switch (S=1) or an appropriate number of switches exceeding 1 (S ≥ 2). Various network topologies such as multidrop (MD), StarNet (voice/data), ACDNet, and Minimal Spanning Tree (MST) with no capacity constraint are considered. Execution times for the optimized DLNet and MDNet topologies achieved by using Sharma's Exchange algorithm (by making SDF parameter Fopt =1) are also plotted (with the suffix OPT).

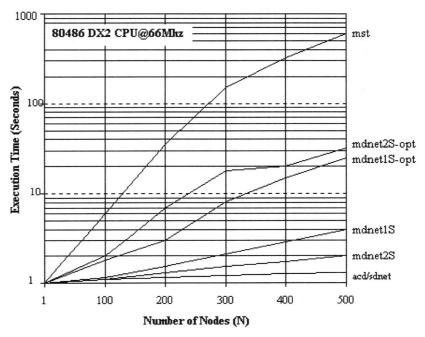

Figure 6.19 Execution times as a function of number of nodes (N) and networking option.

The curves show that most networks (star data, MD data, voice and ACD) require less than 2 seconds to run for 250 nodes or less. Whereas the optimized MD data network topologies (mdnet1S-Opt) take about 13 seconds to run for 250 nodes, an optimized MD datanet with 250 nodes and two or more switches will take only 5 seconds. The uncapacitated MST topology (not run for most useful networks) will take up to 75 seconds to run for 250 nodes. These curves justify the claim that the EcoNets tool is already an interactive tool. The future hardware trends can only point toward additional improvements. It should be mentioned that until a source code becomes available for Windows 95, the execution times of Figure 6.19 will not improve much with faster CPUs. Despite that fact, the present version does run on Windows 95.

The development of EcoNets was started in 1984. The author employed the Future Basic compiler and source code to harness Apple Computer's Macintosh graphic-user interface. Since that code doesn't allow the use of a floating-point coprocessor and it employs a slow linker, the execution times are about 40 times slower than those illustrated in Figure 6.19. An equivalent compiled code is available for a Macintosh and it is useful for only small networks (N ≤ 60).

Random access memory (RAM) requirements have been plotted in Figure 6.20 for some typical network configurations characterized by the number of nodes or locations (N) and number of switches (S). The RAM requirements are for the EcoNets application only. Windows 3.1 or Windows 95 operating systems

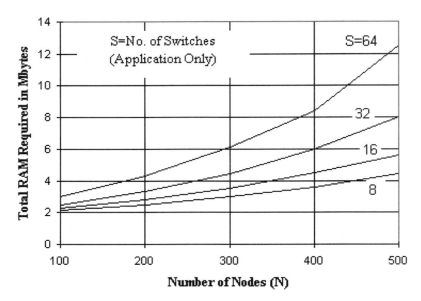

Figure 6.20 Total RAM requirements as a function of number of nodes (N) and switches (S).

create their own demands for RAM. A 250-node application should require about 16 Mbytes for Windows 95 environments. Additional RAM is generally recommended to achieve multitasking.

6.10 Concluding Remarks

An attempt has been made to define (1) inputs required by the EcoNets package; (2) operational capabilities of the EcoNets package such as creating input files and viewing/updating input files in order to model, design, and analyze telecommunications networks in quick succession; and (3) outputs that describe the computational data. Two useful performance metrics (i.e., execution times and random access memory requirements) were also discussed for the EcoNets package. Now we are ready to apply the analytical tools of Chapters 2, 3, and 4, the algorithms of Chapter 5, and the EcoNets software package to obtain some insights related to corporate voice, automatic call distribution (ACD) data, integrated broadband, backbone, personal communication system(PCS), and common channel signaling (CCS) networks. This task will be accomplished in the next eight chapters.

Chapter 7 studies enterprise voice networks and present results obtained by varying several key design parameters as defined in the LINK, NLT, and SDF input files. Chapter 8 deals with the ACD networks that employ one of the many virtual services such as 800-Service or MegaCom 800 Service. Chapter 9 concentrates on data networks with emphasis on the multidrop and star topologies. The dependence of network topologies on the algorithms employed and the number of switches at all levels are studied in detail. Chapter 10 focuses on the implementation considerations and design of integrated broadband networks for achieving significant economies of scale associated with multimedia networks. Chapter 11 focuses on the modeling and optimization of backbone networks. Chapter 12 concentrates on the modeling process for a given network topology through the use of a special FTD input file. Chapter 13 focuses on the design process for a personal communication system (PCS or cellular or mobile systems). Chapter 14 illustrates the design process for a common channel signaling system (CCSS) which is the basis for all advanced intelligent networks (AINs) of today. Each chapter clearly shows the way in which the appropriate design parameters are selected and outputs are interpreted for each network type.

Exercises

6.1 What is the start-up input file? How many input files are named in it? Are these input files named in a strict order? Can one change the name of this start-up file and still manage to run EcoNets?

6.2 Describe the natural structure of each input file. How does a designer define the end of each input file? How can one see the logical structure related to each input file in a printed form?

6.3 Name at least six input file types that are generally most essential to model, analyze, and design a data or voice network.

6.4 Name four input file types that can be prevented to be read by EcoNets. Also name the corresponding four SDF design parameters that inhibit the reading of these special files.

6.5 Name two important uses of a database file (DBF). How is it created? Is it created automatically when a Networking item is executed by EcoNets?

6.6 Name the common and specialized input files required to design or model an ACD network based on the MegaCom800 virtual service.

6.7 Name the input files that define the design parameter Wm (equal to the maximum allowed data rate on a multidrop netlink), the design parameter TGF, and busy hour TCA values of TI, respectively.

6.8 Name the input files that define the link type connecting the CPE to a network switch, the applicable tariff number, virtual service usage tariff for 800-Service, and daily traffic profiles, respectively.

6.9 What are the three major uses of an FTD (or FTF) input file?

6.10 Name the input files required to model a GivenNet topology.

6.11 Name the input files that may be necessary to model and optimize a BBNet.

6.12   Name the input files that may be necessary to model a hierarchical data network.

6.13   Name two methods of designing an integrated voice/data (IV/D) broadband network.

6.14   What happens when an input file named in FILES is not in the working directory? What steps are necessary to correct the matter?

Bibliography

Fike, J. L., and H. D. Jacobsen. "Applying Modeling Techniques in Network Design." Paper presented at the June 3, 1991, Session M13 at the 1991 ICA Conference.

Van Norman, H. J. "WAN Design Tools: the New Generation." *Data Communications*, Oct. 1990, p. 129.

7

Design Process for Voice Networks

7.1 Introductory Remarks

Considering an American household as an enterprise, voice networks can take on any shape and size. The smallest enterprise voice network will then consist of a CPE in the form of one or more telephones and generally a single leased subscriber line. On the other extreme, an enterprise voice network for a large business may consist of many fully owned PABXs located at large branches, some tandem switch nodes collocated with a subset of PABXs, and leased ALs and trunks. The subscriber lines are also generally owned as a part of a premise wiring system (PWS) at each location of the enterprise. Between the two extremes, a myriad of voice networks can be modeled. An economic analysis may prompt the business to access small locations through the public network (PN). According to the concepts introduced in Chapter 1, these small offices and locations are defined as off-net locations. To reach these off-net locations, the enterprise must lease one or more bundles of off-net ALs (also called ONALs) connecting one or more of its network nodes with the public network COs. With a mix of on-net and off-net locations, the private voice network becomes a hybrid network. The hybridization of a voice network can be extended by leasing Centrex services at some locations. This eliminates the need for purchasing PABXs (the other name for voice LANs).

The decision to buy or not to buy a PABX (or a voice LAN) should be made for the same reason as for a data LAN. Availability and benefits of custom calling features, Custom Local Area Signaling Service (CLASS) features or Call Management Service (CMS) features, voice mail, integrated voice/ACD services, integrated voice/data services, call recording capabilities, and cost savings over the life cycle are some of the main reasons that apply to a voice LAN.

Wide Area Networks (WANs) can be synthesized by enhancing the software of certain PABXs to make these also act as tandem switches and connecting them via a desired number of leased-trunk facilities. The remaining PABXs are then connected to the nearest enhanced PABXs via ALs resulting in the multistar, multicenter network topology of Figure 1.8. In some cases, it may be preferable to employ a certain number of standalone tandem switches collocated with a subset of existing PABXs. Still another way to synthesize a WAN is to employ a correct number of intelligent T1/T3 multiplexers that allow the sharing of T1/T3 trunk facilities between competing applications and switching facilities. Chapter 10 presents a methodology for selecting the attributes of T1 or T3 multiplexers for designing an integrated voice/data/image/video broad-band network. In the future, when the enterprise networks begin deploying ATM switches, it will be possible to synthesize a WAN with T3 or OC-N type facilities without intelligent multiplexers. Which approach is ideal for the enterprise at any given time is bound to be affected by the needs of an enterprise and the available switching and transmission facilities. The design process should be primarily concerned with the network topology considerations. The set of available network nodes and transmission facilities will always be changing with time.

Using the network design tools and algorithms of Chapters 3 and 5, and a software tool as described in Chapter 6, one can design enterprise voice networks for either on-net traffic or off-net or any mix of on-net and off-net traffic. The on-net traffic originates from enterprise locations connected to the private network via leased lines. The off-net traffic either terminates into or originates from the off-net locations of the enterprise network. Please consult Figure 1.22 for an illustration of off-net traffic paths. The off-net traffic analysis employs an approximate cost model. An exact analysis can be performed for off-net locations using the ACD network modeling techniques as described in Chapter 8. The methodology presented here can also be used for making purchase/lease decisions for PABXs. The methodology of this chapter can also be employed to optimize the existing network topologies of LANs, MANs, and WANs. Most designers will be surprised by the amount of total savings. The design process described here can also be useful to all enterprises and long distance service (LDS) providers. In particular, the LDS providers can differentiate their services by adding features available from the Equally Accessed Public Intelligent Networks (INs). See Chapter 13 for a discussion of INs and Advanced INs based on the Signaling System 7 (SS7) or Common Channel Signaling (CCS).

A reader interested in getting a good insight into the evolution of voice networks, both public and private, should consult the works by Sharma et al. (Sharma, De Sousa, and Ingle, 1982), Nortel (Nortel, 1994), and Nortel (Nortel, 1995). Some innovative investigations into transmitting voice over IP, Frame Relay, and ATM networks have been reported by Gareiss (Gareiss, 1996), Green (Green, 1996), Pappalardo (Pappalardo, 1997) and a "Voice over ATM" Seminar (TranSwitch Seminar, 1996).

7.2 The Basic Design Process for a Voice Network

7.2.1 A Description of Input Files and Design Parameters

In order to design a voice network topology, the network design package as described in Chapter 6 allows one to execute all the steps defined in Chapter 5, and in particular, Sections 5.4.1 and 5.4.2, using the input design data of input files. The network design software package also allows one to vary a large number of critical design parameters iteratively to arrive at an optimum voice network.

The first step is to prepare the set of input files that must be named in FILES. The required input files for designing a voice network are VHD, LINK, MAP, NLT, TARIFF, SDF, NAME, LATA, FILES, DTP, and SWF. The input files CSABDS, UTBL, WUTBL, MUTBL, and RSTBL as named in FILES are useful for ACD networks only.

The V&H coordinates for each site are derived from the enterprise database (EDB) as reflected in Figure 6.3. If this EDB is not available, the enterprise is not ready for network planning. One will need to buy a private line pricer (PLP) to prepare such an EDB. These tools basically relate the NPA-NNX of the site's telephone number to its V&H coordinates. Large population growth in urban areas is putting great pressure on the nation's numbering plan. This in turn is causing the emergence of new NPAs and/or changes in old telephone numbers. The PLPs are always being updated. The TCA values of TIs handled by a site can be computed using the procedure described in Chapter 2. The number of information workers, type of profession, or number of active PABX terminations may influence the values.

The LINK file is used only if a different access link type must connect each CPE to the nearest tandem switch. In order to use a LINK file, one must set the SDF design parameter F_{lk} to 1. A LINK file can be ignored by setting F_{lk} to 0. For this case, the SDF parameter ALT defines the access link type. The LINK file will be ignored by the modeling examples of this chapter since we employ the same access link types throughout the voice network. Only a single trunk link type is allowed to interconnect the tandem switches for all networks.

The MAP file is required by every networking task since it is used to draw the boundary map. The MAP file is provided under the name MAPUSA. It consists of V&H coordinates for United States boundary. Only 79 points were chosen to get an easy to recognize boundary consisting of straight line segments. Using the procedure of Chapter 6 we can create a MAP file for the boundary of a country other than the United States, Canada, and Mexico.

The next important input file is an NLT file. It defines the set of link types that will be used in modeling the AL and trunk facilities. These link types can be a mix of leased and fully owned facilities. For the voice application, the important link attributes are the maximum number of voice conversations that can be multiplexed on the link (equal to AL/TK multiplexing factor, or simply MF), the corresponding tariff number (TF#), and a multiplying factor for a private facility, F_{pf}. The AL type is either defined by the LINK file (for F_{lk} equal to 1) or by the SDF parameter ALT (for $F_{lk} = 0$). The trunk link type is always defined by the SDF design parameters TKLT. It defines the single link type used for connecting all the tandem switches according to a fully connected mesh topology as defined by Figure 1.2 (a).

We will use a tariff file that defines at least three tariffs: Tariffs 9, 10, and 11 for the voice grade (VG), 56Kbps DDS, and T-1 lines valid for 1987. Such tariffs were defined in Table 2.4. We will also assume that the average monthly cost of the two local loops and termination charges equals an amount of \$102.00 per VG, \$62.76 per DDS and \$0.00 per T1 circuit, respectively. This implies no bypass for the VG circuits and some form of bypass for the DDS and T1 circuits. Although most of the analyses assume some bypass for the DDS and T1 circuits, an attempt is made near the end of this chapter to show the comparison between network cost with and without bypass of the local exchange carrier.

The next important input file that needs to be updated is the SDF file. It defines the values of 56 design parameters. These can be viewed (Mnemonic and corresponding value) by using the ViewUpdateInputFile item of the File menu. Each SDF mnemonic is also provided with a suffix of V, A, D, or C implying Voice, ACD, Data, or Common application, respectively. Only a few of them affect a voice network significantly. These are TGF, ALT, B_{al}, ECC, ECD, TKLT, B_{tk} and BBTF as defined in SDF, and the parameter MF as defined in NLT. The ALT and TKLT parameters define the types of access links (sometimes used to lower the transmission cost by means of multiplexing several voice conversations into each AL or trunk link). The parameter MF defines the number of voice conversations that can be multiplexed on to an AL or a trunk. The choice of ALT, TKLT, and the corresponding MFs (as now defined by the NLT file) allows one to compute the cost of a voice network for any mix of AL and trunk circuits. Design parameters B_{al} and B_{tk} define the GOS required by the AL and TK subnetworks, respectively. Design parameter ECD is used in computing the cost of each call-minute for a topology map. The design parameter, economical call cost (ECC), defines the threshold above which a virtual network (e.g., Sprint's VPN or AT&T's SDN or MCI's Vnet) option becomes cheaper. If the cost of a five-minute call becomes greater than ECC by using an AL bundle, a virtual service (e.g., VPN or SDN or Vnet) is selected. Such an option will make a location as an off-net location to be served by the OCC. The traffic presented to the OCC is switched by their public network to any continental U.S. location. Any value can be chosen to reflect the SDF parameter ECC. By watching the cost of a 5-minute voice call between the United States and other countries, one can force every CPE site to be an on-net location by setting ECC equal to 13.33. In other words, each location is connected to the private voice network via a leased line. It is quite easy to compute the cost of a hybrid network that employs both private leased lines and virtual facilities. One can also vary the level of hybridization by varying ECC. It is recommended to always compare the cost of two options, in other words, a totally private network with leased lines only and a desired hybrid network with economical AL design. This also provides a cost basis for selecting any of the many available virtual network options.

As an example, one can compute the cost of a voice call by assuming 5 busy hours per business day, 20 business days per month, and (3600 / ECD) calls handled per busy hour (BHR) per line. The total cost per call minute (TCPM) on a given network handling "A" busy hour originating Erlangs can be computed as follows:

$$TCPM = TMC / (A*K)/2 \tag{7.1}$$

where

A = Total busy hour Erlangs handled by all customer nodes (e.g., see Table 7.1 for such a value at the end of the table)

TMC = Total monthly cost =AL\$+TK\$

K = (20 business days/mo.)*(5BHRs/day)*(3600 / ECD)

= Expected number of calls per month per unit erlang

= 1200 for a 300-second call duration as an example

ECD = Expected call duration in seconds

The performance of a voice network is greatly determined by the amount of traffic handled. The software design package allows one to vary the traffic volume by changing the traffic growth factor (TGF). Please refer to the curves of Figure 3.7 to understand the role of economies of scale in voice networks. The analyses presented here generally use four values: 1, 5, 10, and 20. The parameter TGF multiplies the traffic intensities as defined by the VHD input file. The traffic profile of a DTP input file determines the equivalent number of busy hours.

A design parameter called BBTF that determines the amount of traffic flowing on the backbone network trunks also influences the cost of a brand new network. The analysis presented in Section 2.5.4 assumes a uniform community of interest among all the locations. We will consider two values for the backbone traffic factor (BBTF) for a symmetric situation. The value of 1 represents a uniform community of interest. The value of 2 for the community of interest creates more intraswitch traffic at the expense of interswitch traffic. Other values (real or integer) of BBTF can also be assumed. According to our model, when BBTF=2, it almost halves the cost of interswitch trunks. The amount of traffic that does not flow on a trunk becomes part of the intraswitch traffic. By using the value of BBTF equal to 1 and 2, one can study the two extreme situations that generally characterize most voice networks.

In the analysis presented here, the backbone network is assumed to handle only the on-net to on-net traffic. Only when the cost of an AL bundle exceeds a certain threshold as determined by ECC, do we employ the VPN or the SDN options to handle that location's traffic. In order to analyze the exact costs associated with the off-net traffic presented on incoming or outgoing virtual facilities, one is advised to study the design process for ACD networks as presented in Chapter 8.

Although the NAME file doesn't influence any analysis, its use does provide a complete site database in the DBF output file. It can be read only when the SDF design parameter F_{nn} is set to 1.

The daily traffic profile named DTP8 will be used to model typical business hours for all enterprise networks modeled in the chapters that follow.

The FTD file is used for modeling and optimizing an existing network for which exact end-to-end traffic flows are known. The examples considered in Chapters 7 and 9 are assumed to be new systems. As a result, no FTD files were required. Readers interested in employing an FTD file should consult Chapters 11 and 12 where FTD file is employed to model either a backbone network or a given network topology.

A separate LATA file was employed for each of the VHD 41 and VHD59 examples. The V&H coordinates and the TCA values of site TIs were obtained through a random number simulation for the VHD100 example. Since a database relating V&H coordinates to LATA values is still unavailable, a LATA file could not be created. For this example, Flt was set to a zero value to prevent the use of a LATA file. That assigned a zero LATA value to each site, thus forcing EcoNets to consider each link to be of the Inter-LATA type. Such an approximation does not compromise a truly WAN environment.

Three SWF files need to be created, one for each case of 1, 2, and 3 switches. The original version of EcoNets required a manual input of switch nodal IDs. Each SWF starts with the number of switches followed by the nodal IDs for switch locations.

Finally, the FILES file is updated with the names of newly created input files, each time an example voice network is modeled and optimized. If an input file is not in the directory, the software warns the designer by printing the input file number (e.g., 1 implies a missing VHD file, 2 implies a missing LINK file, and so on) in a standard dialog window. No task can be accomplished until the missing input file is placed in the directory. In some cases, one can set the SDF parameters such as F_{lk}, F_{lt}, F_{ftd}, F_{nn} to a zero value. But this will make EcoNets ignore the names of corresponding input files listed in FILES. To illustrate, since we will not be using a LINK file for the voice network examples of Chapter 7, one can list any name of *any existing* LINK file (e.g., LINK17) in FILES if $F_{lk} = 0$.

When all the required input files are available in the directory, the designer is ready to model any voice network. Some designers may find that it takes an inordinate amount of time to prepare the input files when compared to the time it takes to model and optimize a network. Experienced telecommunications and DP managers have already made this painful discovery. Some may find it proper to outsource this input file creation process to a workforce in a developing country. But if a designer has the enterprise database in the form of Table 6.3, one doesn't have to outsource to a third party since personal productivity tools are already available to most engineers/designers. The presence of a third party can only add more delays and human errors.

7.2.2 Synthesis of a 41-Node Voice Network

We will now synthesize a voice network serving 41 locations as defined by the VHD41 input file. It is the first file read by the program. See Table 7.1 for a printout of a related database. It represents an actual user data for each site, associated V&H coordinates, and time-consistent, busy hour traffic intensities in Erlangs. The latest EcoNets version lists the BHR traffic intensities (TIs) in milli-Erlangs to allow more accurate entries while still using the integer representation. One can assume that each location is served by a single PABX since these are available in all sizes now. Such a PABX handles not only the local area traffic but also acts as a gateway for the interlocation traffic. Consequently, it also functions as not only a switch but also a traffic concentrator. The objective is to synthesize an economical network topology for the available tariffs. The program then reads the contents of the applicable MAP, NLT, TARIFF, SDF, and LATA files. The program then draws the V&H boundary of the geographical region as defined by the MAP input file.

The printed data of Table 7.1 and other tables of Chapter 7 may look too orderly compared to those appearing on the screen. Additional editing has been performed for tables included in the book. For example, to achieve a desired amount of editing for a proposal, one can use a word processing program to open any TEXT-based output file and perform the desired amount of editing. EcoNets employs a single font type and size for the entire document. Most word processing programs generally allow the use of many font types/sizes within a single document.

Table 7.1 Network Design Database for VHD41 Input File

#	-V-	-H-	LOAD (Erlangs)	LATA	LINK	NAME
1	7027	4203	7	524	0	KANMO
2	9213	7878	7	730	0	LAXCA
3	5015	1430	1	224	0	NWKNJ
4	4687	1373	1	920	0	HFDCT
5	6263	2679	1	922	0	CINOH
6	5972	2555	1	324	0	COLOH
7	4997	1406	9	132	0	NYCNY
8	8492	8719	5	722	0	SFOCA
9	7501	5899	1	656	0	DENCO
10	7489	4520	2	532	0	WICKA
11	7947	4373	1	536	0	OKLOK

Continued

Table 7.1 *Continued*

#	-V-	-H-	LOAD (Erlangs)	LATA	LINK	NAME
12	8436	4034	1	552	0	DALTX
13	8266	5076	1	546	0	AMLTX
14	8938	3536	1	560	0	HOUTX
15	8483	2638	1	490	0	NOLLA
16	8549	5887	3	664	0	ALBNM
17	6807	3482	6	520	0	STLMO
18	7010	2710	1	470	0	NASTN
19	8173	1147	1	952	0	TMPFL
20	8351	527	2	460	0	MIAFL
21	7260	2083	2	438	0	ATLGA
22	9345	6485	1	668	0	TSCAR
23	9135	6748	3	666	0	PHXAR
24	8486	8695	1	722	0	OAKCA
25	8665	7411	1	721	0	LASNE
26	5986	3426	5	358	0	CHIIL
27	7707	4173	1	538	0	TULOK
28	5536	2828	1	340	0	DETMI
29	4422	1249	3	128	0	BOSMA
30	5574	2543	1	320	0	CLEOH
31	6261	4021	2	635	0	CDRIA
32	6272	2992	1	336	0	INDIN
33	6529	2772	1	462	0	LOUKY
34	6657	1698	1	422	0	CHTNC
35	6113	2705	1	328	0	DATOH
36	5251	1458	5	228	0	PHLPA
37	5622	1583	3	236	0	WASDC
38	5510	1575	1	238	0	BALMA

Continued

#	-V-	-H-	LOAD (Erlangs)	LATA	LINK	NAME
39	5621	2185	6	234	0	PITPA
40	5363	1733	1	226	0	HARPA
41	5166	1585	1	228	0	ALNPA

Total Nodal Traffic = 95 Erlangs

NLT#=0	TARIFF#=1
NLT#=1	TARIFF#=1
NLT#=2	TARIFF#=3
NLT#=3	TARIFF#=4

TRAFFIC GROWTH FACTOR (TGF) = 1

Although the tariffs of 1987 (as defined in Table 2.4) were employed to compute the results of Chapter 8, the lessons to be learned remain the same. One of those lessons is that topologies remain the same despite some small perturbations in the tariffs.

The network topologies illustrated in Chapter 7 were derived using an Apple Computer's Macintosh WS. The designer working with an IBM-compatible PC will see some differences, but these are not important.

Using the CreateDB item of the Networking menu, one should first derive the nodal distribution of 41 locations over the Continental United States (CONUS) as shown in Figure 7.1. It is up to the designer to choose the nodal numbers to be printed or not (by setting the common SDF parameter F_{np} to 0 or 1). It shows a concentration of nodes in the northeast corner of the country. That region can be expanded by merely providing the desired V- and H- coordinate limits (Vmin, Vmax, Hmin, and Hmax) for the map boundary. For this study, we can ignore map expansions. A study of the nodal distribution may help the designer find entry errors (e.g., some nodes showing up in the sea). Table 7.1 shows (near its end) that the NLT file allows the selection of "0" link type. The newer version of EcoNets allows only non-zero link types and it should remove much confusion.

Using the FindCOGs item of the Networking menu, we can now study the seven sites for optimal location of switches. We selected the number of decompositions (d) of three and the horizontal decompositions as a starting point (as defined by SDF parameters NDEC and DECT). The older version of EcoNets required a manual selection of NDEC and DECT. The output recommends that the locations (17), (16, 39), (25, 1, 18, and 41) as possible switch sites. See Figure 7.2 for the illustrations of such COGs. The various groupings imply the particular nodal decomposition employed. Node 17 is recommended for a single switch. Nodes 16 and 39 are recommended for two switch locations. One should use nodes 16, 17, and 39 for locating three switches. Nodes 25, 1, 18, and 41 should be selected as sites for four switches. The recommended sites for five switches are at nodes 25, 1, 17, 18, and 41.

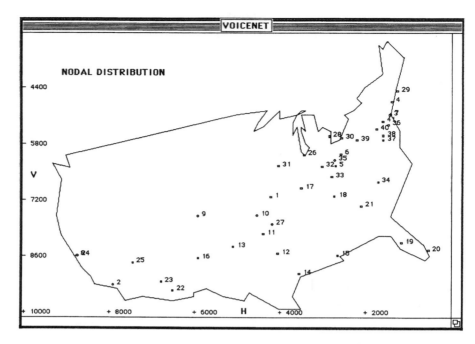

Figure 7.1 Nodal distribution for the VHD41 input file.

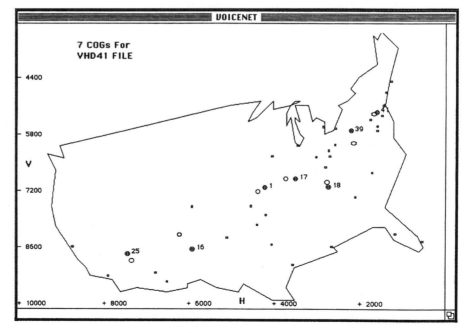

Figure 7.2 Graph showing seven COGs based on three decompositions for VHD41.

The database of Table 7.1 also shows a total of 95 Erlangs (or 95,000 milli-Erlangs) handled by all the sites. Half of these Erlangs are of the originating type. This value also determines the average number of concurrently busy calls in the system during a busy hour. If the expected call duration (ECD) of a voice call is expressed in seconds, then the number of busy hour call completions (equal to N_{bhc}) can be defined as follows:

$$N_{bhc} = (A/2)*(3600 / ECD) \qquad (7.2)$$

Since each call is handled by exactly two switches for the case of a fully connected backbone mesh network, the total Erlangs of Table 7.1 represent the total call intensity handled by the network switches during each busy hour.

The cost of transmission is summarized for both access lines (AL\$) and trunks (TK\$) on each network plot. The output graph also shows the average cost of a voice call-minute. Tables 7.2, 7.3, and 7.4 represent the printouts of AL costs for 1, 2, and 3 switches, respectively. Figures 7.3, 7.4, and 7.5 represent the synthesized network topologies for 1, 2, and 3 switches, respectively. Although the computations were also performed for four and five switches, the results are not included here for brevity. The peak hour Erlangs handled by each site as defined by the 4th column in Table 7.1 was used to compute the number (N) of ALs. We employ a grade-of-service as defined by the blocking probability equal to 0.01 for both access lines and trunks according to the Erlang-B formula. One can compare the values of N with those obtained from Figures 3.1 through 3.6 that plot N required circuits versus A Erlangs and several useful values of blocking probabilities, B. The SDF parameters B_{al} and B_{tk} define the blocking probabilities for ALs and trunks respectively. These parameters can be modified according to the simple technique as defined in Chapter 6.

Table 7.2 COG and Access Line (AL) Connections Printout

A. COG PRINTOUT

NO. OF DECOMPOSITIONS=3
INITIAL DECOMPOSITION FACTOR=1 (0 FOR VERT, 1 FOR HORIZ) COGs
WITH THEIR V-H COORDs & CLOSEST NODES WITH THEIR V-H COORDs

6814	3711	17	6807	3482
8190	6197	16	8549	5887
5936	2125	39	5621	2185
8834	7319	25	8665	7411
7131	4353	1	7027	4203
6896	2746	18	7010	2710
5210	1655	41	5166	1585

Continued

Table 7.2 *Continued*

B. ACCESS LINE CONNECTION ANALYSIS

NO. OF SWITCHES=1 AT 17
AL BLOCKING FACTOR=.01   ECONOMICAL CALL COST=13.31
AL TRUNK LINK TYPE=0   AL MULTIPXG. FACTOR=1
TRAFFIC GROWTH FACTOR=1   EXP.CALL DURATION-SEC=300

#	SW	ALs	MIs	$/AL	$/ALB
1	17	14	238	520	7290
2	17	14	1584	1035	14494
3	17	5	861	803	4019
4	17	5	945	830	4154
5	17	5	306	564	2822
6	17	5	394	620	3103
7	17	17	870	806	13717
8	17	11	1739	1084	11934
9	17	5	795	782	3913
10	17	7	392	619	4336
11	17	5	457	661	3305
12	17	5	543	702	3511
13	17	5	683	746	3734
14	17	5	674	743	3719
15	17	5	593	718	3590
16	17	8	939	828	6629
18	17	5	252	529	2648
19	17	5	855	801	4009
20	17	7	1054	865	6059
21	17	7	465	665	4660
22	17	5	1243	926	4630
23	17	8	1268	934	7472
24	17	5	1731	1082	5412
25	17	5	1374	968	4840

Continued

#	SW	ALs	MIs	$/AL	$/ALB
26	17	11	260	534	5882
27	17	5	358	597	2989
28	17	5	452	657	3287
29	17	8	1033	858	6870
30	17	5	490	681	3409
31	17	7	242	523	3664
32	17	5	229	515	2575
33	17	5	241	522	2612
34	17	5	566	709	3546
35	17	5	329	579	2895
36	17	11	807	786	8652
37	17	8	707	754	6037
38	17	5	729	761	3808
39	17	13	555	706	9178
40	17	5	717	757	3788
41	17	5	793	782	3910

SW.LOC	#VTERMs	SW.COST
1	276	207102

TOTAL ACCESS LINE COST 207102

Table 7.3 Access Line (AL) Connections Printout

NO. OF SWITCHES= 2   AT   16 39

AL BLOCKING FACTOR= .01   ECONOMICAL CALL COST= 13.31

AL TRUNK LINK TYPE= 0   AL MULTIPXG. FACTOR= 1

TRAFFIC GROWTH FACTOR= 1   EXP.CALL DURATION-SEC= 300

#	SW	ALs	MIs	$/AL	$/ALB
1	16	14	717	757	10610
2	16	14	663	740	10368
3	39	5	306	564	2820
4	39	5	391	618	3093
5	39	5	256	532	2660

Continued

Table 7.3 *Continued*

#	SW	ALs	MIs	$/AL	$/ALB
6	39	5	161	471	2357
7	39	17	315	570	9693
8	16	11	895	814	8963
9	16	5	331	580	2901
10	16	7	547	703	4922
11	16	5	515	693	3465
19	16	5	587	716	3580
13	16	5	271	542	2710
14	16	5	753	769	3846
15	39	5	916	821	4107
17	39	13	555	706	9178
18	39	5	469	668	3343
19	39	5	871	807	4035
20	39	7	1010	851	5960
21	39	7	519	694	4860
22	16	5	314	569	2848
23	16	8	329	578	4631
24	16	5	888	812	4062
25	16	5	483	677	3387
26	39	11	409	630	6930
27	16	5	603	721	3607
28	39	5	205	499	2497
29	39	8	48I	676	5408
30	39	5	114	441	2206
31	39	7	614	724	5074
32	39	5	327	578	2890
33	39	5	341	587	2935
34	39	5	362	599	2999

Continued

#	SW	ALs	MIs	$/AL	$/ALB
35	39	5	226	513	2565
36	o9	11	257	533	5866
37	39	8	190	490	3920
38	39	5	196	493	2468
40	39	5	164	473	2367
41	39	5	238	520	2603

SW LOC	#VTERMs	SW COST
1	99	69900
2	169	102834

TOTAL ACCESS LINE COST = $172734

Table 7.4 Access Line (AL) Connection Analysis

NO. OF SWITCHES= 3
AL BLOCKING FACTOR=.01
AL TRUNK LINK TYPE= 0
TRAFFIC GROWTH FACTOR= 1

AT   16 17 39
ECONOMICAL CALL COST= 13.31
AL MULTIPXG. FACTOR= 1
EXP.CALL DURATION-SEC= 300

#	SW	ALs	MIs	$/AL	$/ALB
1	17	14	238	520	7290
2	16	14	663	740	10368
3	39	5	306	564	2820
4	39	5	391	618	3093
5	39	5	256	532	2660
6	39	5	161	471	2357
7	39	17	315	570	9693
8	16	11	895	814	8963
9	16	5	331	580	2901
10	17	7	392	619	4336
11	17	5	457	661	3305
12	17	5	543	702	3511
13	16	5	271	542	2710
14	17	5	674	743	3719

Continued

Table 7.4 *Continued*

#	SW	ALs	Mls	$/AL	$/ALB
15	17	5	593	718	3590
18	17	5	252	529	2648
19	17	5	855	801	4009
20	39	7	1010	851	5960
21	17	7	465	665	4660
22	16	5	314	569	2848
23	16	8	329	578	4631
24	16	5	888	812	4062
25	16	5	483	677	3387
26	17	11	260	534	5882
27	17	5	358	597	2989
28	39	5	205	499	2497
29	39	8	481	676	5408
30	39	5	114	441	2206
31	17	7	242	523	3664
32	17	5	229	515	2575
33	17	5	241	522	2612
34	39	5	362	599	2999
35	39	5	226	513	2565
36	39	11	257	533	5866
37	39	8	190	490	3920
38	39	5	196	493	2468
40	39	5	164	473	2367
41	39	5	238	520	2603

SW LOC	#VTERMs	SW COST
1	58	39870
2	91	54790
3	106	59482

TOTAL ACCESS LINE COST 154142

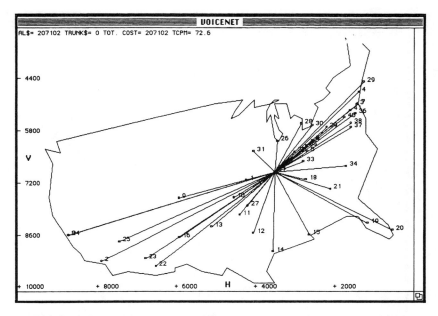

Figure 7.3 A voice network topology with one switch and all on-net locations: VDH41, TGF = 1, and BBTF = 2.

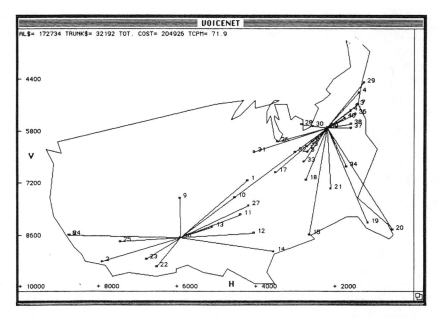

Figure 7.4 A voice network topology with two switches and all on-net locations: VDH41, TGF = 1, and BBTF = 2.

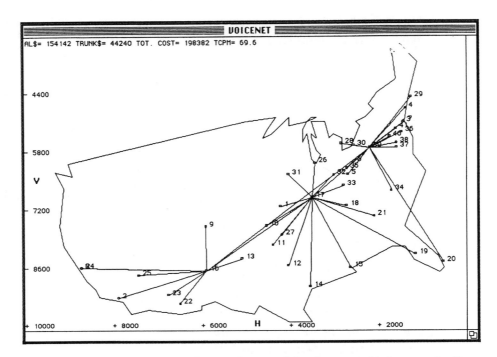

Figure 7.5 A voice network topology with three switches and all on-net locations: VDH41, TGF = 1, and BBTF = 2.

Table 7.5 represents the trunking analysis for trunk bundle dimensioning and transmission costs for two switches. Each one of the first three matrices is of (S x S) size where S is the number of switches. The first such matrix represents the Erlangs flowing between any two switches. For example, the first number at the top left of the matrix implies the Erlangs flowing from one AL bundle to another AL bundle that terminate on the first switch at node 16 (defined hereon as Switch 16) in Table 7.5. The number to its right implies the Erlangs flowing from Switch 16 to Switch 39. The number below the top left number represents the Erlangs flowing from Switch 39 to Switch 16. The remaining number represents the intranodal Erlangs at Switch 39. The second matrix represents a modified erlang (sum of Erlangs flowing in two directions) table that is used to compute the number of trunks between the switches. The third matrix represents the actual number of computed trunks between the switches.

When computing the number of trunks for a given bundle, we assume that a four-wire trunk is required to handle only one conversation. The Erlangs originated in both directions must be added to compute the trunks. If one computes the number of trunks for each direction of traffic separately and then adds them together to create a single bundle, extra costs will be incurred. Such a conclusion has a basis in the charts of Figure 3.7 in Chapter 3. The fourth matrix actually lists each trunk bundle, the associated nodes, the trunk mileage, per-trunk cost, and the total cost of each trunk bundle These tables should familiarize the way computations are tabulated for any number of switches.

Table 7.5 Trunking Analysis

TRUNK LINK/NODE TYPE 0
TRUNK MULTIPLEXING FACTOR 1
BACKBONE TRAFFIC FACTOR=2
TRAFFIC GROWTH FACTOR=1
ECONOMICAL CALL COST=13.31
TRUNK BLOCKING=.01
TOTAL INTER/INTRA NODAL TRAFFIC=95
NO. OF SWITCHES=2 AT 16 39

ORIGINAL TRUNK TRAFFIC MATRIX AS FOLLOW:

24	11
11	47

MODIFIED TRAFFIC MATRIX IS AS FOLLOWS:

0	22
0	0

MODIFIED TRUNK MATRIX IS AS FOLLOWS:

0	32
0	0

TRUNK COST MATRIX IS AS FOLLOWS:

FROM	TO	MILES	COST	TOTALS
16	39	1493	1006	32192

*****SUMMARY OF SYSTEM COSTS*****

TOTAL ACCESS LINE COST ($)=172734
TOTAL TRUNK COST($)=32192
TOTAL SYSTEM COST($)=204926
TOTAL COST PER 5-MIN.CALL($)=3.60

Figures 7.6 and 7.7 plot the cost per call-minute and the number of off-net locations, respectively, as a function of traffic growth factor (TGF) and for two design constraints: only VG lines chosen on the basis of either all on-net ALs (the worst case) or economical AL design with eight VG circuits multiplexed per 56Kbps DDS AL/trunk (the best case), respectively. The results show that the need for a hybrid network is eliminated only for large values of TGF (>10) while using the design option of eight VG circuits multiplexed per 56Kbps AL/TK. More than half of the locations will require a VPN/SDN/Vnet type service even for TGF=20 when only VG circuits are employed on an economical AL basis. Figures 7.8 through 7.10 represent the total transmission costs (AL$+TK$) of the best networks plotted as a function of two values of BBTF and for three values of

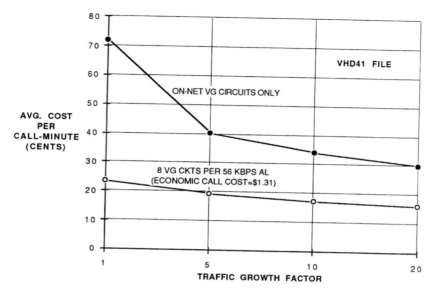

Figure 7.6 Average cost per minute (ACPM) versus traffic growth factor: VHD41 and one switch.

TGF, respectively. The results show that when BBTF=1, it generally makes a single switch network to be the most optimum case. See Figure 7.11 for a plot of the network cost distribution when we add the monthly costs of NMC and hardware. For brevity, we plot the results for only one set of values: TGF=10 and BBTF=2.

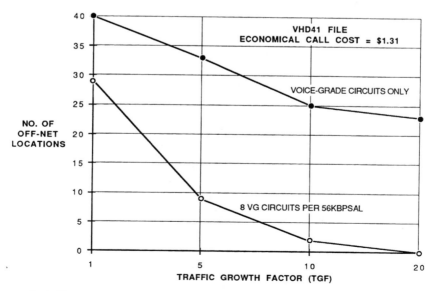

Figure 7.7 Number of off-net locations versus TGF: VHD41, ECC = $1.31, and one switch.

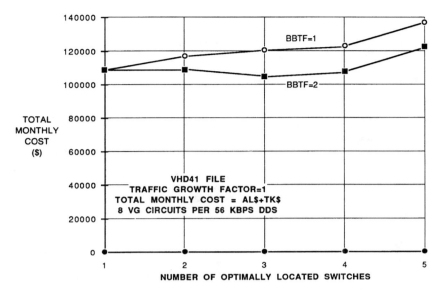

Figure 7.8 Total monthly cost for network transmission versus the number of optimally located switches: VHD41, TGF = 1, BBTF = 1 or 2, and eight VG circuits per economically designed AL/TK.

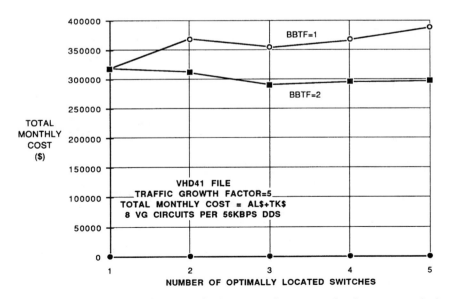

Figure 7.9 Total monthly cost for network transmission versus the number of optimally located switches: VHD41, TGF = 5, BBTF = 1 or 2, and eight VG circuits per economically designed AL/TK.

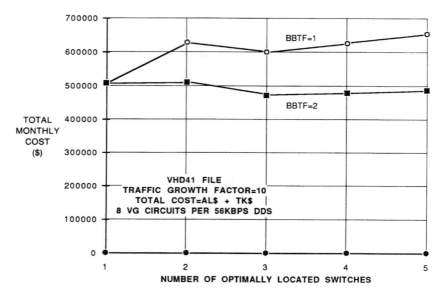

Figure 7.10 Total monthly cost for network transmission versus the number of optimally located switches: VHD41, TGF = 10, BBTF = 1 or 2, and eight VG circuits per economically designed AL/TK.

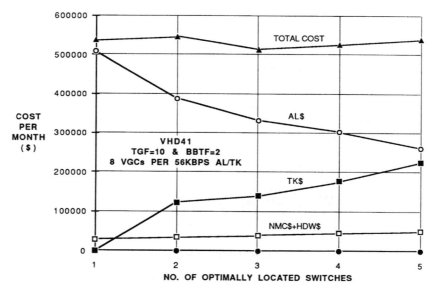

Figure 7.11 Distribution of total monthly cost versus the number of optimally located switches: VHD41, TGF = 10, BBTF = 2, and eight VG circuits per economically designed AL/TK.

The model employed for computing the cost of hardware requires the knowledge of AL and trunk terminations (equal to N_{alt}, N_{tt}, respectively)as shown in Tables 7.2 through 7.5. The total hardware cost can then be approximated as follows:

$$C_{th} = (N_{alt} + N_{tt})*800 \ ..... \ \$ \qquad (7.3)$$

The cost of hardware on a monthly basis can be computed by using the methodology of Section 2.4.4 and assuming a seven-year life cycle and 10 percent interest rate. The resulting value of hardware cost per month as employed in our analysis is as follows:

$$C_{hm} = C_{th} \ / \ 60.237 + 3000*S \ ...\$/MO \qquad (7.4)$$

The monthly cost of network management and control can be determined by considering the cost of network maintenance assuming that the switches have a dual redundancy. We will use a sample formula based on actual practice and it includes the labor and overhead associated with personnel needed to manage a private network with S tandem voice switches.

$$C_{nmc} = 7500 + 4500*S \ .....\$/MO \qquad (7.5)$$

7.2.3 Synthesis of a 59-Node Voice Network

See Table 7.6 for a definition of the VHD59 database. This database also represents an actual user. Such a network handles 140 Erlangs (In and Out) during the peak period when compared to the 95 Erlangs handled by the 41-node network. Therefore, the 59-node network should cost more than that of the 41-node network. Figure 7.12 illustrates the nodal distribution for the VHD59 file. A startling fact is apparent: the nodes are clustered around Atlanta (node #1) which is also the firm's headquarters. Figure 7.13 illustrates the seven recommended COGs. Figures 7.14 through 7.16 represent some useful network topologies. The solution for a single switch at node 1 is cheaper by $5,000 when compared to a single switch at location 37 as recommended by the COG analysis. This can be explained by the fact that node #1 handles higher traffic. If node #1 is used as a switch, the associated heavier AL bundle is eliminated and in its place a thinner AL bundle is implemented. Consequently, the designer must always temper the COG analysis with such alternative solutions. See Tables 7.7 and 7.8 for samples of AL cost analysis for eight VG circuits per 56Kbps AL/TK. A "0" for #ALs signifies the use of VPN/SDN type service. See Figure 7.17 for the plots of call-minute cost versus TGF for two useful set of design parameters. Due to clustering of nodes around the major switch location ID of 1, the call-minute costs of a 59-node network are generally lower than those for a 41-node network. See Figure 7.18 for the plot of off-net locations versus TGF and two useful designs.

Table 7.6 Network Design Database for VHD59 Input File

#	-V-	-H-	LOAD (Erlangs)	LATA	LINK	NAME
1	7260	2083	12	438	0	ATLGA
2	7089	1674	2	442	0	AGSGA
3	6749	2001	1	420	0	AVLNC
4	5510	1575	1	238	0	BALMD
5	7518	2446	2	476	0	BHMAL
5	8806	3298	1	562	0	BPTTX
7	8476	2874	1	492	0	BTRLA
8	6901	1589	2	434	0	CAESC
9	7098	2366	1	472	0	CHATN
10	5986	3426	6	358	0	CHIIL
11	7021	1281	1	436	0	CHSWV
12	6261	4021	4	635	0	CIDIA
13	6657	1698	2	422	0	CLTNC
14	5972	2555	1	324	0	CMHOH
15	7556	2045	2	438	0	CSGGA
16	6263	2679	4	922	0	CVGOH
17	8436	4034	6	552	0	DALTX
18	6113	2705	1	328	0	DAYOH
19	5622	1583	3	236	0	WASDC
20	7501	5899	3	656	0	DENCO
21	5536	2828	2	340	0	DETMI
22	8409	3168	1	486	0	ESFLA
23	6729	3019	1	330	0	EVVIN
24	5942	2982	1	334	0	FWAIN
25	8479	4122	6	552	0	GSWTX
26	7827	3554	1	528	0	HOTAR
27	8938	3536	3	560	0	HOUTX

Continued

#	-V-	-H-	LOAD (Erlangs)	LATA	LINK	NAME
28	6272	2992	2	336	0	INDIN
29	8035	2880	2	482	0	JANMS
30	7649	1276	1	452	0	JAXFL
31	8665	7411	1	721	0	LASNV
32	9213	7878	2	730	0	LAXCA
33	6459	2562	1	466	0	LEXKY
34	7721	3451	1	528	0	LITAR
35	7364	1865	1	446	0	MCNGA
36	7899	2639	1	482	0	MEIMS
37	7471	3125	5	468	0	MEMTN
38	7692	2247	1	478	0	MGMAL
39	8351	527	11	460	0	MIAFL
40	7027	4203	2	524	0	MKCMO
41	8148	3218	1	486	0	MLOLA
42	5781	4525	2	628	0	MSPMN
43	8483	2638	5	490	0	MSYLA
44	4997	1406	8	132	0	NYCNY
45	7954	1031	1	458	0	ORLFL
46	8165	606	1	460	0	PBIFL
47	5251	1458	1	228	0	PHLPA
48	6982	3088	1	468	0	PUKTN
49	7266	1379	2	440	0	SAVGA
50	6529	2772	1	462	0	SDFKY
51	9468	7629	1	732	0	SDOCA
52	8492	8719	2	722	0	SFOCA
53	7310	3836	1	522	0	SGFMO
54	8272	3495	2	486	0	SHVLA
55	7496	1340	1	440	0	SSIGA

Continued

Table 7.6 *Continued*

#	-V-	-H-	LOAD (Erlangs)	LATA	LINK	NAME
56	6807	3482	3	520	0	STLMO
57	5704	2820	1	326	0	TOLOH
58	8173	1147	2	952	0	TPAFL
59	6801	2251	2	474	0	TYSTN

TOTAL NODAL TRAFFIC LOAD=   140 ERLANGS

See Figures 7.19 through 7.21 for the plots of transmission costs as a function of number of switches and two values of BBTF. See Figure 7.22 for the distribution of all network-related costs versus number of switches and two values of BBTF. Due to extreme clustering of nodes around the major switch location of 1, the network rarely demands more than two switches. A single switch seems quite effective even for BBTF=2.

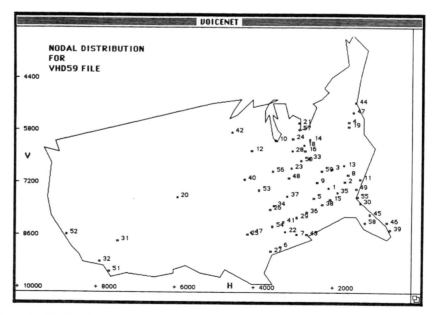

Figure 7.12 Nodal distribution for the VHD59 input file.

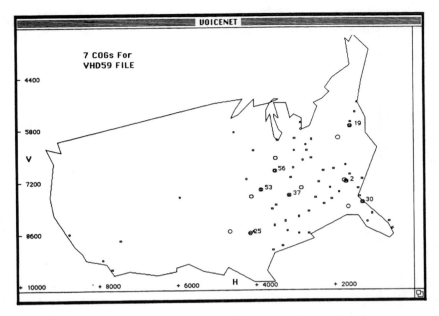

Figure 7.13 Graph showing seven COGs based on three decompositions for VHD59.

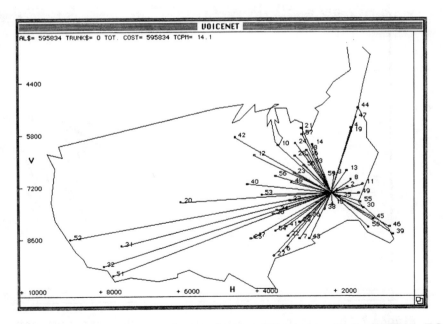

Figure 7.14 A voice network topology with one switch: VHD59, TGF = 10, and eight VG circuits per economically designed 56Kbps AL.

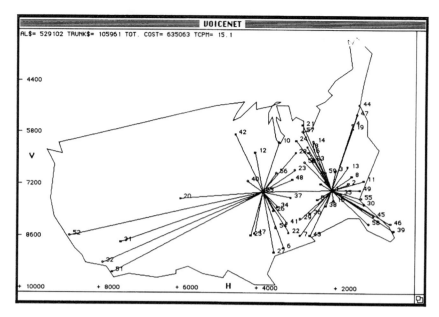

Figure 7.15 A voice network topology with two switches: VHD59, TGF = 10, BBTF = 2, and eight VG circuits per economically designed 56Kbps AL or TK.

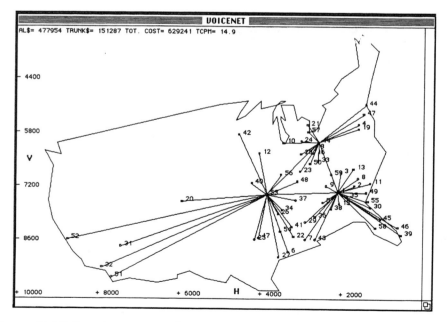

Figure 7.16 A voice network topology with three switches: VHD59, TGF = 10, BBTF = 2, and eight VG circuits per economically designed 56Kbps AL or TK.

Table 7.7 Access Line Connection Analysis

NO. OF SWITCHES= 1   AT   1
AL BLOCKING FACTOR= 01   ECONOMICAL CALL COST= 1.31
AL TRUNK LINK TYPE= 1   AL MULTIPXG. FACTOR= 8
TRAFFIC GROWTH FACTOR= 10   EXP. CALL DURATION-SEC= 300

#	SW	ALs	Mls	$/AL	$/ALB
2	1	4	140	1228	4913
3	1	3	163	1299	3897
4	1	3	576	2444	7332
5	1	4	140	1230	4921
6	1	3	621	2520	7562
7	1	3	458	2190	6571
8	1	4	193	1388	5553
9	1	3	103	1116	3349
10	1	10	585	2459	24594
11	1	3	264	1604	4813
12	1	7	689	2635	18447
13	1	4	226	1488	5953
14	1	3	433	2115	6345
15	1	4	94	1080	4320
16	1	7	367	1914	13401
17	1	10	720	2687	26875
18	1	3	412	2051	6153
19	1	6	541	2385	14312
20	1	6	1209	3513	21081
21	1	4	593	2473	9895
22	1	3	499	2314	6943
23	1	3	340	1832	5498
24	1	3	504	2322	6968
25	1	10	751	2739	27397
26	1	3	498	2310	6932

Continued

Table 7.7 *Continued*

#	SW	ALs	Mls	$/AL	$/ALB
27	1	6	701	2656	15938
28	1	4	424	2087	8349
29	1	4	351	1866	7467
30	1	3	283	1660	4981
31	1	0	1742	0	7860
32	1	0	1933	0	15720
33	1	3	295	1696	5089
34	1	3	456	2183	6551
35	1	3	76	993	2981
36	1	3	267	1614	4842
37	1	9	336	1820	16382
38	1	3	146	1246	3739
39	1	17	600	2485	42258
40	1	4	674	2609	10439
41	1	3	455	2181	6544
42	1	4	902	2995	11983
43	1	9	424	2087	18790
44	1	13	746	2732	35522
45	1	3	398	2008	6026
46	1	3	547	2395	7187
47	1	3	665	2594	7783
48	1	3	329	1800	5402
49	1	4	222	1477	5910
50	1	3	317	1764	5293
51	1	0	1887	0	7860
52	1	0	2134	0	15720
53	1	3	554	2407	7222
54	1	4	549	2398	9594

Continued

#	SW	ALs	Mls	$/AL	$/ALB
55	1	3	246	1549	4649
56	1	6	465	2209	13257
57	1	3	544	2390	7170
58	1	4	413	2053	8214
59	1	4	154	1271	5087

SW LOC	#VTERMs	SW COST
1	1853	595834

TOTAL ACCESS LINE COST 595834

Table 7.8 Access Line Connection Analysis

NO. OF SWITCHES= 2
AL BLOCKING FACTOR= .01
Al TRUNK LINK TYPF= 1
TRAFFIC GROWTH FACTOR= 10

AT 53 1
ECONOMICAL CALL COST= 1.31
AL MULTIPXG. FACTOR= 8
EXP. CALL DURATION-SEC= 300

#	SW	ALs	Mls	$/AL	$/ALB
2	1	4	140	1228	4913
3	1	3	163	1299	3897
4	1	3	576	2444	7332
5	1	4	140	1230	4921
6	53	3	502	2319	6959
7	1	3	458	2190	6571
8	1	4	193	1388	5553
9	1	3	103	1116	3349
10	53	10	438	2128	21289
11	1	3	264	1604	4813
12	53	7	336	1822	12756
13	1	4	226	1488	5953
14	1	3	433	2115	6345
15	1	4	94	1080	4320
16	1	7	367	1914	13401
17	53	10	361	1896	18969
18	1	3	412	2051	6153

Continued

Table 7.8 *Continued*

#	SW	ALs	Mls	$/AL	$/ALB
19	1	6	541	2385	14312
20	53	6	655	2577	15464
21	1	4	593	2473	9895
22	53	3	406	2033	6099
23	53	3	317	1762	5287
24	1	3	504	2322	6968
25	53	10	380	1954	19543
26	53	3	186	1367	4102
27	53	6	523	2354	14128
28	53	4	423	2082	8330
29	1	4	351	1866	7467
30	1	3	283	1660	4981
31	53	0	1208	0	7860
32	53	4	1412	3857	15430
33	1	3	295	1696	5089
34	53	3	178	1342	4028
35	1	3	76	993	2981
36	1	3	267	1614	4842
37	53	9	230	1501	13512
38	1	3	146	1246	3739
39	1	17	600	2485	42258
40	53	4	146	1247	4990
41	53	3	329	1799	5398
42	53	4	530	2366	9465
43	1	9	424	2087	18790
44	1	13	746	2732	35522
45	1	3	398	2008	6026
46	1	3	547	2395	7187

Continued

#	SW	ALs	Mls	$/AL	$/ALB
47	1	3	665	2594	7783
48	53	3	258	1585	4755
49	1	4	222	1477	5910
50	1	3	317	1764	5293
51	53	0	1379	0	7860
52	53	0	1588	0	15720
54	53	4	322	1779	7119
55	1	3	246	1549	4649
56	53	6	194	1392	8355
57	1	3	544	2390	7170
58	1	4	413	2053	8214
59	1	4	154	1271	5087

SW LOC	#VTERMS	SW COST
1	810	237418
2	1025	291684

TOTAL ACCESS LINE COST= 529102

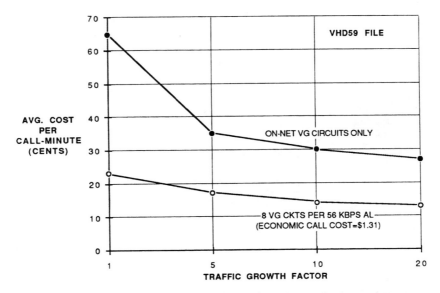

Figure 7.17 Average cost per minute versus the TGF for an optimized network: VHD59, one switch, and economical call costs of $13.31 and $1.31.

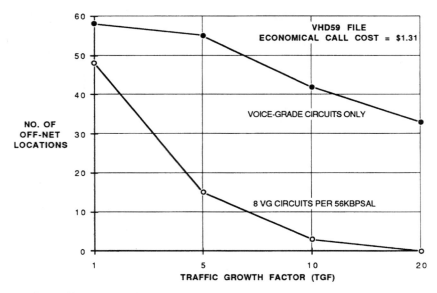

Figure 7.18 Number of off-net locations versus the TGF for an optimized network: VHD59, one switch, and economical call cost of $1.31.

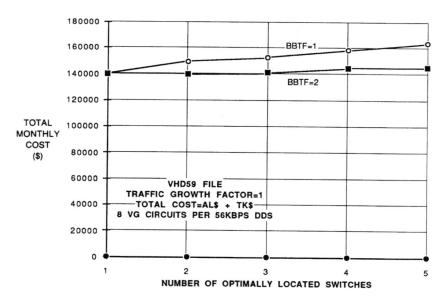

Figure 7.19 Total monthly costs for network transmission versus the number of optimally located switches: VHD59, TGF = 1, BBTF = 1 or 2, and eight VG circuits per economically designed AL/TK.

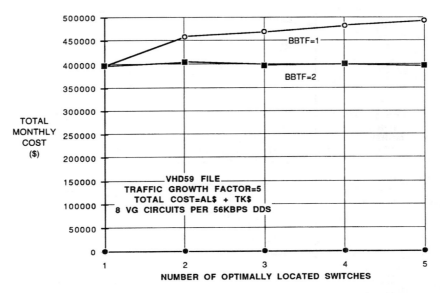

Figure 7.20 Total monthly costs for network transmission versus the number of optimally located switches: VHD59, TGF = 5, BBTF = 1 or 2, and eight VG circuits per economically designed AL/TK.

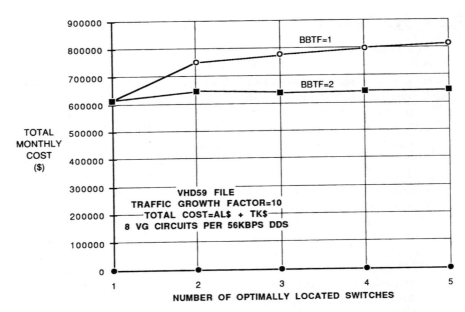

Figure 7.21 Total monthly costs for network transmission versus the number of optimally located switches: VHD59, TGF = 10, BBTF = 1 or 2, and eight VG circuits per economically designed AL/TK.

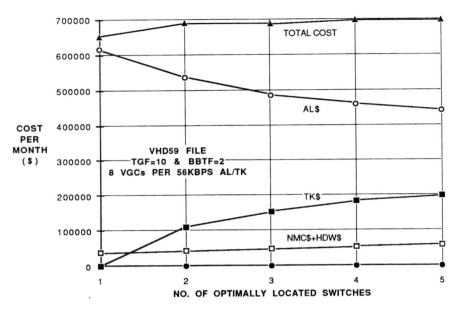

Figure 7.22 Distribution of total monthly costs versus the number of optimally located switches: VHD59, TGF = 10, BBTF = 2, and eight VG circuits per economically designed AL/TK.

7.2.4 Synthesis of a 100-Node Voice Network

See Table 7.9 for a definition of a randomly produced database for 100 nodes. Such a network handles 270 Erlangs (In and Out) during a busy hour for TGF=1. This value is approximately twice that for the 59-node network for a similar TGF. Figure 7.23 shows the nodal distribution for such a network. Figure 7.24 illustrates the seven recommended COGs for such a network. Figures 7.25 through 7.27 represent some useful network topologies. Figure 7.28 plots the call-minute cost as a function of TGF for two sets of design parameters. Figure 7.29 plots the number of off-net locations as a function of TGF for two designs. Figures 7.30, 7.31, and 7.32 plot the network transmission costs as a function of TGF and two values of BBTF. See Figure 7.33 for the distribution of all network-related costs versus number of switches for BBTF=2.

Although a study of the three networks just shown should be adequate to form some conclusions, a network serving 200 nodes was also synthesized for a few judicious design parameters and the results will be discussed to make a point. Table 7.10 defines a randomly produced DB for a 200-node network. Such a network generates 483 total Erlangs (In and Out). Figures 7.34 and 7.35 represent some useful network topologies for a practical design. Figures 7.36 and 7.37 represent total network costs as a function of the number of switches for two values of TGF and BBTF.

Table 7.9 Network Design Database for VHD100 Input File

#	-V-	-H-	LOAD (Erlangs)	LATA	LINK
1	6053	1503	1	0	0
2	6807	1551	4	0	0
3	7262	2762	2	0	0
4	6230	7031	1	0	0
5	7731	8494	4	0	0
6	6189	6192	1	0	0
7	7273	3024	2	0	0
8	5026	4496	3	0	0
9	6660	4249	4	0	0
10	7488	2566	2	0	0
11	8509	4905	1	0	0
12	5587	5771	4	0	0
13	7966	6606	4	0	0
14	6525	1653	3	0	0
15	5414	1499	2	0	0
16	7469	4943	4	0	0
17	6498	8705	4	0	0
18	9159	5733	4	0	0
19	5816	2549	2	0	0
20	8940	3945	2	0	0
21	8617	8046	1	0	0
22	7741	1659	2	0	0
23	5554	5528	3	0	0
24	7705	8603	2	0	0
25	8320	2532	4	0	0
26	6418	1915	1	0	0
27	6153	1601	2	0	0

Continued

Table 7.9 *Continued*

#	-V-	-H-	LOAD (Erlangs)	LATA	LINK
28	6343	6014	3	0	0
29	8173	1147	4	0	0
30	7236	2104	3	0	0
31	6017	6721	3	0	0
32	7345	8087	3	0	0
33	5143	3203	1	0	0
34	5013	2230	4	0	0
35	8618	3012	3	0	0
36	7186	4742	1	0	0
37	7526	6805	2	0	0
38	6801	5433	4	0	0
39	7894	2331	3	0	0
40	8018	6063	2	0	0
41	6167	8590	3	0	0
42	5493	5736	2	0	0
43	7187	6475	1	0	0
44	6161	4713	3	0	0
45	6900	4811	4	0	0
46	5991	2238	3	0	0
47	7987	7595	4	0	0
48	7317	1829	4	0	0
49	5454	2975	3	0	0
50	8337	538	1	0	0
51	5306	3144	4	0	0
52	9420	6266	2	0	0
53	5419	3706	3	0	0
54	7281	8340	1	0	0

Continued

#	-V-	-H-	LOAD (Erlangs)	LATA	LINK
55	7421	3060	2	0	0
56	5260	2880	1	0	0
57	7096	2534	4	0	0
58	5462	5410	3	0	0
59	8075	2360	3	0	0
60	7154	8706	2	0	0
61	6085	5037	4	0	0
62	8521	4680	1	0	0
63	5710	3050	1	0	0
64	5900	7580	3	0	0
65	7406	5950	3	0	0
66	6385	3240	4	0	0
67	7340	4173	2	0	0
68	7002	8292	3	0	0
69	7396	3910	3	0	0
70	8574	4784	2	0	0
71	9011	4392	3	0	0
72	9297	4734	1	0	0
73	6095	7726	4	0	0
74	8782	4000	1	0	0
75	9393	3742	1	0	0
76	6226	9256	1	0	0
77	5597	2282	3	0	0
78	5837	3537	3	0	0
79	7597	4733	3	0	0
80	5533	5945	4	0	0
81	8278	7304	3	0	0
82	7695	4471	2	0	0

Continued

Table 7.9 *Continued*

#	-V-	-H-	LOAD (Erlangs)	LATA	LINK
83	5705	1659	4	0	0
84	7450	4789	1	0	0
85	8866	4387	4	0	0
86	7962	4721	1	0	0
87	8869	6056	3	0	0
88	7396	6516	4	0	0
89	8210	5927	3	0	0
90	5131	2540	4	0	0
91	7267	5263	2	0	0
92	9240	7833	4	0	0
93	6383	3739	4	0	0
94	7237	8655	2	0	0
95	5453	3478	2	0	0
96	6257	6593	4	0	0
97	7701	1956	4	0	0
98	5323	2628	3	0	0
99	7718	6720	4	0	0
100	8692	8268	4	0	0

TOTAL NODAL TRAFFIC LOAD=   270 Erlangs

7.3 Summary of Voice Network Syntheses and Conclusions

The main objective of this chapter was to provide a consistent design methodology for a voice network. The numbers and conclusions are incidental to the example networks selected, design parameters selected, and assumptions employed. To illustrate, 5 busy hours per day were assumed for the data gathered. A different DTP file may have yielded 10 busy hours a day. That would have reduced the cost-per-minute figures by half. Combinatorics play an important part in network modeling. Another example illustrates the previous point. An optimum voice network (as defined by the VHD200 file, enclosed DTP8 file, TGF of 100 and 1992 tariffs of Table 2.5, T3 facilities using ADPCM or MF=1344)

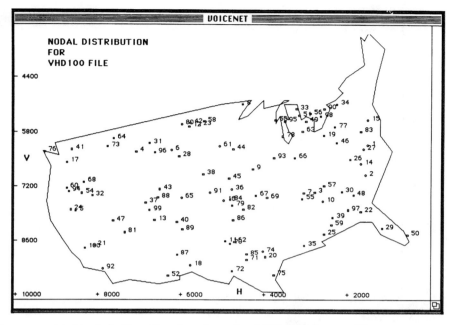

Figure 7.23 Nodal distribution for the VHD100 input file.

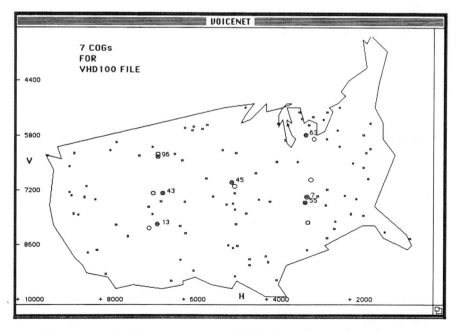

Figure 7.24 Graph showing seven COGs based on three decompositions for VHD100.

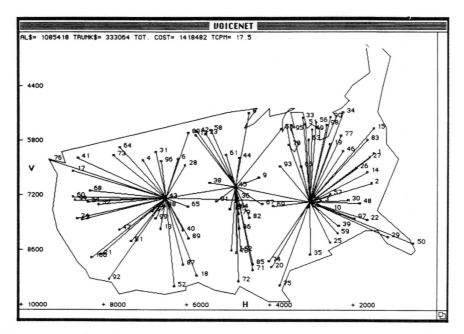

Figure 7.25 A voice network topology with three switches: VHD100, TGF = 10, BBTF = 2, and eight VG circuits per 56Kbps AL or TK.

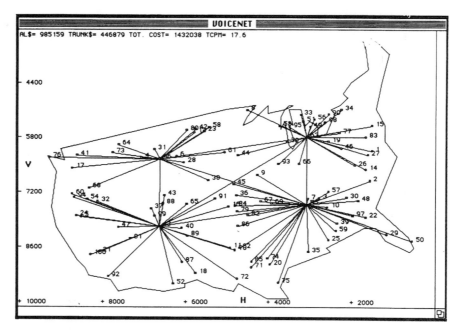

Figure 7.26 A voice network topology with four switches: VHD100, TGF = 10, BBTF = 2, and eight VG circuits per 56Kbps AL or TK.

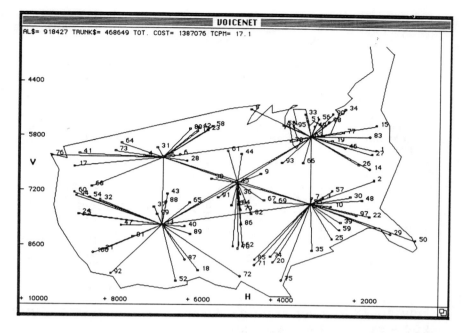

Figure 7.27 A voice network topology with five switches: VHD100, TGF = 10, BBTF = 2, and eight VG circuits per 56Kbps AL or TK.

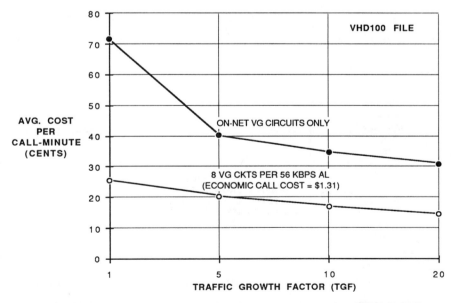

Figure 7.28 Average cost per call minute versus the TGF for an optimized network: VHD100, and economical call costs of $13.31 and $1.31.

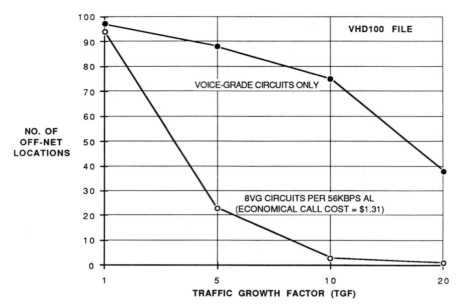

Figure 7.29 Number of off-net locations versus the TGF: VHD100, ECC = $1.31 and one switch.

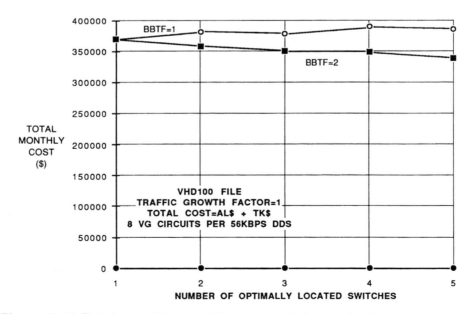

Figure 7.30 Total monthly cost for network transmission versus the number of optimally located switches: VHD100, TGF = 1, BBTF = 1 or 2, and eight VG circuits per economically designed AL/TK.

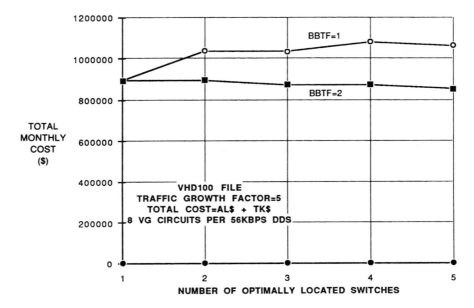

Figure 7.31 Total monthly cost for network transmission versus the number of optimally located switches: VHD100, TGF = 5, BBTF = 1 or 2, and eight VG circuits per economically designed AL/TK.

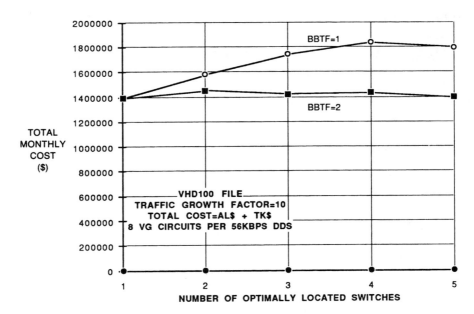

Figure 7.32 Total monthly cost for network transmission versus the number of optimally located switches: VHD100, TGF = 10, BBTF = 1 or 2, and eight VG circuits per economically designed AL/TK.

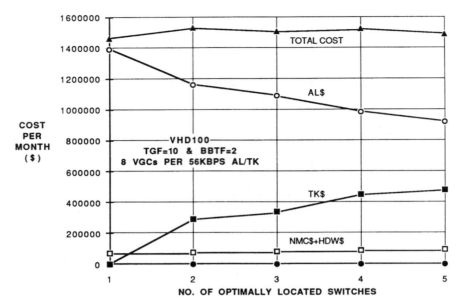

Figure 7.33 Distribution of total monthly costs versus the number of optimally located switches: VHD100, TGF = 10, BBTF = 2, and eight VG circuits per economically designed AL/TK.

Table 7.10 Network Design Database for VHD200 Input File

#	-V-	-H-	LOAD (Erlangs)	LATA	LINK
1	6375	1919	4	0	0
2	9063	7322	1	0	0
3	5550	1786	3	0	0
4	5878	1408	1	0	0
5	6774	8875	1	0	0
6	6859	1711	3	0	0
7	5204	1370	4	0	0
8	6309	6818	4	0	0
9	6410	1731	3	0	0
10	6726	5527	2	0	0
11	6411	2341	4	0	0
12	9301	6816	1	0	0

Continued

#	-V-	-H-	LOAD (Erlangs)	LATA	LINK
13	5237	5454	4	0	0
14	7331	6802	3	0	0
15	7753	2748	3	0	0
16	8267	2327	4	0	0
17	5731	5090	4	0	0
18	7285	5559	1	0	0
19	5367	1996	2	0	0
20	7557	2836	1	0	0
21	6780	3572	4	0	0
22	5982	6680	1	0	0
23	8853	6918	1	0	0
24	6095	7733	4	0	0
25	7597	2240	3	0	0
26	9045	7737	3	0	0
27	8187	2924	4	0	0
28	5736	2809	3	0	0
29	6652	4493	4	0	0
30	6330	2377	4	0	0
31	7348	4859	3	0	0
32	7995	3965	1	0	0
33	7371	5936	1	0	0
34	8100	3190	3	0	0
35	8257	6049	3	0	0
36	6798	3127	3	0	0
37	5938	7411	1	0	0
38	9492	6114	4	0	0
39	8501	8787	4	0	0
40	8636	3780	2	0	0

Continued

Table 7.10 *Continued*

#	-V-	-H-	LOAD (Erlangs)	LATA	LINK
41	8553	8152	2	0	0
42	8229	5933	1	0	0
43	8976	5173	3	0	0
44	8451	4011	3	0	0
45	7730	7076	1	0	0
46	8981	7684	4	0	0
47	5487	2264	3	0	0
48	8556	4222	1	0	0
49	9449	7188	1	0	0
50	7338	8522	3	0	0
51	7389	6223	2	0	0
52	5547	3849	4	0	0
53	6031	3529	2	0	0
54	7533	5603	2	0	0
55	8113	3389	1	0	0
56	6957	1536	2	0	0
57	5100	4099	1	0	0
58	8141	3528	2	0	0
59	6568	7444	1	0	0
60	7766	7118	2	0	0
61	5809	7385	3	0	0
62	6168	5105	2	0	0
63	7065	3947	1	0	0
64	6300	2475	2	0	0
65	8250	580	4	0	0
66	7549	2006	2	0	0
67	6251	6962	2	0	0

Continued

#	-V-	-H-	LOAD (Erlangs)	LATA	LINK
68	8005	1493	2	0	0
69	6354	9224	1	0	0
70	5790	2633	4	0	0
71	7440	3971	3	0	0
72	8751	5813	1	0	0
73	7565	2565	4	0	0
74	9391	5914	3	0	0
75	6394	7889	1	0	0
76	6192	7056	1	0	0
77	7593	3348	4	0	0
78	5182	3265	4	0	0
79	8250	540	4	0	0
80	8083	6676	2	0	0
81	9311	4048	3	0	0
82	6370	4455	1	0	0
83	9094	7703	3	0	0
84	5646	4061	3	0	0
85	7721	8195	4	0	0
86	5626	3606	1	0	0
87	8052	8885	3	0	0
88	7325	1604	2	0	0
89	7208	2733	4	0	0
90	7597	1813	2	0	0
91	7553	5983	1	0	0
92	7627	8627	4	0	0
93	8710	5670	1	0	0
94	5303	1812	3	0	0
95	8006	8934	1	0	0

Table 7.10 *Continued*

#	-V-	-H-	LOAD (Erlangs)	LATA	LINK
96	7615	4964	1	0	0
97	9001	8394	1	0	0
98	8064	5872	1	0	0
99	5575	6063	1	0	0
100	8655	3881	1	0	0
101	7943	3687	1	0	0
102	9262	5267	4	0	0
103	6480	8521	3	0	0
104	7154	7953	1	0	0
105	5687	6496	1	0	0
106	6147	3748	1	0	0
107	5280	4182	2	0	0
108	5913	3851	3	0	0
109	6153	6886	2	0	0
110	6331	8623	3	0	0
111	8978	6181	2	0	0
112	8992	7485	1	0	0
113	8535	7874	3	0	0
114	6366	2377	1	0	0
115	5071	4279	2	0	0
116	7262	2338	3	0	0
117	5934	5384	1	0	0
118	6579	7701	4	0	0
119	7760	4949	2	0	0
120	7302	1711	1	0	0
121	6047	7159	4	0	0
122	6114	5346	3	0	0

Continued

#	-V-	-H-	LOAD (Erlangs)	LATA	LINK
123	7441	8279	2	0	0
124	6427	929	2	0	0
125	6200	2615	3	0	0
126	6556	1313	1	0	0
127	6245	2807	4	0	0
128	6560	4456	1	0	0
129	6016	6584	1	0	0
130	7267	6105	3	0	0
131	7768	9076	4	0	0
132	9310	6425	4	0	0
133	6529	6230	1	0	0
134	8146	725	2	0	0
135	7975	7018	2	0	0
136	5726	3313	2	0	0
137	8363	3830	1	0	0
138	7811	1415	3	0	0
139	8721	4771	2	0	0
140	8185	2225	1	0	0
141	6027	5569	3	0	0
142	8239	3032	3	0	0
143	7968	3485	1	0	0
144	8441	2707	2	0	0
145	6328	7292	4	0	0
146	7012	5775	1	0	0
147	6689	4816	4	0	0
148	7705	4045	4	0	0
149	5168	3177	2	0	0
150	6439	2651	3	0	0

Continued

Table 7.10 *Continued*

#	-V-	-H-	LOAD (Erlangs)	LATA	LINK
151	7694	4000	3	0	3
152	7087	5562	2	0	0
153	6607	3805	1	0	0
154	7434	1924	3	0	0
155	7347	8414	4	0	0
156	9318	6815	2	0	0
157	8526	4247	2	0	3
158	6556	2175	3	0	0
159	7731	1053	2	0	0
160	7370	1747	4	0	0
161	7953	5624	2	0	0
162	9064	7599	3	0	0
163	6786	3129	1	0	0
164	8067	3836	1	0	Q
165	9416	6310	3	0	0
166	6651	8820	1	0	0
167	7321	2156	2	0	0
168	8914	8094	1	0	0
169	8084	7033	3	0	0
170	6772	8464	2	0	0
171	5265	5524	1	0	0
172	6394	3182	2	0	0
173	6725	2120	3	0	0
174	5802	4191	2	0	0
175	7450	2233	4	0	0
176	6059	3909	1	0	0
177	8565	8610	3	0	0

Continued

#	-V-	-H-	LOAD (Erlangs)	LATA	LINK
178	8055	1079	2	0	0
179	7008	7709	3	0	0
180	7026	7878	1	0	0
181	5700	4235	2	0	0
182	6893	1343	3	0	0
183	6667	1681	2	0	0
184	7102	1191	4	0	0
185	6423	8454	3	0	0
186	7064	8251	3	0	0
187	8834	3133	4	0	0
188	7998	1398	4	0	0
189	6179	5937	3	0	0
190	7481	7681	1	0	0
191	5936	6994	4	0	0
192	7862	1268	4	0	0
193	8581	8385	2	0	0
194	6093	5675	4	0	0
195	8315	2999	3	0	0
196	6267	2993	3	0	0
197	6777	7594	2	0	0
198	6541	6837	4	0	0
199	8870	6407	4	0	0
200	8145	8951	2	0	0

TOTAL NODAL TRAFFIC LOAD=   483 Erlangs

employs five switches and costs only 2.1 cents per call minute. The reader is encouraged to model such a network.

Only the general trends shown by some charts should be useful to the designer. For example, the observation that the cost per call minute goes down as TGF increases is generally true for all voice networks. It is almost impossible to predict the network behavior with different mixes of AL and trunk facilities and

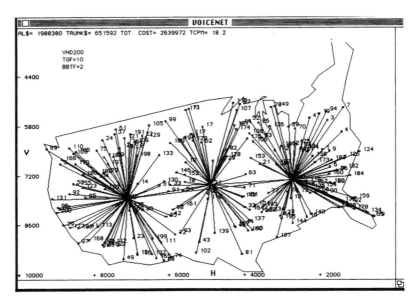

Figure 7.34 A voice network topology with five switches: VHD200, TGF = 10, BBTF = 2, and eight VG circuits per 56Kbps AL or TK.

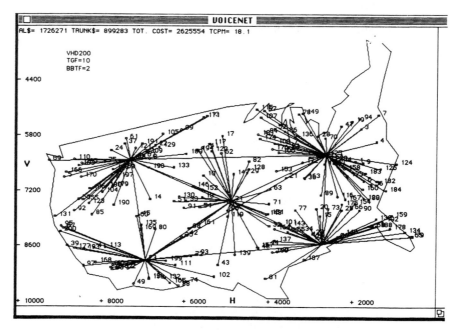

Figure 7.35 A voice network topology with three switches: VHD200, TGF = 10, BBTF = 2, and eight VG circuits per 56Kbps AL or TK.

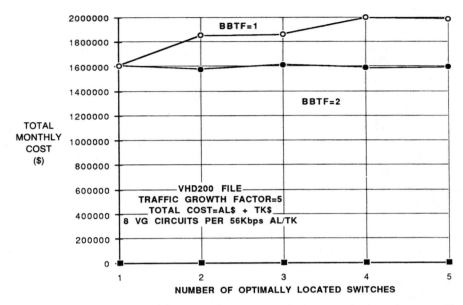

Figure 7.36 Total monthly cost for network transmission versus the number of optimally located switches: VHD200, TGF = 5, BBTF = 1 or 2, and eight VG circuits per economically designed AL/TK.

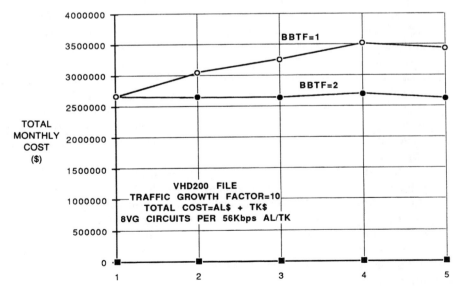

Figure 7.37 Total monthly cost for network transmission versus the number of optimally located switches: VHD200, TGF = 10, BBTF = 1 or 2, and eight VG circuits per economically designed AL/TK.

nodal traffic intensities. *Intuitions are constantly proven wrong and common sense is of no great use when working with network modeling.* The designer is advised to start with an open mind when launching a new network modeling task.

If one had to vary all the design variable such as link types, ALT, TKLT and their corresponding MF parameters, BBTF, TGF, number of switches, AL and trunk blocking, one will end up with an astronomical collection of charts. Only a small fraction of them should provide a useful insight into voice networking. This is where the mind of the designer comes into play. Through insights gathered over the years, the number of studied configurations are reduced to a practical number. The design process applied to the four VHD files using only a judicious combinations of link types, ALMF, TKMF, BBTF, and number of optimally located tandem switches should be quite instructive.

The first observation derived from Figures 7.7, 7.18, and 7.29 implies that a private network must employ VPN- or Vnet- or SDN-like services to be cost effective at low traffic volumes. These figures show that if an economical AL-bundle design is employed using ECC=$1.31, most locations will require some off-net services to remain cost effective. Similar conclusions can be drawn from Figures 7.6, 7.17, and 7.28 that plot average cost of a call-minute as a function of TGF for two designs: one based on all on-net voice circuits and the other based on an optimum mix of on-net and off-net facilities with eight voice conversations multiplexed on a 56Kbps AL or a trunk. The case of multiplexing 48 conversations per T1 AL or a trunk was also considered but such a network was generally more costly than the best case. These two designs yield the same cost per call-minute only for very large values of TKMF. For very large values of TGF(>10), all the networks studied tend to provide full connectivity to all locations via privately leased lines.

Figures 7.8, 7.9, 7.10, 7.19, 7.20, 7.21, 7.30, 7.31, and 7.32 plot total network transmission costs as a function of the number of optimally placed switches (S) and two values of BBTF. The results show that a network with uniform community of interest (BBTF=1) requires only a single tandem switch for a minimum cost criterion. However, most networks show some cyclical variation of network transmission costs as a function of S. The only exception is the 59-node network that was characterized by nodal clustering around node 1. For that case, a single switch generally yields the best solution. Two switches are optimum only for a large value of TGF. For networks with 100 or 200 nodes, five switches seem to be appropriate although further analyses needs to be performed.

In order to select the final design, plots of Figures 7.11, 7.22, and 7.33 are suggested. These plots exhibit costs for not only the ALs and trunks but also for the switching and NMC-related hardware on a monthly basis. An interesting fact emerges: the hardware costs increase the network costs by only 1 or 2 cents per call-minute. Transmission costs are still the major factors. Furthermore, the hardware costs don't influence the shape of the plot relating total cost to the number of optimally located switches. It just moves the plot up and down, without changing the plot minima.

A study of network costs suggests that the costs per call-minute are almost identical for very large values of TGF for the four networks studied. Only the 59-node network is slightly cheaper. It appears that costs per call-minute reach their asymptotic value as TGF reaches about 20 for most properly designed networks for given tariffs. Another approach for reaching major conclusions is to display some of the output data in the form as shown in Table 7.11. The printout relates TGF, ALMF (AL multiplexing factor), and TKMF (trunk multiplexing factor) to the computed values of cents/CM (cost of a call-minute).

It is interesting to note that a 100-node private voice network that employs all on-net VG circuits is very costly. The transmission costs range between 35 and 29 cents per call minute, even when TGF is varied between the high values of 10 and 40. Therefore, the network designer must employ some kind of digital multiplexing on AL and trunk circuits. Another interesting aspect of the data is that ALMF=8 and TKMF=8 yield the cheapest networks. The perceived quality of end-to-end trans-mission in the network must be studied carefully before employing 56Kbps DDS cir-cuits for both ALs and trunks. The data of Table 7.11 also shows that the cost of a call minute is not reduced by employing 56Kbps DDS-ALs with ALMF=8 and T1 trunks with TKMF=48 (valid for ADPCM) when compared to the cheapest network for ALMF and TKMF=8. In fact, the cost of the network goes up somewhat when T1 circuits are used for both ALs and trunks. However, T1 and T3 facilities may become attractive for very high traffic intensities as experienced by large corporations.

Since a 100-node network handles about 8.1 million call-minutes in a month for TGF=10, even a 10 cent saving per call-minute from that for a VPN/SDN/Vnet implementation implies a monthly savings of about $810,000. Such a savings can pay for a large amount of hardware.

Additional savings can be achieved through the use of facilities from other OCCs and bulk-transmission carriers. In some cases, one can now lease a frac-tional T1 with less than 24 64Kbps PCM channels. By using the low bit rate voice compression technique, one can carry the desired number of VGCs on a frac-tional T1. The methodology presented here should be adequate to study any arbitrary tariff as opposed to always being dependent upon a fixed DB.

Table 7.11 Costs per Call-Minute versus TGF and ALMF and TKMF for a 100-Node: AL and Trunk Blocking Factors of .01 and AT&T's Tariffs 9, 10, and 11

TGF	ALMF	TKMF	Cents/CM
10	1(all On-Net VG)	1	34.8
20	1(all On-Net VG)	1	31.0
40	1(all On-Net VG)	1	28.6
10	8(56Kbps AL)	8(56Kbps Trunk)	17.1
20	8(56Kbps AL)	8(56Kbps Trunk)	15.1
10	8(56Kbps AL)	4(T1 Trunk)	17.6
20	8(56Kbps AL)	4(T1 Trunk)	15.1
10	48(T1 AL)	48(T1 Trunk)	18.1
20	48(T1 AL)	48(T1 Trunk)	15.3

One can also achieve a significant savings by lowering the grade-of-service on ALs and trunks. Special results were computed for a 100-node network for TGF=10 and these are shown in Table 7.12 for only the best designs, in other words, for the proper number of optimally located switches.

Since the analyses presented earlier assume bypass of the local BOCs for the 56Kbps and T1 circuits, it should be useful to study the effects of no bypass on 56Kbps and T1 circuits. This can be done by adding $737 and $2,000, respectively, to the "1st" fixed value in Tariff file for the two types of circuits. These values were dictated by the facts that the average length of the DDS and T1 LECs were found to be as 20 and 10 miles, respectively, for VHD41 and VHD59 files. See Figure 7.38 for the plots of costs per call-minute versus the number of network nodes and two multiplexing techniques and for TGF=10. The cost difference is about 2 to 5 cents per call-minute. Networks with 41, 100, and 200 nodes are almost identical in costs. A voice network for the VHD59 file is slightly cheaper due to the fact that nodes are clustered around the switch node #1. Since networks with only voice grade circuits are always designed with no bypass, curves were not plotted. The bypassing of a few VG circuits is generally not worth the trouble.

The results just presented should enable the designer to make some predictions as to the expected costs for their future networks. These results should also help an enterprise assess the performance of their existing networks. The results should never be taken as a substitute for an in-depth analysis of their existing or future needs since each user has a unique situation.

Exercises

7.1 Create new VHD17, LATA17, NAME17 input files using the design data of Table E7.1 for a 17-site corporation engaged in manufacturing specialized electronic equipment.

7.2 Using the VHD17, LATA17, NAME17, the enclosed DTP8 file, and the 1987 tariff input files (Table 2.4), model voice networks using the B_{al} and B_{tk} equal to 0.1, ECC equal to $13.31 or $1.0, for one, two, and three switches optimally located and three sets of leased lines Voice Grade

Table 7.12 Cost per Call-Minute versus AL and Trunk Blocking Factor (VHD100), TGF = 10, ALMF = 8, TKMF = 8, and AT&T's Tariffs 9, 10, and 11)

AL Blocking	Trunk Blocking	Cost per Call-Minute
.01	.01	17.1 cents
.01	.1	15.1 cents
.1	.01	16.1 cents
.1	.1	14.0 cents

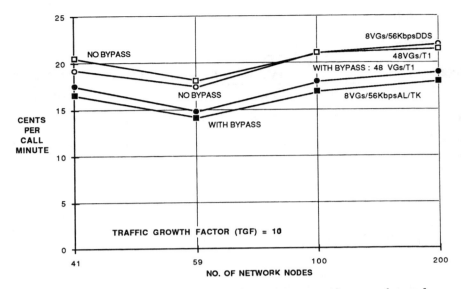

Figure 7.38 Average cost per call-minute versus the number of network nodes: TGF = 10, BBTF = 2 for bypass or no-bypass conditions, and two types of access lines and trunks.

Table E7.1 A voice network design database for 17 sites.

*************** NODAL DEFINITION DATA ***************

NODE#	-V-	-H-	LOAD (MErl)	LATA (Region#)	LINK	NAME (CCCCST)
1	8438	4061	40000	552	0	LCLNTX
2	8436	4034	5000	552	0	DALLTX
3	8296	1094	1300	952	0	SRSTFL
4	8360	906	1300	939	0	FTMYFL
5	6421	8907	1300	674	0	TACMWA
6	6336	8596	1300	676	0	BELVWA
7	4410	1248	1400	128	0	DANVMA
8	6479	2598	1300	466	0	VERSKY
9	9258	7896	1300	730	0	TOAKCA
10	9233	7841	1400	730	0	NORWCA
11	9210	7885	1400	730	0	WLAXCA

Continued

Table E7.1 *Continued*

*************** NODAL DEFINITION DATA ***************

NODE#	-V-	-H-	LOAD (MErl)	LATA (Region#)	LINK	NAME (CCCCST)
12	7292	5925	1400	656	0	DENVCO
13	7731	4025	1300	538	0	TULSOK
14	7235	2069	1300	438	0	NORCGA
15	5972	2555	2500	324	0	COLMOH
16	9228	7920	2500	730	0	STMNCA
17	8173	1147	2500	952	0	TMPAFL

TOTAL TRAFFIC LOAD = 68500

(MF=1), 56Kbps (MF=8), and T1 (MF=24) for both ALs and trunks. Tabulate the results for a total of 18 topologies and plot system costs as a function of the number of optimally located switches (see Figure 5.1). Using these charts, find the optimum voice network topology.

7.3 Repeat Exercise 7.2 for the 1991/92 AT&T tariffs of Table 2.5 and the 1996 MCI tariffs of Table 2.6. Tabulate the differences in the results for the optimum network topologies.

Bibliography

Gareiss, R. 1996. "Voice over the Internet." *Data Communications*, September 1996, pp. 93–100.

Green, T. 1996. "IP Voice Vendor Looking for Carriers." *Network World*, August 26, 1996, p.14.

Nortel Document. 1994. "Telephony 101." http://www.nortel.com

Nortel Document. 1995. "Long Distance 101." http://www.nortel.com

Pappalardo, D. 1997. "Frame Relay Gets a New Set of Priorities." *Network World*, February 3, 1997, p. 21.

Sharma, R., P. De Sousa, and A. Ingle. 1982. *Network Systems*. New York: Van Nostrand and Reinhold.

TranSwitch Seminar. 1996. "Voice Over ATM." Participated by Nortel, Nyborg & Company, Mitel Corporation and Fujitsu.

8

Design Process for ACD Networks

8.1 Introductory Remarks

An Automatic Call Distribution (ACD) system is employed to manage incoming and/or outgoing voice calls. Some businesses such as airlines, car rentals, and hotel chains derive their income primarily through the mechanism of telemarketing. Some other businesses allow homeowners to call their sales agents and purchase products being advertised on television. Some telephone companies and large department stores employ an ACD to manage their outgoing calls to their customers.

The so-called 800 service is generally employed to handle the incoming calls from the general public. The incoming traffic can also employ the Wide Area Telecommunications Service (WATS or INWATS) lines. MegaCom 800 Services provide the same capabilities via T1 lines that connect the ACD center with the OCC's POP. In some special cases, Foreign Exchange (FX) and local business lines may also be used to handle both the incoming and outgoing traffic and reduce the transmission cost. Most of the OCCs such as AT&T, MCI, and Sprint provide such services and their tariffs are widely advertised and frequently revised.

The ACD node provides circuit switched paths between the reservation/sales agents of the business and the terminating trunks (these should be called access lines but the word *trunks* has survived). The ACD switching node allows these incoming calls to be queued whenever all agents are busy. These calls are queued in the trunks. Consequently, the call holding times on trunks are extended by the call waiting times. A model of the ACD switching node is illustrated in Figure 3.10.

See Chapter 3 for a discussion of ACD traffic analysis tools. One can study two papers by Sharma (Sharma, 1985) and (Sharma, 1986) to get a deeper insight into the workings of an ACD system.

The so-called computer-telephone-integration (CTI) technology is adding dimensions to the basic ACD technology. Some call it the CTI revolution. However, the ACD still dominates the so-called call center market due primarily to ACD's rich features and capabilities. The promises of (1) ACD's integration with a host and (2) lower per-termination costs of a CTI-based system still intrigue many innovators, small customer service centers, catalog/mail-order/database marketing operations, service bureaus, and messaging services. See the paper by MacPherson and Cleveland (MacPherson and Cleveland, 1990) to get an idea of (CTI) technology trends.

The ACD node(s) can be either purchased or leased as a Centrex service. Another solution is to add the ACD features on an existing digital PABX(s). A financial justification for any of these situations can only be made after a careful evaluation of associated ACD networks considering identical GOS, class-of-service, service features, special incremental costs, and cost of money during the life cycle. The design process of this chapter should be of interest to all enterprises (e.g., airlines, car rentals, and hotel reservation companies), virtual service (VS), or the so-called 800 service, or even the new digital 800-service providers who can differentiate their services by adding features through the use of public intelligent networks.

8.2 The Basic Design Process for an ACD Network

The process of traffic data gathering is identical to that for voice networks. Time-consistent averages of traffic intensities for each applicable hour must be obtained for computing not only the number of transmission facilities (that connect the ACD node to the OCC's POP) to be leased, but also the number of agents required to handle the incoming calls for all hours. See Figures 8.1(a) and 8.1(b) for two traffic patterns: one for a typical airline reservation, and one for a home merchandising business. The traffic pattern of Figure 8.1(a) implies a predictable level of traffic intensities. Consequently, the task of agent scheduling is simplified. The traffic pattern of Figure 8.1(b) will always be tied to the product being advertised on television. If the product has a great appeal to the public and it is attractively priced, the volume of incoming calls may exceed all expectations of even the most astute student of human psychology or behavior. As a result, a large fraction of the traffic may get blocked by not only the public telephone network of the virtual service provider but also the private ACD trunk network leased by the customer. Furthermore, the task of agent scheduling becomes quite complex. Some hard decisions need to be made for reaching a compromise between the cost of transmission facilities and the desired sales volumes and customer satisfaction levels.

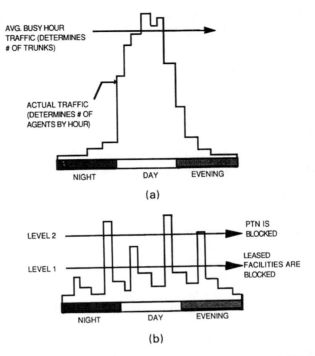

Figure 8.1 Typical 24-hour traffic patterns for two ACDs employed for airline reservation and home merchandising applications.

8.2.1 A Description of Input Files and Design Parameters

In order to design an ACD network topology, the network design package as described in Chapter 6 allows one to execute all the steps defined in Chapter 5, and in particular, Sections 5.4.1 and 5.4.2 using the input design data of input files. The network design software package also allows one to vary a large number of critical design parameters iteratively to arrive at an optimum ACD network.

The first step is to prepare the set of input files that must be named in FILES. The required input files for designing an ACD network are VHD, LINK, MAP, NLT, TARIFF, SDF, NAME, LATA, FILES, DTP, SWF, and a subset of CSABDS, UTBL, WUTBL, MUTBL, and RSTBL as named in FILES.

We will start with the VHD file. The V&H coordinates for each region are derived from the enterprise database (EDB) as reflected in Figure 6.3. Only the area code (i.e., NPA) is employed to approximate the V&H coordinates since calls can come from any part of the region. The TCA values of TIs originating from the region can be computed using the procedure described in Chapter 2. If it is a brand-new ACD system, the TIs can be simulated as a function of the calling populations within each region. If an existing ACD network has no capability for collecting traffic data, one can derive the traffic patterns from judicious samples of calls received on a typical day. In most cases, the existing provider of virtual services can provide the detailed call traffic patterns for a fee.

The LINK file is used only if each ACD switch employs a mix of virtual service lines, FX, or local business lines to serve traffic from different regions. The LINK file will be ignored by the modeling examples of this chapter since we employ only a single type of virtual facilities throughout the ACD network.

The MAP file is required by every networking task since it is used to draw the boundary map. The MAP file (i.e., list of V&H coordinates for boundaries of countries other than the United States, Canada, and Mexico) can be obtained by using the procedure described in Chapter 6.

The next important input file is an NLT file. It defines the set of link types that will be used in modeling the AL bundle (between the POP and ACD) and trunk bundles. The link types can now represent a mix of leased and fully owned facilities. For the ACD application, the important link attributes are link type number and the corresponding tariff number (TF#) and facility (leased/privately owned) factor F_{pf}.

We will use a tariff file that defines at least three tariffs: Tariffs 9, 10, and 11 for the voice grade (VG), 56Kbps DDS, and T1 lines. Such tariffs were defined in Table 2.4. We will also assume that the average monthly cost of the two local loops and termination charges equals an amount of \$102.00 per VG, \$62.76 per DDS, and \$0.00 per T1 circuit, respectively. This implies no bypass for the VG circuits and some form of bypass for the DDS and T1 circuits. Although most of the analyses assume some bypass for the DDS and T1 circuits, an attempt is made near the end of this chapter to show the comparison between network cost with and without bypass of the local exchange carrier.

The next important input file that needs to be updated is the SDF file. It defines the values of 56 design parameters. These can be viewed (Mnemonic and corresponding value) by using the ViewUpdateInputFile item of the File menu. Each SDF mnemonic is also provided with a suffix of V, A, D, or C that implies Voice, ACD, Data, or Common application, respectively. Only a few of them affect an ACD network significantly. These are TGF, ALT, B_{al}, ECD, DREQ, PEXD, C_{lbr}, F_{rst}, ACD, TKLT, F_{fdx}, MTKU, BBTF, Fvc6 (for simulating actual days in a month for computing costs), and DPM (days per month for computing cost summaries on network map). The ALT parameter defines the types of access links connecting POP to an ACD Node. The TKLT parameter defines the data trunks used to handle ACD DB messages between the primary ACD node and secondary ACD nodes. The primary ACD node is always listed as the first ACD switch in the SWF. Only a single trunk link type must be employed to interconnect the ACD switches in an ACD network. The choice of ALT and TKLT allows one to compute the cost of an ACD network for any mix of AL and trunk circuits. Design parameter B_{al} defines the GOS required by the AL bundle. Design parameter ECD is used in computing the cost of each all-minute for a topology map.

Although the NAME file doesn't influence any analysis, its use does provide a complete site database in the DBF output file. In order to read the NAME file, one must set F_{nn} to 1.

The LATA file represents the region numbers as required by the CSABDS file for computing the CSA Band numbers.

The FTD file is never used for modeling an ACD network since all calls are destined for ACD nodes only. The fourth column of a VHD file will always be employed by the ACD network to define the incoming TCA traffic intensity (TIs) originating from the region.

The daily traffic profile input files DTP and DTP4 will be used to model typical business hours for the ACD network modeled in Chapter 8. These list the fraction daily traffic volumes occurring during each hour of the day (see Table 8.2).

The SWF files are required to define the number switches and their nodal IDs. The original version of EcoNets required a manual input of switch nodal IDs.

Several specialized databases defining virtual tariffs must be employed to compute the cost of incoming calls. The first table common to all virtual network services defines the Customer Service Area (CSA) Bands, represented by numbers varying from 1 to 6, for each source-destination combination of regions (e.g., Northern California to Tennessee). To illustrate, a call coming from Northern California and destined to an ACD switch in Memphis, Tennessee will use a CSA band of 4. Table 8.3 illustrates CSA bands (CSABDS) for all combinations of states/regions. Each of the originating 60 state/region is provided with 62 values of CSA bands arranged in two rows of values. Two destination regions, namely Virgin Islands (VI) and Puerto Rico (PR), are added to the 60 states/regions shown on the leftmost column. A close observation will show that a CSABDS table is not symmetrical in terms of its values. The input file, CSABDS, reflects such an extensive database.

The second table computes a rate step for a given source state/region and a CSA band. Such a table is called the Rate Step Table (RSTBL) as illustrated in Table 8.4. It is used to compute the tariffs for 800- and the WATS virtual services. RSTBL is not required to compute the tariff for the MegaCom 800 virtual service. There are a total of 26 separate rate steps. For each originating home state/region (e.g., North and West Texas), there are six (6) unique rate steps for each of the CSA bands. For example, the rate steps for the home or originating states of Iowa and California are 14 and 18, respectively, considering only the CSA band of 6. The input file, RSTBL, defines such a database.

The monthly cost of virtual service on 800-lines can then be computed as a function of hours utilized on 800 virtual lines, the rate step and time of day (e.g., business hours, evenings, and nights, or weekends). The Use Rate Table (URTBL or simply UTBL) that defines such hourly costs for using 800-virtual lines is illustrated in Table 8.5(a). The WATS Use Rate Table WUTBL that defines such hourly costs for using the INWATS virtual lines is illustrated in Table 8.5(b). The hourly cost of a virtual service using MegaCom800 lines can be computed as a function of the CSA Band number and the time of day the calls arrive. The MegaCom 800 Use Rate Table (MUTBL) that defines such hourly costs for using MegaCom virtual lines is illustrated in Table 8.5(c). Tables 8.5(a), (b), and (c) reflect the 1987 tariffs. Some users are still being grandfathered with these services. The designer should check with the various 800-service providers about the latest tariffs and make the proper changes before modeling the monthly costs. Whenever the type of current service must be changed, the customer is required to accept the new simplified tariffs that look similar to Table 8.5(c) but with upper limits on mileage bands replacing the leftmost column. Setting ACDT equal to 4 will handle that situation.

All of the specialized input files, CSABDS, UTBL, WUTBL, MUTBL, and RSTBL must be named in FILES. The design parameter, ACDT, determines which set of specialized input files must be read to synthesize an ACD network.

The designer is always interested in factors that influence the network performance and monthly costs significantly. In particular, the designer is also interested to know the important input files that will influence the outputs of an ACD network design process.

The cost and performance of an ACD network is mainly determined by the amount of traffic handled. The software design package can enable one to vary the traffic volume by changing the traffic growth factor (TGF). The analyses presented here use three TGF values: 1, 5, and 10. The parameter TGF multiplies the traffic intensities as defined by the VHD input file. Traffic volumes influence the cost of labor in an ACD network in a major way. The higher the traffic handled by an ACD switch, the lower the unit cost of a labor minute due to higher economies of scale. When one increases the number of ACD switches, one loses such economies of scale related to labor costs. Since the cost of virtual service is generally directly proportional to traffic intensity, economies of scale don't play any part in determining the communications cost (as these did for enterprise voice networks employing private leased lines).

Finally, the FILES file must be updated with the names of newly created input files each time an example ACD network is modeled and optimized. If an input file is not in the directory, the software warns the designer by printing the input file number (e.g., 1 implies a missing VHD file, 2 implies a missing LINK file, and so on) in a standard dialog window. No task can be accomplished until the missing input file is placed in the directory. For some situations, one may have to set the SDF parameters such as F_{lk}, F_{ftd}, and F_{nn} to a zero value. This will make EcoNets ignore the names of corresponding names of input files listed in FILES. To illustrate, since we will not be using a LINK file for the voice network examples of Chapter 8, one can list any name of a missing LINK file (e.g., LINK35) in FILES if Flk = 0. The parameter, F_{lt} can't be set to zero since a LATA file is very essential to ACD modeling. As pointed out earlier, the LATA file defines the region numbers as defined in Table 8.3. Its values are needed to compute the CSABDS numbers that are used to compute the monthly costs. The design parameter, BBTF, determines the amount of traffic flowing on the backbone network trunks. It may influence the cost of the backbone network. Since an ACD network generally employs a small number of switches, BBTF should not be a very significant factor as it was for the voice networks.

When all the required input files are available in the directory, the designer is ready to model any ACD network. Some designers may find that it takes an inordinate amount of time to prepare the input files when compared to the time it takes to model and optimize an ACD network. Experienced telecommunications and DP managers have already made this painful discovery. The field of network design is a very complex one. Only a few stubborn souls have the stomach to face the challenge. Some may find it proper to outsource this input file creation process to a work force in a developing country. But if a designer had the raw data in the form of Table 6.3, he or she doesn't have to outsource to a third party since personal productivity tools are already available to most engineers/designers. The presence of a third party can only add more delays and human errors.

8.2.2 The Computational Model

A VHD49 file was employed to model ACD networks that handle incoming calls from 49 regions of CONUS. A single LATA file was employed for the VHD49 case. It defines the correct geographical region number (varying from 1 to 60) for each area of originating calls. The data of a LATA also influences the network costs. When the regions are very close to the ACD node, the cost of virtual services tends to be lower. Three SWF files were created, one for each case of 1, 2, and 3 switches. The original version of EcoNets required a manual input of switch nodal IDs. Each SWF starts with the number of switches, followed by the nodal IDs for switch locations. Table 8.1 illustrates the database (DB) for creating the VHD49, LATA49, and NAME49 files.

In order to compute the cost of handling off-net traffic, one must first obtain the percent distribution of call volumes by each hour of the day. If such data is not available (e.g., for a new ACD system), one must construct such a distribution through the process of consultations with the key players of the business. Such data will not only pinpoint the applicable traffic pattern but also the number of hours observed during a month for each applicable rate step. Two Daily Traffic

Table 8.1 ACD Network Design Database (49 States/Regions)

#	-V-	-H-	LOAD (MERLs)	LATA	LINK	NAME
1	8173	1173	526	11	0	TMPFL
2	4422	1249	540	24	0	BOSMA
3	5986	3426	876	15	0	CHIIL
4	5574	2543	506	41	0	CLEOH
5	5510	1575	268	23	0	BALMD
6	8938	3536	358	52	0	HOUTX
7	5251	1458	582	46	0	PHLPA
8	4997	1406	1125	37	0	NYCNY
9	8492	8719	401	5	0	SFOCA
10	9213	7878	923	6	0	LAXCA
11	7954	1031	419	11	0	ORLFL
12	8669	8239	407	5	0	FRSCA

Continued

Table 8.1 *Continued*

#	-V-	-H-	LOAD (MERLs)	LATA	LINK	NAME
13	7574	7066	449	54	0	SLKUT
14	9250	7810	573	6	0	ANHCA
15	7260	2083	586	12	0	ATLGA
16	5076	2327	486	38	0	BUFNY
17	6263	2680	506	41	0	CINOH
18	5972	2554	468	42	0	COLOH
19	8432	4033	366	51	0	DFWTX
20	5536	2829	537	25	0	DETMI
21	8282	557	459	11	0	FLDFL
22	6272	2992	477	17	0	INDIN
23	7029	4202	497	29	0	KANMO
24	7937	1538	533	11	0	MIAFL
25	5788	3588	505	59	0	MILWI
26	5780	4526	260	27	0	MNPMN
27	4960	1354	321	26	0	NSANY
28	5016	1430	2426	34	0	NWKNJ
29	8486	2638	479	21	0	NORLA
30	9133	6748	519	3	0	PHOAZ
31	5619	2184	613	46	0	PITPA
32	6799	8915	486	44	0	PTLOR
33	9202	7718	526	6	0	RIVCA
34	8303	8581	459	5	0	SACCA
35	6807	3483	293	29	0	STLMO
36	8436	4034	466	52	0	SANTX
37	9462	7632	566	6	0	SNDCA
38	8583	8619	494	5	0	SJSCA
39	6337	8896	533	57	0	SEAWA

Continued

#	-V-	-H-	LOAD (MERLs)	LATA	LINK	NAME
40	5623	1578	376	10	0	WASDC
41	6233	1823	419	56	0	SALVA
42	8111	8340	358	32	0	RENNV
43	6206	4167	383	18	0	WTRIA
44	5703	2820	433	41	0	TOLOH
45	6529	2773	444	20	0	LOKKY
46	4913	2195	556	38	0	ROCNY
47	5917	1223	439	39	0	NFKNC
48	4550	1290	444	47	0	PRORI
49	7501	5899	533	7	0	DENCO

TOTAL TRAFFIC LOAD=25999 MERLs

Profiles (DTPs) were considered for analysis. A DTP file was created to model the reservation-related traffics for every day of the week or month. It should be easy to make a DTP file for each day of the week to reflect a particular traffic pattern. The DTP4 file was created to model the home merchandising traffic. Table 8.2 illustrates these two traffic profiles.

A subroutine was created for computing the cost of any virtual service on a monthly basis. Such a subroutine executes the following tasks:

Table 8.2 DTP and DTP4 Files

Hour	DTP Value	DTP4 Value
1	.01	.041667
2	.01	.041667
3	.01	.041667
4	.01	.041667
5	.02	.041667
6	.02	.041667
7	.02	.041667
8	.03	.041667
9	.04	.041667
10	.06	.041667

Continued

Table 8.2 *Continued*

Hour	DTP Value	DTP4 Value
11	.08	.041667
12	.12	.041667
13	.10	.041667
14	.12	.041667
15	.10	.041667
16	.08	.041667
17	.04	.041667
18	.03	.041667
19	.02	.041667
20	.02	.041667
21	.02	.041667
22	.02	.041667
23	.01	.041667
24	.01	.041667

Note: DTP and DTP4 columns represent typical airline and home merchandising applications respectively.

1. Compute the total number of originating calls handled by the system per month = NCPM.

2. Compute the time consistent average (TCA) of busy-hour erlangs (BHRERL) for each ACD switch.

3. Compute the number of leased lines required to handle the traffic for a given AL blocking factor.

4. Using the applicable DTP file, compute the total number of handled erlangs during the night or weekend, business and evening hours of a day for each leased line.

5. Using the correct combination of CSABDS, RSTBL, URTBL,WUTBL, and MUTBL, compute the total cost for each leased line.

6. Compute the total cost of all leased lines in a virtual private network (VPN).

7. Using the Erlang-C formulae (e.g., Equation 3.11 of Chapter 3), compute the number of agents required for each hourly period as a function of traffic intensity and the specified PEXD (equal to the probability of delay exceeding a specified delay = DREQ) as defined in SDF file.

8. Compute the total cost of agent labor for the entire month.

9. Compute the cost of data trunks required to retrieve data from the database.

10. Compute the total cost of a call-minute.

Table 8.3 62 CSA Band Values (varying from 1 to 6) for Each State/Region

1 AK   0 5 5 5 5 5 5 5 5 5 5 5 5 5 5 5 5 5 5 5 5 5 5 5 5 5 5 5 5 5 5
5 5 5 5 5 5 5 5 5 5 5 5 5 5 5 5 5 5 5 5 5 5 5 5 5 5 5 5 5 5 5

2 AL   6 0 5 2 5 5 5 4 4 5 1 1 5 5 3 2 2 3 4 1 1 5 5 5 4 4 4 1 2 5 4
5 5 4 5 4 4 4 2 5 3 2 3 5 4 4 4 1 5 1 3 3 3 5 5 3 5 3 4 5 5 5

3 AZ   6 4 0 3 2 1 1 5 5 5 5 5 5 2 4 4 4 4 3 4 4 5 5 5 5 5 4 4 4 3 3
1 5 5 1 5 5 5 5 3 5 5 3 3 5 5 5 5 3 4 3 3 2 1 5 5 3 5 4 2 5 5

4 AR   6 2 5 0 5 5 4 5 4 4 3 3 5 5 2 2 3 3 2 2 1 5 4 5 4 4 4 1 1 5 3
5 5 5 4 5 5 5 3 4 3 3 1 5 4 4 5 3 4 1 1 3 3 5 5 4 5 3 3 4 5 5

5 NCA   6 4 1 4 0 0 2 5 5 5 5 5 5 1 4 4 4 3 3 4 4 5 5 5 4 4 3 4 3 2 2
1 5 5 2 5 5 5 5 3 5 5 3 1 5 5 5 5 3 4 3 3 3 1 5 5 1 5 4 2 5 5

6 SCA   6 4 1 4 0 0 2 5 5 5 5 5 5 1 4 4 4 3 3 4 4 5 5 5 4 4 3 4 3 3 3
1 5 5 1 5 5 5 5 3 5 5 3 1 5 5 5 5 3 4 3 3 2 1 5 5 2 5 4 2 5 5

7 CO   6 5 1 3 3 3 0 5 5 5 5 5 5 2 4 4 4 3 1 5 4 5 5 5 4 4 3 4 3 2 1
2 5 5 1 5 5 5 5 3 5 5 1 4 5 5 5 5 2 4 3 4 2 1 5 5 4 5 4 1 5 5

8 CT   6 3 5 4 5 5 5 0 2 2 4 3 5 5 3 3 3 4 4 3 4 2 2 1 3 3 4 4 4 5 4
5 1 1 5 1 1 2 3 4 2 2 4 5 1 2 1 3 4 3 5 5 5 5 1 2 5 2 3 5 5 5

9 DE   6 3 5 4 5 5 5 2 0 1 3 3 5 5 3 3 3 4 4 3 4 3 1 2 3 3 4 4 4 5 4
5 2 1 5 2 1 2 2 4 3 2 4 5 1 2 2 3 4 3 5 5 5 5 2 1 5 2 4 5 5 5

10 DC   6 3 5 4 5 5 5 2 1 0 3 3 5 5 5 3 3 4 4 3 4 3 1 3 3 3 4 4 4 5 5
5 3 1 5 2 2 2 2 4 2 2 4 5 1 1 2 3 4 3 5 5 5 5 3 1 5 1 4 5 5 5

11 FL   6 1 5 2 5 5 5 4 2 2 0 1 5 5 3 3 2 4 4 2 1 5 2 4 4 4 5 1 3 5 4
5 4 3 5 4 4 4 1 5 3 2 3 5 3 3 3 4 1 5 1 3 3 3 5 4 2 5 2 4 5 4 4

12 GA   6 1 5 2 5 5 5 4 3 3 1 0 5 5 3 2 2 4 4 1 2 5 3 4 3 3 5 1 3 5 4
5 4 3 5 3 3 3 1 5 3 2 4 5 3 3 4 1 5 1 4 4 4 5 4 2 5 2 4 5 5 5

13 HI   3 5 2 4 1 1 3 5 5 5 5 5 0 2 4 4 5 4 4 5 4 5 5 5 5 5 4 5 4 2 3
2 5 5 3 5 5 5 5 4 5 5 4 1 5 5 5 3 5 4 4 4 2 5 5 2 5 4 3 5 5

14 ID   6 5 3 4 2 2 2 5 5 5 5 5 5 0 4 4 4 3 3 4 4 5 5 5 4 4 3 4 3 1 3
1 5 5 3 5 5 5 5 3 5 5 3 1 5 5 5 5 3 5 4 4 4 1 5 5 1 5 3 1 5 5

15 NIL   6 3 5 2 5 5 4 5 4 4 4 5 3 5 5 0 0 1 1 3 2 3 5 4 5 2 1 2 3 1 5 3
5 5 4 5 4 4 4 3 4 2 2 3 5 3 3 5 3 4 2 4 4 4 5 5 3 5 2 1 5 5 5

Continued

Table 8.3 *Continued*

16 SIL	6 2 5 2 5 5 4 5 4 4 5 3 5 5 0 0 1 1 3 1 3 5 4 5 3 2 3 2 1 5 3
	5 5 4 5 4 4 4 3 4 2 1 3 5 3 3 5 3 4 1 4 4 4 5 5 3 5 3 2 5 5 5
17 1N	6 3 5 3 5 5 5 4 3 3 4 3 5 5 1 1 0 2 4 1 4 5 3 4 3 1 3 3 2 5 4
	5 4 4 5 3 3 3 3 4 1 1 4 5 3 2 4 3 4 2 5 5 5 5 4 3 5 2 2 5 5 5
18 IA	6 3 5 3 5 5 3 5 5 4 5 4 5 5 1 1 2 0 2 3 3 5 4 5 2 2 1 3 1 4 1
	5 5 5 4 5 5 5 4 3 3 3 2 5 4 4 5 4 1 3 4 4 4 5 5 4 5 3 1 3 5 5
19 KS	6 3 4 2 5 5 1 5 5 5 5 4 5 4 3 2 3 1 0 3 3 5 5 5 4 4 3 3 1 4 1
	4 5 5 2 5 5 5 4 3 4 4 1 5 5 5 5 4 1 3 2 3 2 3 5 4 5 4 3 3 5 5
20 KY	6 2 5 3 5 5 5 4 3 3 4 2 5 5 2 1 1 3 4 0 3 5 3 4 3 3 4 3 1 5 4
	5 4 3 5 4 4 4 2 5 2 1 4 5 3 3 4 2 5 1 4 4 4 5 4 1 5 1 3 5 5 5
21 LA	6 1 5 1 5 5 4 5 4 4 3 2 5 5 3 2 3 3 3 2 0 5 4 5 4 4 4 1 2 5 3
	5 5 4 3 5 5 5 3 5 3 3 1 5 4 4 5 3 4 2 1 1 2 5 5 4 5 3 4 5 5 5
22 ME	6 4 5 4 5 5 5 1 2 2 4 3 5 5 3 3 3 4 4 3 4 0 2 1 3 3 3 4 4 5 4
	5 1 1 5 1 1 2 3 4 2 3 4 5 2 2 1 3 4 3 5 5 5 5 1 2 5 2 3 5 5 5
23 MD	6 3 5 4 5 5 5 2 1 1 3 3 5 5 3 3 3 4 4 3 4 3 0 2 3 3 4 4 4 5 4
	5 3 1 5 2 1 2 2 4 2 2 4 5 1 1 2 3 4 3 5 5 5 5 3 1 5 1 4 5 5 5
24 MA	6 3 5 4 5 5 5 1 2 2 4 3 5 5 3 3 3 3 4 3 4 1 2 0 3 3 4 4 4 5 4
	5 1 1 5 1 1 2 3 4 2 2 4 5 1 2 1 3 4 3 5 5 5 5 1 2 5 2 3 5 5 5
25 NMI	6 4 5 4 5 5 4 3 3 3 4 4 5 5 1 2 2 1 4 2 4 4 3 3 0 0 1 4 2 5 4
	5 3 3 5 3 3 2 3 2 1 2 4 5 3 2 3 4 3 3 5 5 5 5 3 3 5 3 1 5 5 5
26 SMI	6 4 5 4 5 5 4 3 3 2 4 4 5 5 1 2 1 2 4 2 4 4 3 3 0 0 3 4 3 5 4
	5 3 3 5 2 3 1 3 4 1 1 4 5 2 2 3 4 3 3 5 5 5 5 3 3 5 2 1 5 5 5
27 MN	6 4 5 3 5 5 3 5 5 4 5 5 5 4 1 2 2 1 2 3 4 5 4 5 1 2 0 4 2 3 1
	5 5 5 4 4 4 4 5 1 3 3 3 5 3 3 5 5 1 3 4 4 4 4 5 4 5 3 1 3 5 5
28 MS	6 1 5 1 5 5 4 5 4 4 2 1 5 5 3 2 3 3 3 2 1 5 4 5 4 4 4 0 2 5 4
	5 5 4 4 5 5 5 3 5 3 3 2 5 4 4 5 2 4 1 1 3 3 5 5 3 5 3 3 5 5 5
29 MO	6 3 5 1 5 5 4 5 4 4 4 3 5 5 1 1 2 1 1 1 2 5 4 5 3 3 3 2 0 5 1
	5 5 5 4 5 5 5 4 4 3 3 1 5 4 4 5 4 3 1 3 3 3 5 5 4 5 3 2 4 5 5
30 MT	6 5 3 4 2 3 2 5 5 5 5 5 5 1 4 4 4 3 3 4 5 5 5 5 4 4 2 5 3 0 2
	2 5 5 3 5 5 5 5 1 4 4 3 1 5 5 5 5 1 5 4 4 4 1 5 5 1 5 3 1 5 5
31 NE	6 4 4 3 5 5 1 5 5 5 5 4 5 3 2 3 3 1 1 4 4 5 5 5 3 3 2 4 1 3 0
	4 5 5 3 5 5 5 5 2 4 4 2 4 5 5 5 5 1 4 3 3 3 3 5 5 5 4 2 1 5 5
32 NV	6 5 1 4 1 1 2 5 5 5 5 5 5 1 4 4 4 3 3 5 4 5 5 5 4 4 4 4 4 3 3
	0 5 5 3 5 5 5 5 3 5 5 3 1 5 5 5 5 3 4 3 3 3 1 5 5 2 5 4 2 5 5

Continued

Table 8.3 *Continued*

```
33 NH     6 4 5 4 5 5 5 1 2 2 4 3 5 5 3 3 3 3 4 3 4 1 2 1 3 3 4 4 4 5 4
          5 0 1 5 1 1 2 3 4 2 2 4 5 2 2 1 3 4 3 5 5 5 5 1 2 5 2 3 5 5 5

3J NJ     6 3 5 4 5 5 5 1 1 1 4 3 5 5 3 3 3 3 4 4 3 4 3 1 1 3 3 4 4 4 5 4
          5 2 0 5 2 1 2 2 4 2 2 4 5 1 2 1 3 4 3 5 5 5 5 2 2 5 2 3 5 5 5

35 NM     6 4 1 3 3 3 1 5 5 5 5 5 5 5 3 4 4 4 3 2 4 3 5 5 5 5 5 4 4 3 3 2
          2 5 5 0 5 5 5 5 4 5 5 1 4 5 5 5 5 3 4 2 2 1 1 5 5 4 5 4 2 5 5

36 NENY   6 3 5 4 5 5 5 1 1 2 4 3 5 5 3 3 3 3 4 3 4 2 1 1 3 2 4 4 3 5 4
          5 1 1 5 0 0 0 3 4 2 2 4 5 1 1 1 3 4 3 5 5 5 5 1 2 5 2 3 5 5 5

37 SENY   6 3 5 4 5 5 5 1 1 1 4 3 5 5 3 3 3 3 4 3 4 2 1 1 3 3 4 4 3 5 4
          5 1 1 5 0 0 0 2 4 2 2 4 5 1 2 1 3 4 3 5 5 5 5 1 2 5 2 3 5 5 5

38 WNY    6 3 5 4 5 5 5 2 1 1 4 3 5 5 3 3 3 3 4 3 4 2 1 2 3 2 4 4 3 5 4
          5 2 1 5 0 0 0 3 4 1 2 4 5 1 1 2 3 4 3 5 5 5 5 2 2 5 1 3 5 5 5

39 NC     6 2 5 4 5 5 5 3 1 1 3 1 5 5 3 3 3 4 4 2 4 4 1 3 3 3 4 3 4 5 4
          5 4 2 5 3 3 3 0 5 2 1 4 5 2 2 3 1 5 1 4 4 4 5 3 1 5 1 4 5 5 5

40 ND     6 5 4 3 4 4 2 5 5 5 5 5 5 2 2 3 3 1 2 4 5 5 5 5 2 3 1 4 3 1 1
          3 5 5 3 5 5 5 5 0 4 4 3 3 5 5 5 5 1 4 4 4 4 3 5 5 3 4 2 1 5 5

41 NOH    6 3 5 4 5 5 5 3 2 2 4 3 5 5 2 2 1 3 4 1 4 4 2 3 3 1 4 4 3 5 4
          5 4 3 5 3 3 1 3 4 0 0 4 5 2 1 4 3 4 2 5 5 5 5 3 1 5 1 3 5 5 5

42 SOH    6 3 5 4 5 5 5 3 3 2 4 3 5 5 2 2 1 3 4 1 4 4 2 3 3 1 4 4 3 5 4
          5 4 3 5 3 3 1 2 4 0 0 4 5 2 1 4 3 4 2 5 5 5 5 3 1 5 1 3 5 5 5

43 OK     6 3 4 1 5 5 1 5 5 5 4 4 5 5 3 3 3 3 1 3 2 5 5 5 4 4 3 2 1 4 2
          5 5 5 1 5 5 5 4 4 4 4 0 5 5 5 5 4 3 3 1 2 1 3 5 4 5 4 3 3 5 5

44 0R     6 5 2 4 1 2 3 5 5 5 5 5 5 1 4 4 4 3 5 4 5 5 5 4 4 3 4 4 2 3
          1 5 5 3 5 5 5 5 3 5 5 3 0 5 5 5 5 3 5 3 3 3 1 5 5 1 5 4 2 5 5

45 EPA    6 3 5 4 5 5 5 1 1 1 4 3 5 5 3 3 3 3 4 4 3 4 3 1 2 3 3 4 4 3 5 4
          5 2 1 5 1 1 1 2 4 2 2 4 5 0 0 2 3 4 3 5 5 5 5 2 2 5 2 3 5 5 5

46 WPA    6 3 5 4 5 5 5 2 1 1 4 3 5 5 3 3 3 3 4 4 3 4 3 1 2 3 2 4 4 3 5 4
          5 3 2 5 1 2 1 2 4 1 1 4 5 0 0 3 3 4 3 5 5 5 5 3 1 5 1 3 5 5 5

47 RI     6 3 5 4 5 5 5 1 2 2 4 3 5 5 3 3 3 3 4 4 3 4 2 2 1 3 3 4 4 4 5 4
          5 1 1 5 1 1 2 3 4 2 3 4 5 2 2 0 3 4 3 5 5 5 5 1 2 5 2 3 5 4 4

48 SC     6 1 5 3 5 5 5 3 2 2 2 1 5 5 3 3 2 4 4 1 3 4 2 4 4 4 4 5 3 3 5 5
          5 4 3 5 3 3 3 1 5 2 2 4 5 3 2 4 0 5 1 4 4 4 5 4 1 5 1 4 5 5 5

49 SD     6 4 4 3 5 5 2 5 5 5 5 5 5 3 2 3 3 1 2 4 4 5 5 5 3 3 1 4 2 1 1
          4 5 5 3 5 5 5 5 1 4 4 3 4 5 5 5 5 0 4 3 3 3 3 5 5 4 4 2 1 5 5
```

Continued

Table 8.3 *Continued*

```
50 TN    6 1 5 1 5 5 5 4 4 3 3 1 5 5 2 2 2 3 4 1 3 5 3 5 3 3 4 1 1 5 4
         5 5 4 5 4 4 4 1 5 2 2 3 5 3 3 4 2 5 0 4 4 4 5 5 1 5 2 3 5 5 5

51 ETX   6 2 3 1 5 5 2 5 5 5 3 3 5 4 3 2 3 3 1 3 1 5 5 5 4 4 4 1 1 4 2
         4 5 5 2 5 5 5 4 4 4 1 5 5 5 5 4 3 2 0 0 0 3 5 5 5 4 4 3 5 5

52 STX   6 2 2 1 5 5 2 5 5 5 3 3 5 4 3 3 3 3 1 3 1 5 5 5 4 4 4 1 2 4 3
         4 5 5 1 5 5 5 4 4 4 4 1 5 5 5 5 4 3 2 0 0 0 3 5 5 5 4 3 3 5 5

53 WTX   6 3 2 1 5 5 1 5 5 5 3 3 5 4 3 3 3 3 1 3 2 5 5 5 4 4 4 2 2 4 2
         4 5 5 1 5 5 5 4 4 4 4 1 5 5 5 5 4 3 3 0 0 0 2 5 5 5 4 4 3 5 5

54 UT    6 5 1 4 2 2 5 1 5 5 5 5 5 1 4 4 4 3 3 5 4 5 5 5 4 4 4 4 4 2 3
         1 5 5 1 5 5 5 5 3 5 5 3 3 5 5 5 5 3 4 3 3 3 0 5 5 3 5 4 1 5 5

55 VT    6 4 5 4 5 5 5 1 2 2 4 3 5 5 3 3 3 3 4 3 4 1 2 1 3 3 4 4 4 5 4
         5 1 1 5 1 1 2 3 4 2 2 4 5 2 2 1 3 4 3 5 5 5 5 0 2 5 2 3 5 5 5

56 VA    6 3 5 4 5 5 5 3 1 1 3 3 5 5 3 3 3 4 4 1 4 4 1 3 3 3 4 3 4 5 4
         5 3 2 5 3 2 2 1 5 2 1 4 5 2 1 3 2 4 1 5 5 5 5 3 0 5 1 4 5 5 5

57 WA    4 5 3 4 1 2 3 5 5 5 5 5 5 1 4 4 5 3 3 5 5 5 5 5 4 4 3 5 3 1 3
         1 5 5 3 5 5 5 5 2 5 5 3 1 5 5 5 5 3 5 4 4 4 2 5 5 0 5 3 2 5 5

58 WV    6 3 5 4 5 5 5 3 2 1 4 3 5 5 3 3 2 4 4 1 4 4 1 3 3 3 4 4 3 5 4
         5 3 2 5 3 3 2 1 5 1 1 4 5 2 1 3 2 4 2 5 5 5 5 3 1 5 0 3 5 5 5

59 WI    6 4 5 3 5 5 4 4 4 3 5 4 5 5 1 2 2 1 3 3 4 5 3 4 1 1 1 4 2 4 2
         5 4 4 5 3 3 3 4 3 2 2 3 5 3 3 4 4 2 3 5 5 5 5 4 3 5 3 0 4 5 5

60 WY    6 5 2 4 3 3 1 5 5 5 5 5 5 1 4 4 4 3 2 4 4 5 5 5 4 4 3 4 3 1 1
         2 5 5 2 5 5 5 5 2 5 5 3 2 5 5 5 5 1 5 4 4 4 1 5 5 3 5 3 0 5 5
```

Note: The standard 2-letter state designation terminates the region designation (e.g., Northern California = NCA or Northeast New York = NENY). The Number Associated with the state/region is shown on the left. There are 62 CSA band values for each region, i.e., the 60 as shown as the leftmost column plus Virgin Island (VI) and Puerto Rico (PR). The "0" CSA band numbers are for two identical states/regions forming a pair. Examples: The CSA band for the (58-10) state/region pair = 1, for the (55-35) pair = 5 and so on.

8.2.3 A Discussion of Output Files

Figures 8.2 to 8.4 illustrate the ACD network topologies for 1, 2, and 3 ACD centers using the 800 Service for TGF=1. The corresponding results are tabulated in Tables 8.6 through 8.8. The part that computes the cost of 800 Service for each location inserts a zero for a number of access lines (ALs) and cost of ALs since we are dealing with virtual networks. The part that describes the cost of labor lists the traffic intensity (in milliErlangs) handled by each ACD location/site during each hour of the day and the number of agents required for handling the traffic while providing the desired GOS. The desired GOS is defined at the start of each printout. The part that analyzes trunking describes each data trunk that connects an ACD center to the ACD database computer collocated at the first switch location of SWF. Only the star-type topology is employed for data trunks. Four types of ACDs can be analyzed. The SDF design parameter ACDT defines the type of an ACD system. ACDT=1 models the 800 Service. ACDT=2 models the MegaCom 800 Service. ACDT=3 models the INWATS deployment. ACDT=4 can model the new distance-dependent virtual tariffs. The computed costs per month or costs per minute are plotted in Figures 8.5 to 8.10 for several ACD types, TGF values, and number of optimally located ACD nodes for identical daily profiles.

Table 8.4 Rate Steps for Each CSA Band and State/Region

	CSAI	CSA2	CSA3	CSA4	CSAS	CSA6
1 AK	4	7	9	11	17	17
2 AL	4	7	9	11	17	17
3 AZ	6	9	12	15	18	18
4 AR	4	7	9	11	15	15
5 NCA	8	12	15	17	18	18
6 SCA	7	11	15	17	18	18
7 CO	7	8	10	12	16	16
8 CT	1	7	10	14	18	18
9 DE	1	5	9	13	18	18
10 DC	1	4	8	12	18	18
11 FL	7	10	12	13	18	18
12 GA	4	7	10	12	18	18
13 HI	23	23	24	25	26	26
14 ID	5	9	13	15	18	18
15 NIL	3	6	8	10	15	15
16 SIL	3	6	8	10	15	15
17 IN	3	6	8	10	16	16
18 IA	4	7	9	11	14	14
19 KS	5	7	9	12	14	14
20 KY	3	5	8	10	17	17
21 LA	5	8	10	13	16	16
22 ME	6	9	12	16	18	18

Continued

Table 8.4 *Continued*

	CSAI	CSA2	CSA3	CSA4	CSAS	CSA6
23 MD	2	5	9	12	18	18
24 MA	2	7	11	14	18	18
25 NMI	5	8	9	12	17	17
26 SMI	4	7	9	12	17	17
27 MN	6	8	10	12	15	15
28 MS	5	7	9	11	16	16
29 MO	5	7	8	10	15	15
30 MT	7	10	12	14	17	17
31 NE	5	8	9	12	14	14
32 NV	5	8	13	16	18	18
33 NH	2	7	11	15	18	18
34 NJ	1	5	9	13	18	18
35 NM	6	8	10	13	17	17
36 NENY	3	7	10	14	18	18
37 SENY	1	7	10	14	18	18
38 WNY	3	5	10	14	18	18
39 NC	4	7	8	12	18	18
40 ND	6	9	11	14	15	15
41 NOH	3	5	7	10	17	17
42 SOH	3	5	8	10	17	17
43 OK	5	7	9	12	15	15
44 OR	5	9	15	17	18	18
45 EPA	I	5	8	12	18	18
46 WPA	3	5	8	12	18	18
47 RI	1	6	11	14	18	18
48 SC	4	7	9	12	18	18
49 SD	5	8	10	12	15	15
50 TN	5	6	8	10	17	17

Continued

	CSAI	CSA2	CSA3	CSA4	CSAS	CSA6
51 ETX	6	9	11	14	16	16
52 STX	8	11	12	!4	16	16
53 WTX	7	9	11	14	16	16
54 UT	6	7	11	14	18	18
55 VT	2	7	11	14	18	18
56 VA	3	5	8	11	18	18
57 WA	8	11	15	17	18	18
58 WV	2	5	7	11	18	18
59 WI	3	7	9	11	16	16
60 WY	5	9	10	13	16	16

NOTES: CSA = Customer Service Area

CSAI implies that CSA Band is equal to 1,

CSA2 implies that CSA Band is equal to 2 and so on.

WTX = West Texas, NENY = North East New York and so on

Table 8.5(a) The Use-Rate Table for AT&T's 800 Service (1/1/87)

	PER HOUR OF USE, First 25 Hours/mo		PER RATE PERIOD, Next 75 Hours/mo		PER ACCESS LINE >100 Hours/mo		
							All Hours Late Night & Weekends
RATE STEP	Business Day	Evening	Business Day	Evening	Business Day	Evening	
1	12.06	8.07	11.45	7.68	10.85	7.26	5.41
2	12.68	8.49	12.04	8.06	11.41	7.64	5.69
3	12.99	8.70	12.32	8.26	11.69	7.84	5.83
4	13.21	8.85	12.55	8.40	11.90	7.98	5.92
5	13.41	8.99	12.74	8.53	12.08	8.09	6.00
6	13.57	9.11	12.91	8.64	12.23	8.18	6.07
7	13.81	9.24	13.13	8.79	12.42	8.33	6.19
8	14.07	9.43	13.36	8.96	12.66	8.48	6.31
9	14.28	9.56	13.55	9.09	12.86	8.61	6.39
10	14.45	9.68	13.73	9.19	13.01	8.72	6.51

Continued

Table 8.5(a) *Continued*

RATE STEP	Business Day	Evening	Business Day	Evening	Business Day	Evening	All Hours Late Night & Weekends
11	14.62	9.79	13.89	9.31	13.16	8.81	6.56
12	14.78	9.90	14.05	9.41	13.31	8.92	6.62
13	14.90	9.98	14.17	9.48	13.41	8.99	6.69
14	15.05	10.09	14.30	9.57	13.54	9.08	6.76
15	15.22	10.19	14.45	9.68	13.69	9.17	6.83
16	15.41	10.32	14.63	9.80	13.85	9.28	6.90
17	15.56	10.42	14.79	9.91	14.01	9.39	6.96
18	15.86	10.64	15.07	10.10	14.28	9.56	7.11
19							
20							
21							
22							
23	16.24	10.88	15.43	10.34	14.62	9.79	7.33
24	16.54	11.08	15.71	10.52	14.88	9.97	7.45
25	16.80	11.26	15.96	10.69	15.12	10.13	7.59
26	16.99	11.39	16.15	10.82	15.31	10.25	7.68

NOTE: The Use-Rate Table (URTBL) employs the rate steps obtained from Tables 8.3 & 8.4 for computing the total monthly cost of WATS Service. These rate change with time. To handle uniform increases or special discounts, one can use the SDF Parameter #28.

Table 8.5(b) The WATS Use-Rate Table (WUTBL) for AT&T's WATS Service (1/1/87)

RATE STEP	PER HOUR OF USE, First 25 Hours/mo		PER RATE PERIOD, Next 75 Hours/mo		PER ACCESS LINE >100 Hours/mo		All Hours Late Night & Weekends
	Business Day	Evening	Business Day	Evening	Business Day	Evening	
1	12.06	8.07	11.45	7.68	10.85	7.26	5.41
2	12.68	8.49	12.04	8.06	11.41	7.64	5.69

Continued

RATE STEP	Business Day	Evening	Business Day	Evening	Business Day	Evening	All Hours Late Night & Weekends
3	12.99	8.70	12.32	8.26	11.69	7.84	5.83
4	13.21	8.85	12.55	8.40	11.90	7.98	5.92
5	13.41	8.99	12.74	8.53	12.08	8.09	6.00
6	13.57	9.11	12.91	8.64	12.23	8.18	6.07
7	13.81	9.24	13.13	8.79	12.42	8.33	6.19
8	14.07	9.43	13.36	8.96	12.66	8.48	6.31
9	14.28	9.56	13.55	9.09	12.86	8.61	6.39
10	14.45	9.68	13.73	9.19	13.01	8.72	6.51
11	14.62	9.79	13.89	9.31	13.16	8.81	6.56
12	14.78	9.90	14.05	9.41	13.31	8.92	6.62
13	14.90	9.98	14.17	9.48	13.41	8.99	6.69
14	15.05	10.09	14.30	9.57	13.54	9.08	6.76
15	15.22	10.19	14.45	9.68	13.69	9.17	6.83
16	15.41	10.32	14.63	9.80	13.85	9.28	6.90
17	15.56	10.42	14.79	9.91	14.01	9.39	6.96
18	15.86	10.64	15.07	10.10	14.28	9.56	7.11
19							
20							
21							
22							
23	16.24	10.88	15.43	10.34	14.62	9.79	7.33
24	16.54	11.08	15.71	10.52	14.88	9.97	7.45
25	16.80	11.26	15.96	10.69	15.12	10.13	7.59
26	16.99	11.39	16.15	10.82	15.31	10.25	7.68

NOTE: The Use-Rate Table (URTBL) employs the rate steps obtained from Tables 7.15 & 7.16 for computing the total monthly cost of WATS Service. These rate change with time. To handle uniform increases or special discounts, one can use the SDF Parameter #28.NOTE: The Use-Rate Table (URTBL) employs the customer service area value obtained from Table 7.15 for computing the total monthly cost of AT&T's MegaComService. These rates change with time. In case, the increase is uniform over all values, one can use the SDF Parameter #28 to reflect the change. This parameter can also be used to reflect any special discounts.

Table 8.5(c) The MegaCom-Use-Rate Table (MUTBL) for AT&T's MegaCom Service (1/1/87)

Rate Schedule Per House of Use, Per Rate Period, Per Access Line

Service Area	Business Day	Evening	Night/Weekend
1	10.36	8.07	6.74
2	10.93	8.52	7.11
3	11.22	8.75	7.29
4	11.80	9.20	7.66
5	12.07	9.43	7.86
6	13.81	10.76	8.97

Note: The MegaCom-Use-Rate Table (MUTBL) employs the customer service areas obtained from Table 8.3 for computing the monthly cost of MegaCom Service. To handle special discounts or uniform changes, one can use the SDF design parameter, F_{rst} (Parameter # 29).

The preceding analyses employ a simplified analytical model as described in Section 8.3. Communications costs can be reduced by using foreign exchange (FX) and business local lines. Two improved methods for computing the required number of trunks and agents are discussed next.

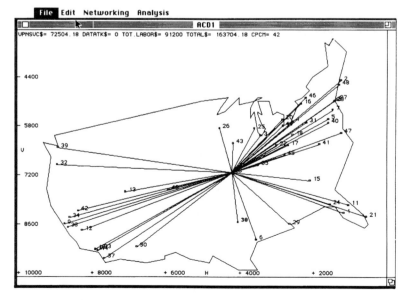

Figure 8.2 An ACD network topology with one switch: 49 regions and AT&T's 800 service.

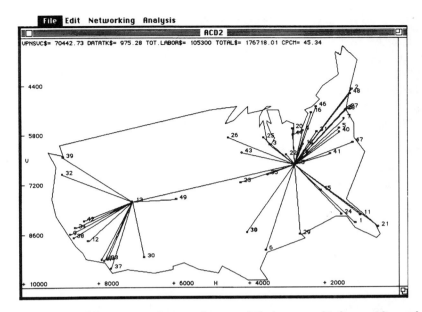

Figure 8.3 An ACD network topology with two switches: 49 regions and AT&T's 800 service.

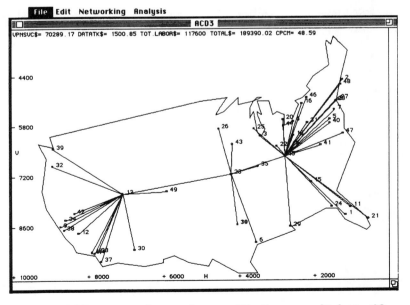

Figure 8.4 An ACD network topology with three switches: 49 regions and AT&T's 800 service.

Table 8.6 ACD Network Design Data

NO. OF SWITCHES= 1   AT   23
AL BLOCKING FACTOR= .01   AL TRUNK LINK TYPE= 0   #PKHRS/DAY (Fph)= 8.333
ACD-TYPE= 1   DREQ(SEC)= 30   PROB(DELAY>DREQ)= .15   LABOR COST ($/HR)= 10
TRAFFIC GROWTH FACTOR (TGF)= 1   EXP.CALL DURATION-SEC= 300
(A) ACCESS LINE ANALYSIS

#	SW	ALs	10*MIs	$/AL	$/ALB
1	23	0	10238	0	1466.87
2	23	0	12456	0	948.16
3	23	0	4110	0	2442.92
4	23	0	6978	0	1411.09
5	23	0	9596	0	747.37
6	23	0	6393	0	998.36
7	23	0	10339	0	1623.04
8	23	0	10930	0	3137.32
9	23	0	15014	0	1118.28
10	23	0	13521	0	2573.99
11	23	0	10445	0	1168.47
12	23	0	13779	0	1135.01
13	23	0	9219	0	1252.13
14	23	0	13397	0	1597.94
15	23	0	6740	0	1634.19
16	23	0	8561	0	1355.32
17	23	0	5388	0	1411.09
18	23	0	6191	0	1305.12
19	23	0	4468	0	1020.67
20	23	0	6414	0	1497.54
21	23	0	12188	0	1280.02
22	23	0	4513	0	1330.22
23	23	0	0	0	1385.99

Continued

#	SW	ALs	10*MIs	$/AL	$/ALB
24	23	0	8900	0	1486.39
25	23	0	4378	0	1408.3
26	23	0	4080	0	725.06
27	23	0	11131	0	895.18
28	23	0	10833	0	6765.45
29	23	0	6759	0	1335.8
30	23	0	10444	0	1447.35
31	23	0	7784	0	1709.49
32	23	0	14921	0	1355.32
33	23	0	13070	0	1466.87
34	23	0	14421	0	1280.02
35	23	0	2379	0	817.09
36	23	0	4480	0	1299.54
37	23	0	13298	0	1578.42
38	23	0	14807	0	1377.63
39	23	0	15004	0	1486.39
40	23	0	9413	0	1048.56
41	23	0	7933	0	1168.47
42	23	0	13525	0	998.36
43	23	0	2604	0	1068.08
44	23	0	6056	0	1207.51
45	23	0	4787	0	1238.19
46	23	0	9222	0	1550.53

8.3 Improved Models for Computing ACD System Resources

Usage level of a given number of trunks (T) and agent positions (P) will determine the total monthly cost of any ACD. Consequently it is necessary to compute

Table 8.7 ACD Network Design Data

NO. OF SWITCHES= 2   AT   45   13

AL BLOCKING FACTOR= .01   AL TRUNK LINK TYPE= 0   #PKHRS/DAY (Fph)= 8.333

ACD-TYPE= 1   DREQ(SEC)= 30   PROB(DELAY>DREQ)= .15   LABOR COST ($/HR)= 10

TRAFFIC GROWTH FACTOR (TGF)= 1   EXP.CALL DURATION-SEC= 300

(A) ACCESS LINE ANALYSIS

#	SW	ALs	10*MIs	$/AL	$/ALB
1	45	0	7254	0	1425.16
2	45	0	8223	0	921.2
3	45	0	2685	0	2373.46
4	45	0	3106	0	1370.97
5	45	0	4973	0	726.12
6	45	0	7990	0	969.97
7	45	0	5798	0	1576.89
8	45	0	6492	0	3048.11
9	13	0	5979	0	1086.48
10	13	0	5784	0	2500.81
11	45	0	7117	0	1135.25
12	13	0	5074	0	1102.74
13	13	0	0	0	1216.53
14	13	0	5798	0	1552.5
15	45	0	3178	0	1587.73
16	45	0	4806	0	1316.78
17	45	0	891	0	1370.97
18	45	0	1892	0	1268.01
19	45	0	7217	0	991.65
20	45	0	3145	0	1454.96
21	45	0	8935	0	1243.63
22	45	0	1067	0	1292.4

Continued

#	SW	ALs	10*MIs	$/AL	$/ALB
23	45	0	4787	0	1346.59
24	45	0	5922	0	1444.13
25	45	0	3483	0	1368.26
26	45	0	6028	0	704.45
27	45	0	6689	0	869.73
28	45	0	6397	0	6573.1
29	45	0	6203	0	1297.82
30	13	0	5031	0	1406.19
31	45	0	3427	0	1660.88
32	13	0	6339	0	1316.78
33	13	0	5545	0	1425.16
34	13	0	5316	0	1243.63
35	45	0	2411	0	793.86
36	45	0	7229	0	1262.59
37	13	0	6232	0	1533.54
38	13	0	5856	0	1338.46
39	13	0	6985	0	1444.13
40	45	0	4742	0	1018.74
41	45	0	3146	0	1135.25
42	13	0	4372	0	969.97
43	45	0	4525	0	1037.71
44	45	0	2616	0	1173.18
45	45	0	0	0	1202.99
46	45	0	5427	0	1506.44
47	45	0	5269	0	1189.44
48	45	0	7820	0	1202.99
49	13	0	3697	0	1444.13

Continued

Table 8.7 *Continued*

(B) ACCESS LINE ANALYSIS SUMMARY

SW LOC	#VTERMs	BHRERL	NCH/MO	SW COST(S)
45	29	18.772	56281	51025.8616289
13	14	7.227	21667	19416.9372638

(C) ACD AGENT POSITION ANALYSIS

W#	HOUR	YMERLs	#AP REQD	TMNHRS
45	1	1564	4	4
45	2	1564	4	8
45	3	1564	4	12
45	4	1564	4	16
45	5	3128	6	22
45	6	3128	6	28
45	7	3128	6	34
45	8	4693	8	42
45	9	6257	10	52
45	10	9386	13	65
45	11	12514	17	82
45	12	18772	24	106
45	13	15643	20	126
45	14	18772	24	150
45	15	15643	20	170
45	16	12514	17	187
45	17	6257	10	197
45	18	4693	8	205
45	19	3128	6	211
45	20	3128	6	217
45	21	3128	6	223
45	22	3128	6	229

Continued

W#	HOUR	YMERLs	#AP REQD	TMNHRS
45	23	1564	4	233
45	24	1564	4	237
13	1	602	2	239
13	2	602	2	241
13	3	602	2	243
13	4	602	2	245
13	5	1204	3	248
13	6	1204	3	251
13	7	1204	3	254
13	8	1806	4	258
13	9	2409	5	263
13	10	3613	7	270
13	11	4818	8	278
13	12	7227	11	289
13	13	6022	9	298
13	14	7227	11	309
13	15	6022	9	318
13	16	4818	8	326
13	17	2409	5	331
13	18	1806	4	335
13	19	1204	3	338
13	20	1204	3	341
13	21	1204	3	344
13	22	1204	3	347
13	23	602	2	349
13	24	602	2	351

TOTAL VIRTUAL NETWORK COST (S)= 70442.73   FOR 43 VIRTUAL LINES
TOTAL COST OF LABOR(S)= 105300
TOTAL #CALLS PER MO. 6 BHR ERLANGS = 77948   25.999

Table 8.8 ACD Network Design Data

NO. OF SWITCHES= 3   AT   23   45   13

AL BLOCKING FACTOR= .01   AL TRUNK LINK TYPE= 0   #PKHRS/DAY (Fph)= 8.333

ACD-TYPE= 1   DREQ(SEC)= 30   PROB(DELAY>DREQ)= .15   LABOR COST ($/HR)= 10

TRAFFIC GROWTH FACTOR (TGF)= 1   EXP.CALL DURATION-SEC= 300

(A) ACCESS LINE ANALYSIS

#	SW	ALs	10*MIs	5/AL	S/ALB
1	45	0	7254	0	1422.05
2	45	0	8223	0	919.2
3	45	0	2685	0	2368.29
4	45	0	3106	0	1367.98
5	45	0	4973	0	724.54
6	23	0	6393	0	967.86
7	45	0	5798	0	1573.45
8	45	0	6492	0	3041.47
9	13	0	5979	0	1084.11
10	13	0	5784	0	2495.36
11	45	0	7117	0	1132.78
12	13	0	5074	0	1100.33
13	13	0	0	0	1213.88
14	13	0	5798	0	1549.12
15	45	0	3178	0	1584.27
16	45	0	4806	0	1313.91
17	45	0	891	0	1367.98
18	45	0	1892	0	1265.25
19	23	0	4468	0	989.49
20	45	0	3145	0	1451.79
21	45	0	8935	0	1240.92
22	45	0	1067	0	1289.58

Continued

#	SW	ALs	10*MIs	5/AL	S/ALB
23	23	0	0	0	1343.65
24	45	0	5922	0	1440.98
25	45	0	3483	0	1365.28
26	23	0	4080	0	702.91
27	45	0	6689	0	867.83
28	45	0	6397	0	6558.77
29	45	0	6203	0	1294.99
30	13	0	5031	0	1403.13
31	45	0	3427	0	1667.26
32	13	0	6339	0	1313.91
33	13	0	5545	0	1422.05
34	13	0	5316	0	1240.92
35	23	0	2379	0	792.13
36	23	0	4480	0	1259.84
37	13	0	6232	0	1530.2
38	13	0	5856	0	1335.54
39	13	0	6985	0	1440.98
40	45	0	4742	0	1016.52
41	45	0	3146	0	1132.78
42	13	0	4372	0	967.86
43	23	0	2604	0	1035.45
44	45	0	2616	0	1170.63
45	45	0	0	0	1200.36
46	45	0	5427	0	1503.16
47	45	0	5269	0	1186.85
48	45	0	7820	0	1200.36
49	13	0	3697	0	1440.98

Continued

Table 8.8 *Continued*

(B) ACCESS LINE ANALYSIS SUMMARY

SW LOC	#VTERMs	BHRERL	NCH/MO	SW COST($)
23	8	2.623	7862	7119.48042122
95	26	16.149	48419	43752.8521305
13	14	7.227	21667	19416.9372638

(C) ACD AGENT POSITION ANALYSIS

SWAB	HOUR	MERLs	SAP REQD	TMNHRS
23	1	218	2	2
23	2	218	2	4
23	3	218	2	6
23	4	218	2	8
23	5	437	2	10
23	6	437	2	12
23	7	437	2	14
23	8	655	2	16
23	9	874	3	19
23	10	1311	3	22
23	11	1748	4	26
23	12	2623	5	31
23	13	2185	5	36
23	14	2623	5	41
23	15	2185	5	46
23	16	1748	4	50
23	17	874	3	53
23	18	655	2	55
23	19	437	2	57
23	20	437	2	59
23	21	437	2	61
23	22	437	2	63

Continued

SWAB	HOUR	MERLs	SAP REQD	TMNHRS
23	23	218	2	65
23	24	218	2	67
45	1	1345	4	71
45	2	1345	4	75
45	3	1345	4	79
45	4	1345	4	83
45	5	2691	5	88
45	6	2691	5	93
45	7	2691	5	98
45	8	4037	7	105
45	9	5383	9	114
45	10	8074	12	126
45	11	10766	15	141
45	12	16149	21	162
45	13	13457	18	180
45	14	16149	21	201
45	15	13457	18	219
45	16	10766	15	234
45	17	5383	9	243
45	18	4037	7	250
45	19	2691	5	255
45	20	2691	5	260
45	21	2691	5	265
45	22	2691	5	270
45	23	1345	4	274
45	24	1345	4	278
13	1	602	2	280
13	2	602	2	282

Continued

Table 8.8 *Continued*

SWAB	HOUR	MERLs	SAP REQD	TMNHRS
13	3	602	2	284
13	4	602	2	286
13	5	1204	3	289
13	6	1204	3	292
13	7	1204	3	295
13	8	1806	4	299
13	9	2409	5	304
13	10	3613	7	311
13	11	4818	8	319
13	12	7227	11	330
13	13	6022	9	339
13	14	7227	11	350
13	15	6022	9	359
13	16	4818	8	367
13	17	2409	5	372
13	18	1806	4	376
13	19	1204	3	379
13	20	1204	3	382
13	21	1204	3	385
13	22	1204	3	388
13	23	602	2	390
13	24	602	2	392

TOTAL VIRTUAL NETWORK COST ($) = 70289.17   FOR 48 VIRTUAL LINES
TOTAL COST OF LABOR($) = 117600
TOTAL #CALLS PER MO. & BHR ERLANGS = 77948   25.999

(D) TRUNKING ANALYSIS

TRUNK LINK/NODE TYPE 0 (MAX LINK CAPACITY@ 9600BPS)
FDX FACTOR = 1 ,   BBTF=2
MAX TRUNK LOAD = .8 ,   TKMF=1
TOTAL INTER/INTRA NODAL TRAFFIC= 80.09 BPS

Continued

Table 8.8 *Continued*

ORIGINAL TRUNK TRAFFIC MATRIX AS FOLLOWS

0	55.33	24.76
0	0	0
0	0	0

MODIFIED TRUNK TRAFFIC MATRIX IS AS FOLLOWS

0	55.33	24.76
0	0	0
0	0	0

MODIFIED TRUNK MATRIX AS FOLLOWS

0	1	1
0	0	0
0	0	0

TRUNK COST MATRIX IS AS FOLLOWS

FROM	TO	MILES	COST	TOTALS
23	45	479	677.57	677.57
23	13	922	823.28	823.28

*****SUMMARY OF SYSTEM COSTS*****
VIRTUAL LINE COST= 70289.17
TOTAL TRUNK COST= 1500.85
TOTAL SYSTEM COST= 71790.02

the optimum values of T and P for a particular GOS specification as correctly as possible. Unfortunately, the state-of-the-art of teletraffic theory does not yield a closed form solution at this time for all practical situations.

The analytical models described earlier assumed two stochastically independent subsystems for computing the number of trunks and agents. The Erlang-B and Erlang-C formulas used for computing the number of trunks and number of agents, respectively, assume infinite sources and a negative exponential service time distribution. The real situations differ substantially from the ideal conditions assumed for mathematical ease.

The two subsystems are quite interdependent upon one another. The number of required agents should consider the fact that the number of calls waiting to be served cannot exceed the number of trunks employed and it is never infinite. Furthermore, the service time distributions are becoming more and more of the constant-value variety, especially through the use of voice response units. In most cases, the upper limit to service time is generally not much larger than the expected call duration (ECD). In any case, closed form solutions do not exist for all cases. Fortunately, the number of agents required should be less than predicted by the simplified theory. The assumption of finite sources for the Erlang-B formula may be employed to estimate the number of agents by using the curves of Figure 3.9. The assumption of infinite callers holds good for computing the number of trunks.

The assumption that the call disappears from the system if it is blocked may not apply to some systems. A percentage of callers always redial immediately after experiencing blocking. Similarly, the assumption that when a call enters the ACD queue it will remain there forever until it is answered may also not hold in some cases. A percentage of callers always abandon the queue after waiting a certain number of seconds and, in some cases, redial again. This will complicate the analysis measurably.

The presence of many input queues dealing with separate trunk bundles (e.g., FX line bundles from different cities and 800-service bundles for special sales groups) also complicates the analysis involving a common pool of agents. Whittaker (Whittaker, 1975) provides an insight into the complexities involved with the analysis of a multiqueue and a shared multiserver system.

One can make the following refinements to improve the quality of results described earlier:

1. With the availability of newer 800-service features, one may prefer to create only one input queue by eliminating different customer service area bands and/or separate FX-line bundles. The results show that costs do not change substantially, while the computations are simplified.

2. One should compute the average delay for all handled calls (as expressed by Equation 3.11) and add this to the service time for trunks. The ring time should also be added to the trunk service time in case the system does not provide a delay recording after the first ring.

3. One may employ an improved value of the traffic offered to the agents (=the offered traffic to the trunks minus the blocked traffic). In some cases, a certain percentage of customers may try to redial immediately after experiencing a call blockage. See Chapter 3 for a discussion of the effects of retrials on a pure Loss System.

4. One may employ an improved value of traffic offered to the agents by considering the effects of call abandonments and/or after-call work times. See a published paper by Hoffman (Hoffman, 1983) for a technique to consider the effects of call abandonments.

A software program based on a complex theory developed by Smith and Smith (Smith and Smith, 1962) was employed to compute an *exact* value of (1) blocking probability experienced by trunks, and (2) probability of exceeding a given delay limit and for any expected call duration for a given number of trunks (T) and agents positions (P). The values of P (number of agent positions) and T (number of trunks) were varied using the trial and error approach until the desired GOS was achieved. The Analysis item "CompositeACD" of EcoNets software represents such a capability (i.e., an exact method requiring iterations).

The results were also computed using a simplified and a refined method. The *simplified* method assumes two independent subsystems with identical negative exponential service time distributions. The *refined* method assumes no retrials or call abandonments but assumes lower traffic for agents and higher ser-

Table 8.9 Comparison of Various Methods for Computing ACD Performance

	ACD Design Parameters		
	A=3 erlangs	A=30 erlangs	A=300 erlangs
Simplified Method:			
Number of Agents (P)	6	36	311
Number of Trunks (T)	8	42	324
Refined Method:			
Number of Agents (P)	6	35	309
Number of Trunks (T)	10	46	338
Exact Method:			
Number of Agents (P)	6	35	304
Number of Trunks (T)	9	45	347

vice times for trunks as discussed earlier. The results are summarized in Table 8.9 for a system handling calls of average duration 300 seconds each with a trunk GOS of 0.01 blocking probability and the probability of delay exceeding DREQ of 30 seconds equaling 0.15 (i.e., PEXD=0.15).

The results show that the refined method yields a value of T (equal to number of trunks) close to the exact solution. The difference between the computed values of P obtained from the refined and exact methods becomes larger with increasing values of erlangs. The values of P obtained from the exact method are always lower than those obtained from the refined method for the reasons discussed earlier. By comparing the monthly costs of a trunk with those of an agent, one can compute the penalty of not using the exact method. The exact method employed here does not allow a closed form solution suitable for a computer. It appears that Whittaker (Whittaker, 1975) has developed a closed form solution but the analytical complexity of his approach discouraged the author to develop a suitable software program. Since the exact method assumes a state of statistical equilibrium to derive the answers, and since the number of agents are altered every half-hour or hour, such a state is impossible to attain. Therefore, the results obtained from the exact method will never be quite meaningful. Furthermore, the number of trunks available is always higher than the exact value of trunks required during non-busy hours. Theory shows that such a state will always demand a higher number of manned agents. This can be explained by higher traffic loads resulting from lower blocking. The values of P during busy hours can be reduced gradually if the observed GOS is better than what is required. The process of reducing P should be continued until the desired GOS is obtained. The reader is advised to use the refined or exact method and perform sample calculations to test the validity of computed results using the simple method.

8.4 Summary of ACD Network Syntheses and Conclusions

The plots of Figure 8.5 clearly show that the costs for MegaCom 800 Service are approximately 15 percent lower than those for the pure 800 services for all values of TGF. The cost of WATS is approximately 2 percent lower than that for the pure 800 service for all values of TGF. Furthermore, one obtains only a 2 percent gain when considering a unit cost per call as the value of TGF is increased from 1 to 10. This is explained by the fact that the cost of virtual service increases at the same rate as the traffic volume, and cost of a labor minute decreases as TGF increases due to economies of scale inherent in Erlang-C (see Figure 8.6). According to the plots of Figure 8.7, costs of virtual services are slightly reduced (approximately 5 to 7 percent) as the number of ACD centers is increased from one to three. This results from the use of smaller customer service bands as defined by Table 8.3 as the ACD center becomes closer to the traffic sources/destinations. However, the cost of a labor-minute rises as the number of ACD switches is increased since the economies of -scale is lost (see Figure 8.8).

According to Figure 8.6, the cost of agent-hours rises with TGF equally for all types of virtual services employed. Only fixed labor costs were assumed for all locations in the study. It is possible to locate the switches at places with low labor costs. Another important fact to remember is that the labor costs for a system employing OUTWATS tariffs may not require as many agents as for the ACD systems handling only incoming calls. The database generates the outgoing calls as a function of the agent availability. Consequently, the labor cost is then directly determined by the total agent erlangs during the month. Such a quantity is

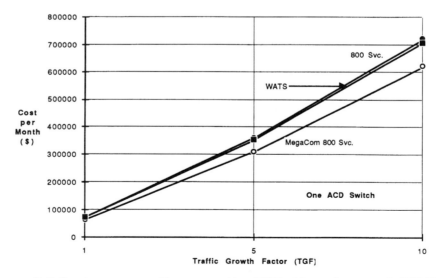

Figure 8.5 Costs per month versus the TGF: 49 regions and AT&T's 800-, MegaCom 800, and WATS services.

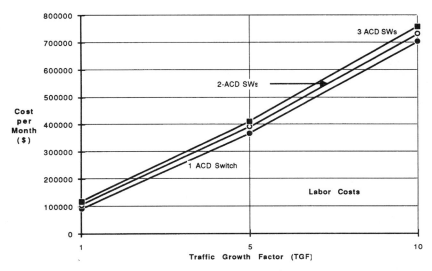

Figure 8.6 Monthly costs per month versus the TGF: 49 regions, one to three switches, and any virtual private network service.

printed in the results. In some cases, the OUTWATS leased lines are employed for handling only the outgoing calls for the employees. For that case, the agent costs can be disregarded.

The results shown in this chapter assume a single bundle of VPN lines. Such a bundle will represent the largest service-area band experienced by the calls. It is possible to lease several VPN line bundles, each reflecting a unique service-area band.

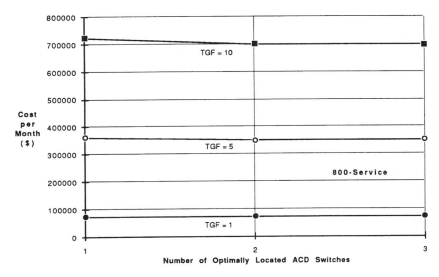

Figure 8.7 Monthly costs of AT&T's 800 service versus the number of optimally located ACD switches: 49 regions and TGF = 1, 5, or 10.

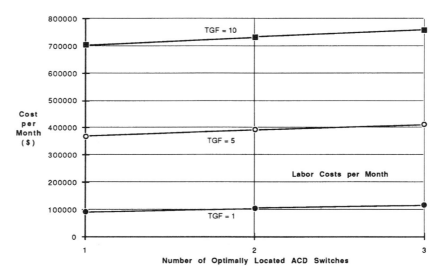

Figure 8.8 Labor costs per month versus the number of optimally located ACD switches: 49 regions and TGF = 1, 5, or 10.

But any resulting savings realized from reduced leasing costs would be neutralized by an increase in the total number of required VPN lines. See Figure 3.7 for a justification of this fact when one must maintain the same GOS for each separate bundle.

Figures 8.9 and 8.10 plot costs per call-minute for the VPN service (800 Service and MegaCom 800 Service) and agent-labor as a function of TGF. The results are plotted for a single optimally located ACD center. The results show

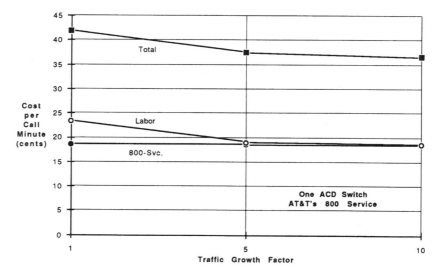

Figure 8.9 Costs per call-minute versus the TGF: one ACD switch, and AT&T's 800 service.

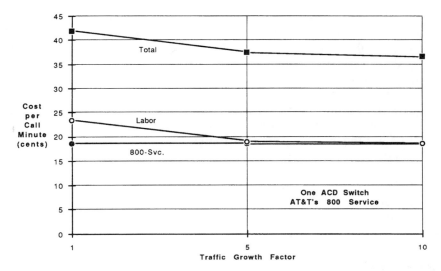

Figure 8.10 Costs per call-minute versus the TGF: one ACD switch, and AT&T's MegaCom 800 Service.

that whereas MegaCom 800 Service costs about 16 cents per call-minute, the 800 Service costs about 18 cents per call-minute. The labor costs range between 24 and 18 cents per call-minute as TGF is increased from 1 to 10. The labor costs were computed using the assumption that an agent-hour costs $10. (it may seem too low with those valid today) including overheads dealing with employee benefits and break periods. The results show that one can reduce the costs of an ACD system drastically by employing voice response units (VRUs) instead of live agents. However, this step should only be taken after much planning and study. While cost savings are important, the perception of an indifferent service associated with VRUs must always be kept in mind.

Comparing with the costs of a call-minute for private voice networks (e.g., see Figures. 7.6, 7.17, and 7.28), the virtual service for ACD yields identical costs for large TGF values. For small values of TGF, the cost of virtual services for ACD operations (about 15–21 cents per call minute as shown in Figures 8.9 and 8.10) is lower than that for the virtual services for plain voice applications (e.g., ECC of $1.31 divided by 5 or equal to 26 cents per call-minute) even for a low monthly call volume of about 78,000 (corresponding to TGF=1). This fact shows the need for selecting the correct VPN/SDN/Vnet type service for voice networks with small traffic volumes.

One can also observe that the total cost of a call-minute does not vary with the number of ACD centers. However, one may decide to realize a multicenter ACD network in order to increase the network availability or survivability. AT&T offers several advanced 800 Services such as (1) single number service for both intrastate and interstate calls, (2) 800 Area Code Routing for several ACD centers, (3) 800 Call Allocator for distributing the calls according to the capability of an ACD center, (4) 800 Command Routing to distribute the calls only to functioning ACD centers, (5) 800 Time Manager to allow only one (or as few as desired) ACD center during low traffic periods and so forth for achieving flexibility and/or survivability of any given multicenter ACD network. Other OCCs offer similar services.

The design of an ACD network handling only outgoing calls has not been discussed here. Such systems are hard to model since the concept of delay is not related to random arrivals of calls but instead influenced by the way new calls are launched. In some systems, no new calls are launched during periods characterized by finite queues. In other systems, new calls are launched as a function of the queue length. In any case, observed delays are invariably much smaller than those observed for an ACD system handling only incoming calls. When an ACD system handles both incoming and outgoing calls concurrently, the problem can get complicated for analysis. But one can use the modeling methodologies presented in this chapter to get meaningful designs.

Exercises

8.1 Using the design data of Table E8.1, create VHD45, LATA45, and NAME45 input files for a corporation that answers incoming calls from 45 regions of CONUS. The incoming calls deal with the specialized electronic equipment that is marketed, sold, and supported via an ACD system.

8.2 Using the VHD45, LATA45, NAME45, DTP8 files, and the ACD tariffs as defined in UTBL, WUTBL, MUTBL, and ACD support data of CSABDS and RSTBL input files, model ACD networks using 800 Service, INWATS, and MegaCom 800 Service. Find the optimum network topology considering the number of optimally located ACD centers and the type of virtual service.

Table E8.1 An ACD Network Design Database for 45 Regions

**************** NODAL DEFINITION DATA ***************

NODE#	-V-	-H-	LOAD (MErl)	LATA (Reg#)	LINK	NAME (NPA-ST)
1	5300	1250	1500	34	0	201-NJ
2	5601	1600	500	10	0	202-DC
3	4700	1380	900	8	0	203-CT
4	7600	2500	800	2	0	205-AL
5	6300	8900	700	57	0	206-WA
6	3700	1500	600	22	0	207-ME
7	6700	7600	450	14	0	208-ID

Continued

NODE#	-V-	-H-	LOAD (MErl)	LATA (Reg#)	LINK	NAME (NPA-ST)
8	8700	8250	950	5	0	209NCA
9	5000	1400	2500	37	0	212SNY
10	9230	7860	3500	6	0	213SCA
11	8400	4000	850	51	0	214ETX
12	5260	1475	680	45	0	215EPA
13	5580	2550	800	41	0	216NOH
14	6500	3430	850	16	0	217SIL
15	5540	4900	650	27	0	218-MN
16	6000	2900	870	17	0	219-IN
17	5510	1580	860	23	0	301-MD
18	5350	1320	250	9	0	302-DE
19	7520	5915	950	7	0	303-CO
20	6180	2100	500	58	0	304-WV
21	8200	590	1500	11	0	305-FL
22	7000	6200	650	60	0	307-WY
23	7000	5000	1200	31	0	308-NE
24	5600	2900	950	26	0	313SMI
25	7100	3400	600	29	0	314-MO
26	6300	2900	2500	21	0	318-LA
27	8500	3200	700	18	0	319-IA
28	7300	1800	1500	12	0	404-GA
29	8800	8500	2500	5	0	408NCA
30	4550	1450	1500	24	0	413-MA
31	5600	3600	800	59	0	414-WI
32	7700	3500	750	4	0	501-AR
33	6800	2800	620	20	0	502-KY
34	6800	8800	950	44	0	503-OR
35	8500	5500	860	35	0	505-NM

Continued

Table E.1 *Continued*

NODE#	-V-	-H-	LOAD (MErl)	LATA (Reg#)	LINK	NAME (NPA-ST)
36	6500	8500	600	57	0	509-WA
37	9400	4100	2500	52	0	512STX
38	7800	2700	800	28	0	601-MS
39	6300	5500	550	49	0	605-SD
40	7000	2500	650	50	0	615-TN
41	5700	5500	250	40	0	701-ND
42	8300	7500	650	32	0	702-NV
43	9000	8000	750	6	0	805SCA
44	8438	4061	4000	51	0	LCLNTX
45	8436	4034	5000	51	0	DALLTX

TOTAL TRAFFIC LOAD = 52990

8.3 Create an input file MUTBL1 representing the new 800-Service as defined in Table E8.2. Model the cost of new 800-Service with many advanced features using the newly created input file and ACDT=2 for an identical database as defined in Exercise 8.1.

8.4 Create an input file MUTBL2 identical to MUTBL1 but with the first column now representing upperbounds for 6 mileage bands defined as follows: 55, 292, 430, 925, 1910 and 3000 miles respectively. Model the cost of the new virtual 800-Service for ACDT=4 for an identical database as defined in Exercise 8.1.

Table E8.2. New 800-Service Tariff with Advanced Features (Effective 3-7-92)
(Cost ($) of 1-Hour of Virtual Facility Use)

CSABAND	BUSINESS	EVENING	NIGHT
1	13.12	10.82	8.70
2	13.57	11.18	9.00
3	13.79	11.35	9.15
4	14.23	11.72	9.44
5	14.46	11.91	9.58
6	15.80	13.01	10.46

Bibliography

Hoffman, H. 1983. "Extended Erlang C: Traffic Engineering for Queuing with Overflow." *Business Communications Review*, July–August 1983.

MacPherson, G., and B. Cleveland. 1990. "Features, Capabilities and Roles Changes for ACDs." *Business Communications Review*, December 1990, pp. 36–41.

Sharma, R. 1985. "Incoming Call Management: Past, Present and Future." *Communication News*, February 1985, pp. 52–55.

Sharma, R. 1986. "ACD Traffic Engineering: A Current Status." *Communication News*, July 1986.

Smith, R., and J. Smith. 1962. "Loss and Delay in Telephone Call Queuing Systems." *ATE Journal*, vol. 18, no. 1, pp. 18–30.

Whittaker, B. 1975. "Analysis and Design of a Multiserver, Multiqueue System with Finite Waiting Space in Each Queue." *Bell Systems Technical Journal*, vol. 54, no. 3, pp. 595–623.

9

Design Process for Data Networks

9.1 Introductory Remarks

Enterprise data networks exist to provide communication paths between data terminals, PCs, WSs, expensive shared peripherals (e.g., laser printers, databases, and servers) and host computers. The population of intelligent data communication devices, also called workstations (WSs), in use at this time is approaching half of the information worker population. About 80 percent of these devices are served by LANs and that number is increasing rapidly. The remaining devices are served by either proprietary networks (e.g., IBM's SNA) or a set of point-to-point connections composed of modems and/or multiplexers.

The data LAN population is now growing at the compound annual rate of about 20 percent. It is very interesting to observe that a digital PABX failed miserably in an attempt to capture the data LAN market through the use of a CS technology. An Ethernet LAN (employing CSMA/CD) succeeded admirably where a digital PABX failed since it provided (1) a high-speed data communications service among a large number of PCs/WSs, expensive peripherals and servers, all operating within the rules of a complicated network operating system (NOS); and (2) a simple LAN connection made possible by a low-cost network interface card (NIC). Currently, the installed LAN base is mostly composed of Ethernets and token ring (TRs). See Appendix B for a discussion of related standards. Although a token ring LAN performs better than an Ethernet (see Chapter 4 for a discussion), it is losing the battle for the installed base to Ethernet for three reasons: (1) the cost of a TR-NIC is much higher than its counterpart for an Ethernet, (2) the evolving standards for Ethernet may soon handle media capacities as high as 1Gbps versus 25Mbps for a TR, and (3) the port capacity on a TR hardly exceeds 32 when compared to a port capacity of over 100 for some Ethernets.

A physical Ethernet bus and/or a physical token ring are almost never used today due to the difficulties of connections in an environment of constant changes and moves. Instead, both Ethernets and token rings now employ intelligent hubs (IHs) that perform the required IEEE 802.3 (for Ethernet) and 802.5 (for TR) protocols for the LAN devices connected to the IH according to a pure star topology. The evolution to such a star topology took some 10 years. Most of the devices are generally connected to the IH through a pair of either unshielded twisted pair (UTP) or shielded twisted pair (STP). The UTP is by far the most commonly used. The specification for running a 10Mbps Ethernet on UTP is called 10BaseT. There are three other specifications for Ethernet: 10Base5 for a thick coaxial cable, 10Base2 for a thin cable, and 10BaseF for a fiber optic cable. Additional specifications for a 100Mbps Ethernet (e.g., 100BaseT, 100Base VG AnyLAN) have been established by IEEE and are being offered by vendors.

Some characteristics of data LANs are seldom emphasized in published literature despite their importance in LAN performance and internetworking. The intelligent hubs (IHs) always broadcast each packet/message to every device served by the LAN (or IH) as part of the original shared media design constraint. Consequently, data LANs are still afflicted with collisions even though these could have been eliminated in the pure star topology served by an IH. It would have required only a small increase in buffering and/or intelligence of a hub. Furthermore, most of the data LANs employ a connectionless PS technology whereby each packet uses a full 48-bit address for destination. In contrast, the FPS, Frame Relay, and ATM technologies employ a virtual connection (PVC or SVC) based PS technology and use only an abbreviated address (VC Identifier as part of the 5-byte cell overhead) to reduce network nodal latencies. These differences between the older LAN-based PS and modern WAN-related FPS technologies have prevented popular data LANs from reaping the benefits of (1) available very high bandwidths in public networks, (2) low latency associated with FPS/ATM, and (3) high availabilities/restorabilities of meshed-ring-based public/private broadband network architectures. Due to the large installed base (we all have heard this story before) of data LANs, it is now too late to correct the old problems. Instead, we are busy creating strange new standards to retrofit these older data LANs into the fabric of modern public/private broadband networks for the purpose of creating enterprise data networks.

We must not forget other types of useful data LANs and WANs. SNA networks handle data traffic according to either a pure PS technique (e.g., SDLC) or an S/F MS-type protocol (e.g., BSC). Matrix type switches handle both low- and high-speed data traffic by circuit switching (CS) techniques that are protocol free. Many data LANs are based on the virtual cell/packet technologies as marketed by AT&T (e.g., ISN), Stratacom (e.g., FastPacket), and Ascom-Timeplex (as based on Doelz's ESR). Some of the low-speed data traffic is also handled by voice networks through the mechanism of circuit switched data calls. A data call is established just like an ordinary voice call. Each established path is generally characterized by short bursts of data separated by long periods of inaction on both directions of traffic flow. This technique creates much waste of available bandwidth. Consequently, most network managers discourage the use of data calls on an enterprise voice network.

Data communications WANs can be synthesized in several ways. One can synthesize a WAN by simply interconnecting widely dispersed data LANs via bridges, routers, or gateways. Bridges link data LANs at the media access control layer, routers link data LANs at the network layer, and gateways link dissimilar data LANs at higher layers for protocol conversions. See Appendix B for further discussions on related standards. Current technology allows bridges and routers to provide high-speed connections with no preferred topology. Such internetworks generally employ point-to-point connections. But when an enterprise data network gets too large and widely dispersed, or when one needs broadband T3 and OC-N facilities, a solution based on bridges and routers becomes either prohibitively expensive or completely nonresponsive. A solution based on either LAN switches or ATM switches for a shared backbone network becomes necessary for MANs and WANs. Lack of standards still prevents the development of a clean solution. With the availability of ATM products employing switched virtual circuits (SVCs), economical interfaces to all existing LANs, and a mix of constant/variable bit rate (CBR/VBR) capabilities, a solution based on a cohesive network architecture should emerge. Standards dealing with LAN Emulation (LANE) for replacing bridges and multiprotocol-on-ATM (MPOA) for replacing routers are needed to create very high-speed end-to-end ATM data paths. A Bay Networks' marketing brochure (Bay Networks, 1995), a paper by Klessig (Klessig, 1995), and a paper by Callon et al. (Callon, Halpern, Drake, Sandick, and Jeffords 1996) should be consulted to get a deeper insight into LAN internetworks.

In many cases, a low-cost FPS node can serve many WSs at an enterprise branch to create a data LAN; several such FPS nodes can then be connected to one another to form a netlink (running at speeds of 56Kbps or 1.5Mbps); and many netlinks can then be served by a wide area backbone network of tandem FP switches (frame relay or ATM type). This will also allow a full sharing of high-speed switches, transmission facilities, and integrated network management and control. In such a network, each FPS is equivalent to an intelligent switching hub (ISH), each netlink is equivalent to a MAN/WAN MDNet, and the entire network becomes an enterprise MAN/WAN data network. Such a solution can be quite attractive for enterprises like transportation companies, banks, or libraries that don't have a large installed base of Ethernets or token rings.

An interesting solution for creating a broadband data network architecture is to employ Switched Multimegabit Data Service (SMDS). SMDS employs a *connectionless* protocol based on the mature IEEE 802.6 MAN specifications. However, SMDS is not a true MAN since it is only a subset (as developed by Bellcore) of IEEE 802.6. It employs 3-level switches (that may or may not be 53-byte-cell-based) interconnected by T1 and T3 circuits. The SMDS Interface Protocol (SIP) allows the handling of two types of packets, one at the network level (up to 9.188 Kbytes), and one at the subscriber access level (53-byte cell). SMDS offers a series of Sustained Information Rates (SIR) ranging between 1.17 and 34Mbps. It is being offered by many RBHCs and one IEC named MCI. Others are slow in offering. The hesitancy is caused by the popularity of connection-based Frame Relay and ATM networks.

Another solution for designing data networks is based on the IP switching technology. It is already implemented on the ubiquitous Internet and many existing data LANs. Data, voice, and video files can be freely exchanged between communicating persons on such networks. Many enterprises are building Intranets and connecting these to the Internets through carefully constructed "firewall." Why not employ IP switches instead of ATM switches? Uncontrolled, variable delay is a good deterrent. A new technology (Roberts, 1997) called Cells in Frames (CIF) may prove a good compromise. Other proprietary solutions are destined to appear.

The question as to which WAN approach will dominate the future marketplace can only be answered by the forces of free enterprise and technology. Some experts believe that all existing LAN and MAN technologies will eventually become seamless interfaces to a hierarchy of ATM networks. No matter what approach is employed, special network synthesis techniques will be required to handle unique topological and response time constraints imposed by the available packet switched network architectures. To illustrate, IBM's SNA architecture employs the multidrop topology and is generally characterized by low media utilization and high sensitivity to upper bound on delays. CCITT's X.25/X.75 architecture employs a mixed star/multidrop topology for the access network and a mesh topology for the backbone network as illustrated in Figures 1.8 and 1.9. AT&T's ISN architecture is characterized by the use of virtual packets and fast packet switching at the backbone network but generally constrained by the star topology at the access line level. The multistar, multicenter topology of Figure 1.8 will generally apply. Doelz's ESR architecture is characterized by protocol-independent, virtual (point-to-point and multipoint) packet switching, and high media utilization but constrained by the directed-link (DL) topology of Figure 1.10. The backbone networks of any data network will always be modeled as a fully connected mesh topology as a start. EcoNets has the capability of optimizing such topologies from the viewpoint of either cost or reliability. Each network topology/technology must be modeled with identical traffic intensities and tariffs before selecting a better solution. To illustrate, a recent paper (Nolle, 1997) shows a dark side of not modeling the entire network topology before selecting it.

It should now be obvious to the reader that each WAN architecture will always require the use of specialized algorithms for computing the relevant response times and optimizing the associated network topologies. The problem is that most of the existing internetworks do not submit to any architectural discipline. Their existing topologies can only be handled by the GivenNet item of EcoNets. A new ATM-based architecture is now emerging that will allow the use of EcoNets capabilities on a wider scale. In the following pages, we present a judicious amount of examples that illustrate the use of network synthesis algorithms (as described in Chapter 5) using both narrowband and broadband facilities. Chapter 10 shows how an *integrated* broadband network can employ most of the algorithms discussed in this chapter. Chapter 13 shows how a modern PCS network requires the use of algorithms for DLNet, MDNet, and StarDataNet topologies using high-speed broadband facilities.

9.2 The Design Process for a Data Network

9.2.1 A Description of Input Files and Design Parameters

In order to design a data network topology, the network design package as described in Chapter 6 allows one to execute all the steps defined in Chapter 5, and in particular, Sections 5.4.1 and 5.4.2, using the input design data of input files. The network design software package also allows one to vary a large number of critical design parameters iteratively to arrive at an optimum voice network.

The first step is to prepare the set of input files that must be named in FILES. The required input files for designing a data network are VHD, LINK, MAP, NLT, TARIFF, SDF, NAME, LATA, FILES, DTP, and SWF. The input files CSABDS, UTBL, WUTBL, MUTBL, and RSTBL as named in FILES are useful for ACD networks only.

The V&H coordinates for each site are derived from the enterprise database (EDB) as reflected in Figure 6.3. If this EDB is not available, the enterprise is not ready for network planning. One will need to buy a private line pricer (PLP) to prepare such an EDB. These tools basically relate the NPA-NNX of the site's telephone number to its V&H coordinates. Population growths in urban areas are putting great pressure on the nation's numbering plan. This is causing the emergence of new NPAs and changes in the old telephone numbers. The PLPs are constantly being updated. The TCA values of T1s handled by a site can be computed using the procedure described in Chapter 2. The number of information workers, type of profession or type, and number of active WSs may influence the values.

The LINK file is used for two design situations: (1) multilevel switches are employed for designing data networks, and (2) each CPE must be connected to the nearest tandem switch by a different access link type. For the first situation, one must set F_{lk} equal to 1. The nodes corresponding to the lowest link types are connected to the closest among those nodes corresponding to the next higher link types with the lowest link type. This process is repeated until no more AL levels are found. At that time, the tandem switches as defined in SWF are connected to one another. In case a LINK file provides a single link type for all sites, the only switches are of the tandem type. For the second situation, one must set the parameters F_{lk} and F_{vc0} equal to 1. For this situation, each site is connected to the closest tandem switch by a link type as defined by the corresponding link type in the LINK file. The second situation is only possible for a StarDataNet, since the DLNet and MDNet topologies allow the sharing of a single link (i.e., netlink) among many nodes. The tandem switches as defined by the input file SWF are always connected to one another by a single trunk link type as defined by the SDF design parameter, TKLT.

The MAP file is shared by every networking task since it is used to draw the boundary map. The MAP files (i.e., list of V&H coordinates for boundaries of countries other than the United States, Canada, and Mexico) can be obtained by using the procedure of Chapter 6.

The next important input file is an NLT file. It defines the set of link types that will be used in modeling the AL and trunk bundles. These link types can be a mix of leased and fully owned facilities. For the data application, the important link attributes are link type number, link capacity C, the maximum allowed data rate (equal to W_m) on the AL, the corresponding tariff number (TF#), and facility (leased/privately owned) factor F_{pf}.

The AL link type can be selected by the parameter ALT of SDF file when LINK file is not used (i.e., $F_{lk}=0$). The older version of EcoNets allowed the use of a "0" ALT when $F_{lk}=0$. One can now use non-zero AL link types (i.e., ALT parameter of SDF) to avoid any confusion. The AL link types are defined by the LINK file for $F_{lk}=1$. The new LINK file structure allows the use of only non-zero numbers to represent AL link types at each network level. For trunk link types, the SDF design parameters TKLT, F_{fdx}, and MTKU are employed. TKLT defines the trunk link type, F_{fdx} defines the full duplex (=1) or simplex nature (=0), and MTKU defines the maximum allowed fraction of link capacity for useful purposes. It should be stressed again that the new version of EcoNets allows the use of only non-zero trunk link types. The older version of EcoNets allowed the use of a "0" for TKLT. That caused some confusion.

We will use a tariff file that defines at least three tariffs: the Tariffs 9, 10, and 11 for the 9600bps, 56Kbps DDS and T1 lines. Such tariffs are defined in Table 2.4. We will also assume that the average monthly cost of the two local loops and termination charges equals an amount of $102.00 per VG, $62.76 per DDS, and $0.00 per T1 circuit. This implies no bypass for the VG circuits and some form of bypass for the DDS and T1 circuits. Although most of the analyses assume some bypass for the DDS and T1 circuits, an attempt is made near the end of this chapter to show the comparison between network cost with and without bypass of the local exchange carrier.

The next important input file that needs to be updated is the SDF file. It defines the values of 56 design parameters. These can be viewed (Mnemonic and corresponding value) by using the ViewUpdateInputFile item of the File menu. Each SDF mnemonic is also provided with a suffix of V, A, D, or C that implies Voice, ACD, Data, or Common application, respectively. Only a few of them may affect a data network significantly. These are ATP, UPR, HPR, IML, RML, N_{cu}, R_{mph}, HTT, F_{opt}, T_{np}, T_{hm}, K_{pg}, BKL, ICPB, TGF,ALT, TKLT, F_{fdx}, MTKU, and BBTF. The ALT parameter defines a single type of access link only when the LINK file is not used (i.e., $F_{lk}=0$). The design parameter TKLT is always required to define the link type for all trunks in a fully connected backbone network.

The trunk link type and the maximum allowable utilization for the backbone trunks are specified by the SDF parameters labeled as TKLT and MTKU, respectively. The design parameter F_{fdx} (equal to 1 for FDX and 0 for HDX) specifies whether the trunk can handle FDX data flows. The design parameter BBTF can be used to determine the amount of traffic flowing on the backbone trunks when no FTD file is used for modeling exact end-to-end traffic flows. BBTF=1 corresponds to a uniformly distributed traffic from every node. BBTF=2 reduces the backbone traffic by a factor of 2. Higher values of BBTF will reduce the cost of trunking even further. The FTD file is used for modeling and optimizing an existing network for which exact end-to-end traffic flows are known. Readers interested in employing an FTD file should consult Chapter 11, which describes the use of the FTD file in modeling a backbone network.

Special design parameters must be specified for computing the response times for each netlink. Design parameter ATP defines analysis type (1 for FPS/Frame Relay/ATM, 2 for Doelz VMPT [it can be also used for VLANs], 3 for SNA multidrop, and 4 for X.25 PS). Therefore, it has a significant influence on the modeled response time. User port and host port rates are specified by SDF parameters UPR and HPR. These also determine response times in all applications. The design parameters IML and RML define the length (bytes) of input message and response message, respectively, for most data networks. The other design parameters N_{cu} and R_{mph} are required for the Doelz's VMPT network modeling. Design parameters T_{np} (nodal processing time per transaction), T_{hm} (half modem time), and HTT (host think time in seconds) are very important for computing the values of turnaround times. The designer must consult the hardware specifications to specify these values. In some cases, the value of T_{hm} may disappear altogether. The value of K_{pg} (propagation constant = seconds/mile) is a fixed value determined by the velocity of light. The design parameters, BLKL, and ICPB specify the average block/packet length in bytes and information characters per block/Packet in PS, FPS, Frame Relay, and ATM applications. The design parameter, F_{opt} (0 for no optimization, 1 for optimization) determines whether the first feasible DLNet/MDNet is subjected to additional optimization process based on the EXCHANGE algorithm. A special design parameter called the F_{dis} is employed to obtain correct distances in miles or kilometers for a country not situated in North America.

Although the NAME file doesn't influence any analysis, its use does provide a complete site database in the DBF output file when $F_{nn}=1$.

The daily traffic profile named DTP8 will be used to model typical business hours for the voice and data networks modeled in this chapter.

In the analysis presented later, each network is analyzed for on-net to on-net traffic flows only. We will obtain the cost of netlinks, backbone trunks, and the entire network for each synthesis job. These values will appear on the associated network graph. The netlinks and the backbone network are completely described in terms of connections and corresponding costs on the printouts that are stored on the disk for later printout if desired. The total number of bits handled by the network during a month will be computed by assuming 5 busy hours per business day and 20 business days per month. We will use this value to derive the cost of transmitting one million bits of user data by the network. This value is part of a detailed output. Such a unit cost figure should be quite meaningful for data networks since several value-added network (VAN) vendors also charge by the number of packets handled. For voice networks, we employed the unit cost of a voice call-minute for network evaluation purposes.

We can use the StarDataNet, DLNet, and MDNet items of the Networking menu of the EcoNets package to synthesize networks based on the star, directed-link, and multidrop topologies using the algorithms developed by the author during the last two decades. Networks based on the star topology should be considered for LANs and MANs for handling high traffic volumes, reliability, and low response times, and WANs for low response times and high reliability. For the

WAN applications, the cost of a multistar, multicenter data network topology will be generally higher than that for a multidrop, multicenter topology network. Networking subroutines EWNET1.0, EWNET1.1, and EWNET1.2 (not part of the enclosed EcoNets version) will be employed to synthesize multidrop topologies based on three popular variations of the Esau-Williams algorithm as described in Chapter 5. The results will show that the MDNet capability of EcoNets generally provides cheaper data networks than the Esau-Williams algorithm. We will also vary the values of TGF, BBTF, W_m (= maximum allowable traffic load on a netlink), and ATP or analysis type to study the network costs and response times. The maximum and average values of system response times will also be studied and compared to the total and unit network costs as discussed earlier.

9.2.2 A Study of Narrowband Data Network Topologies with a Single Switch

We will employ the previously defined VHD41 and VHD59 files with some changes to reflect data applications. The nodal traffic load, which represented milliErlangs for voice, will now represent bits per second loads during the busy hour. The resulting values will be assumed to reflect a value of TGF=1. See Tables 9.1 and 9.2 for database definitions. Furthermore, we will employ two tariffs, the simple 1\$/mile tariff and the set of FCC Tariffs 9, 10, and 11 for computing the monthly costs. The simple tariff should also enable the designer to study the total circuit miles. The nodal distribution graphs of Figures 7.1 and 7.12 and the COG analysis graphs of Figures 7.2 and 7.13 still apply. Since the so-called exact solution as reported by Chandy and Lo (Chandy and Lo, 1973) suffers from (1) lack of design flexibility and (2) exponential growth of computation time with the number of nodes, only the algorithms based on heuristic principles will be considered. Three popular variations of the Esau-Williams (EW) algorithm and Sharma's DLNet and MDNet algorithms will be compared. The three variations of the EW algorithm can be defined as follows:

1. EWNET1.0 is a simplified version of the EW algorithm that prevents the merging of two components or netlinks, each with more than one node. This is supposed to provide good solutions with reduced execution time. An earlier study by Sharma (Sharma, 1983) was based on this variation and the simple 1\$/mile tariff.

2. EWNET1.1 is identical to the original EW algorithm as described in Chapter 5.

3. EWNET1.2 is identical to EWNET1.1 except that the computation of Max (T) function is based on distances alone. Consequently, EWNET1.1 and EWNET1.2 provide identical results for the 1\$/mile tariff.

See Figures 9.1 to 9.12 for the plots that relate monthly costs to the number of EXCHANGE passes for the two VHD data files, two previously described tariffs, TGF=1, and three capacity constraints (or W_m): 3900bps, 5100bps, and 6300bps.

Table 9.1 EOSN/Network Design Database for the VHD41 Input File

TOTAL NUMBER OF NODES = 41

#	-V-	-H-	LOAD (BPS)	LATA	LINK	NAME
1	7027	4203	2100	524	0	KANMO
2	9213	7878	2100	730	0	LAXCA
3	5015	1430	300	224	0	NWKNJ
4	4687	1373	300	920	0	HFDCT
5	6263	2679	300	922	0	CINOH
6	5972	2555	300	324	0	COLOH
7	4997	1406	2700	132	0	NYCNY
8	8492	8719	1500	722	0	SFOCA
9	7501	5899	300	656	0	DENCO
10	7489	4520	600	532	0	WICKA
11	7947	4373	300	536	0	OKLOK
12	8436	4034	300	552	0	DALTX
13	8266	5076	300	546	0	AMLTX
14	8938	3536	300	560	0	HOUTX
15	8483	2638	300	490	0	NOLLA
16	8549	5887	900	664	0	ALBNM
17	6807	3482	1800	520	0	STLMO
18	7010	2710	300	470	0	NASTN
19	8173	1147	300	952	0	TMPFL
20	8351	527	600	460	0	MIAFL
21	7260	2083	600	438	0	ATLGA
22	9345	6485	300	668	0	TCSAR
23	9135	6748	900	666	0	PHXAR
24	8486	8695	300	722	0	OAKCA
25	8665	7411	300	721	0	LASNE
26	5986	3426	1500	358	0	CHIIL

Continued

Table 9.1 *Continued*

#	-V-	-H-	LOAD (BPS)	LATA	LINK	NAME
27	7707	4173	300	538	0	TULOK
28	5536	2828	300	340	0	DETMI
29	4422	1249	900	128	0	BOSMA
30	5574	2543	300	320	0	CLEOH
31	6261	4021	600	635	0	CDRIA
32	6272	2992	300	336	0	INDIN
33	6529	2772	300	462	0	LOUKY
34	6657	1698	300	422	0	CHTNC
35	6113	2705	300	328	0	DATOH
36	5251	1458	1500	228	0	PHILPA
37	5622	1583	900	236	0	WASDC
38	5510	1575	300	238	0	BALMA
39	5621	2185	1800	234	0	PITPA
40	5363	1733	300	226	0	HARPA
41	5166	1585	300	228	0	ALNPA

TOTAL TRAFFIC LOAD= 28500 BPS

Table 9.2 EOSN/Network Design Database for VHD59

TOTAL NUMBER OF NODES = 59

#	-V-	-H-	LOAD (BPS)	LATA	LINK	NAME
1	7260	2083	3600	438	0	ATLGA
2	7089	1674	600	442	0	AGSGA
3	6749	2001	300	420	0	AVLNC
4	5510	1575	300	238	0	BALMD
5	7518	2446	600	476	0	BHMAL
6	8806	3298	300	562	0	BPTTX
7	8476	2874	300	492	0	BTRLA
8	6901	1589	600	434	0	CAESC

Continued

#	-V-	-H-	LOAD (BPS)	LATA	LINK	NAME
9	7098	2366	300	472	0	CHATN
10	5986	3426	1800	358	0	CHIIL
11	7021	1281	300	436	0	CHSWV
12	6261	4021	1200	635	0	CIDIA
13	6657	1698	600	422	0	CLTNC
14	5972	2555	300	324	0	CMHOH
15	7556	2045	600	438	0	CSGGA
16	6263	2679	1200	922	0	CVGOH
17	8436	4034	1800	552	0	DALTX
18	6113	2705	300	328	0	DAYOH
19	5622	1583	900	236	0	WASDC
20	7501	5899	900	656	0	DENCO
21	5536	2828	600	340	0	DETMI
22	8409	3168	300	486	0	ESFLA
23	6729	3019	300	330	0	EVVIN
24	5942	2982	300	334	0	FWAIN
25	8479	4122	1800	552	0	GSWTX
26	7827	3554	300	528	0	HOTAR
27	8938	3536	900	560	0	HOUTX
28	6272	2992	600	336	0	INDIN
29	8035	2880	600	482	0	JANMS
30	7649	1276	300	452	0	JAXFL
31	8665	7411	300	721	0	LASNV
32	9213	7878	600	730	0	LAXCA
33	6459	2562	300	466	0	LEXKY
34	7721	3451	300	528	0	LITAR
35	7364	1865	300	446	0	MCNGA
36	7899	2639	300	482	0	MEIMS

Continued

Table 9.2 *Continued*

#	-V-	-H-	LOAD (BPS)	LATA	LINK	NAME
37	7471	3125	1500	468	0	MEMTN
38	7692	2247	300	478	0	MGMAL
39	8351	527	3300	460	0	MIAFL
40	7027	4203	600	524	0	MKCMO
41	8148	3218	300	486	0	MLOLA
42	5781	4525	600	628	0	MSPMN
43	8483	2638	1500	490	0	MSYLA
44	4997	1406	2400	132	0	NYCNY
45	7954	1031	300	458	0	ORLFL
46	8165	606	300	460	0	PBIFL
47	5251	1458	300	228	0	PHLPA
48	6982	3088	300	468	0	PUKTN
49	7266	1379	600	440	0	SAVGA
50	6529	2772	300	462	0	SDFKY
51	9468	7629	300	732	0	SDOCA
52	8492	8719	600	722	0	SFOCA
53	7310	3836	300	522	0	SGFMO
54	8272	3495	600	486	0	SHVLA
55	7496	1340	300	440	0	SSIGA
56	6807	3482	900	520	0	STLMO
57	5704	2820	300	326	0	TOLOH
58	8173	1147	600	952	0	TPAFL
59	6801	2251	600	474	0	TYSTN

TOTAL TRAFFIC LOAD= 42000 BPS

The maximum line capacity (or C) of 9600bps is assumed for all cases. The EXCHANGE algorithm is identical to that described in Chapter 5 and it is always required for Sharma's MDNet algorithm. The three popular variations of EWNet never employ the exchange algorithm for further cost reductions. The cost values for the zero EXCHANGE pass should therefore reflect the traditional EW algorithms. The results described in this chapter show that additional savings are possible through the use of Sharma's Exchange algorithm. The results also reflect both the actual and the simple tariff of 1$/mile. Kerschenbaum and Chu (1974) have wrongly stated that the Exchange algorithm costs an enormous amount of computation time. The fact is that each Exchange pass costs only one to three times the computation time required for the normal EWNET1.1 algorithm. This is much less than the 70-fold increase in the computation time experienced in work with the algorithm developed by Karnaugh (Karnaugh, 1972) and based on elaborate line cost tradeoffs performed at every stage of the EW algorithm. Karnaugh (Karnaugh, 1972) achieved a cost improvement of only about 1 percent. The Exchange algorithm is shown here to yield better cost improvements over Karnaugh's algorithm with very little increase in computational times.

The cost improvements in percent are drastically different for the two tariffs. Whereas the cost improvements for the simple tariff (1$/mile) can range up to 10 percent, the cost improvements for the realistic set of tariffs (FCC Tariffs 9, 10, 11 as defined by Table 2.4) are less than 2 percent. This points to a need for utmost care when considering claims made by several authors. The results presented here clearly show that no heuristic algorithm can claim to be the least costly in all cases. One can only recommend one solution over another on a statistical basis. A quick study of Figures 9.1 to 9.12 clearly shows that the MDNet algorithm yields a lower cost network after 1 Exchange pass compared to a network based on the EWNet1.1 algorithm with no exchange in almost all cases. The MDNet algorithm is generally more efficient than the EWNet1.1 algorithm even when the exchange algorithm is applied to the EW algorithm. The EWNet1.0 algorithm is generally inferior to EWNet1.1 and MDNet algorithms. But in some rare cases (see Figure 9.10), it is shown to be better than the EWNet1.1 algorithm. In general, the EWNet1.2 algorithm is only slightly inferior to the EWNet1.1 algorithm. The DLNet algorithm is generally inferior to other algorithms but it yields much lower response times as shown later. Only in one case (see Figure 9.5) was it found to be better than all variations of EW algorithm without the Exchange technique. That was quite a surprise. A close observation of the DL topology will show that some netlinks are characterized by loops. Although the human eye can eliminate such loops and hence reduce the network cost somewhat, it is difficult to design an equivalent computer algorithm. This should be a challenge to a keen student of network science.

One can also study the network topologies yielded by the various algorithms. Figures 9.13 and 9.14 represent the network topologies achieved through EWNet1.0 and EWNet1.1/MDNet when one lets W_m go to infinity. It shows that while EWNet1.1 and MDNet yield a single minimal spanning tree without any capacity constraint, the EWNet1.0 yields an inferior minimal spanning tree due to design constraints as described earlier. Figures 9.15 to 9.35 represent some typical network topologies for the two tariffs and a capacity constraint of 5100bps for the two previously defined VHD41 and VHD59 data files using EWNet1.1, MDNet, and DLNet algorithms with and without the Exchange only. It is startling to observe that network topologies practically remain unchanged with varying tariffs (i.e., 1$/mile or Tariffs 9, 10, and 11).

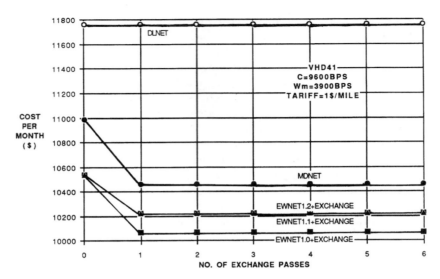

Figure 9.1 Costs per month versus the number of exchange passes for VHD41, 1$ per mile tariff, link capacity (C) = 9600bps, maximum allowable link rate (W_m) = 3900bps, and for directed path (DL), three variations of Esau-Williams (EW), and multidrop (MD) network topologies.

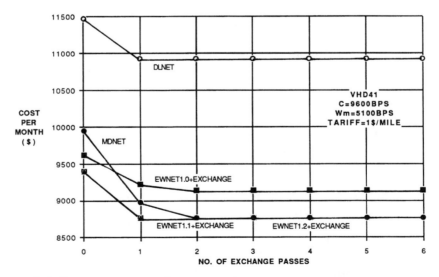

Figure 9.2 Costs per month versus the number of exchange passes for VHD41, 1$ per mile tariff, link capacity (C) = 9600bps, maximum allowable link rate (W_m) = 5100bps, and for directed path (DL), three variations of Esau-Williams (EW), and multidrop (MD) network topologies.

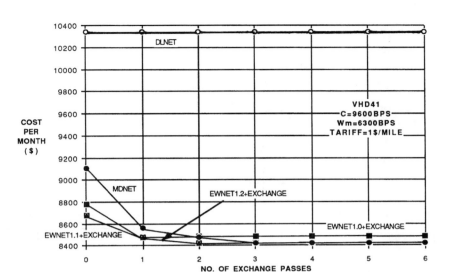

Figure 9.3 Costs per month versus the number of exchange passes for VHD41, 1$ per mile tariff, link capacity (C) = 9600bps, maximum allowable link rate (W_m) = 6300bps, and for directed path (DL), three variations of Esau-Williams (EW), and multidrop (MD) network topologies.

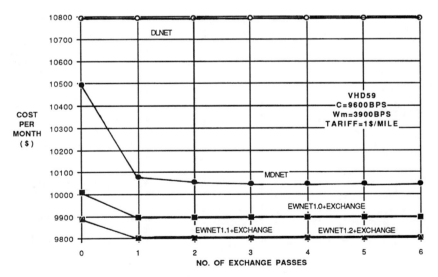

Figure 9.4 Costs per month versus the number of exchange passes for VHD59, 1$ per mile tariff, link capacity (C) = 9600bps, maximum allowable link rate (W_m) = 3900bps, and for directed path (DL), three variations of Esau-Williams (EW), and multidrop (MD) network topologies.

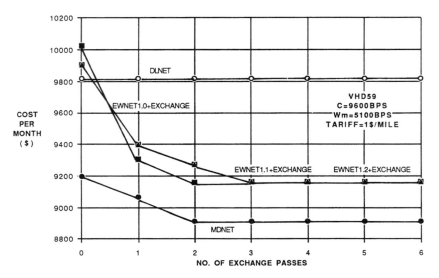

Figure 9.5 Costs per month versus the number of exchange passes for VHD59, 1$ per mile tariff, link capacity (C) = 9600bps, maximum allowable link rate (W$_m$) = 5100bps, and for directed path (DL), three variations of Esau-Williams (EW), and multidrop (MD) network topologies.

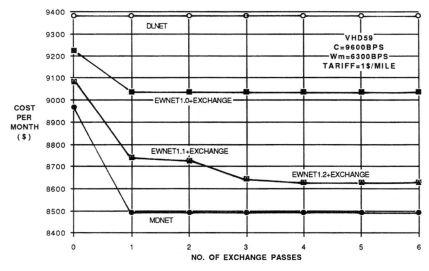

Figure 9.6 Costs per month versus the number of exchange passes for VHD59, 1$ per mile tariff, link capacity (C) = 9600bps, maximum allowable link rate (W$_m$) = 6300bps, and for directed path (DL), three variations of Esau-Williams (EW), and multidrop (MD) network topologies.

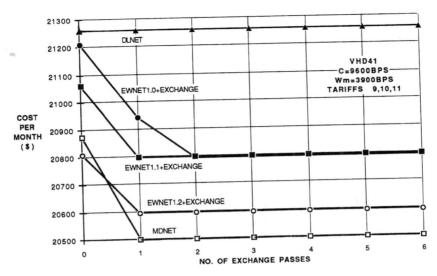

Figure 9.7 Costs per month versus the number of exchange passes for VHD41, Tariffs 9, 10, and 11, link capacity (C) = 9600bps, maximum allowable link rate (W_m) = 3900bps, and for directed path (DL), three variations of Esau-Williams (EW), and multidrop (MD) network topologies.

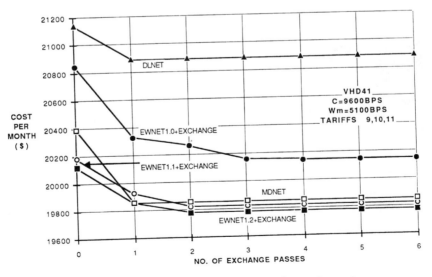

Figure 9.8 Costs per month versus the number of exchange passes for VHD41, Tariffs 9, 10, and 11, link capacity (C) = 9600bps, maximum allowable link rate (W_m) = 5100bps, and for directed path (DL), three variations of Esau-Williams (EW), and multidrop (MD) network topologies.

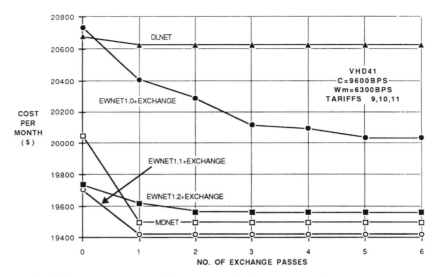

Figure 9.9 Costs per month versus the number of exchange passes for VHD41, Tariffs 9, 10, and 11, link capacity (C) = 9600bps, maximum allowable link rate (W_m) = 6300bps, and for directed path (DL), three variations of Esau-Williams (EW), and multidrop (MD) network topologies.

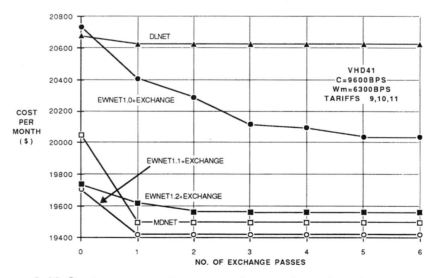

Figure 9.10 Costs per month versus the number of exchange passes for VHD59, Tariffs 9, 10, and 11, link capacity (C) = 9600bps, maximum allowable link rate (W_m) = 3900bps, and for directed path (DL), three variations of Esau-Williams (EW), and multidrop (MD) network topologies.

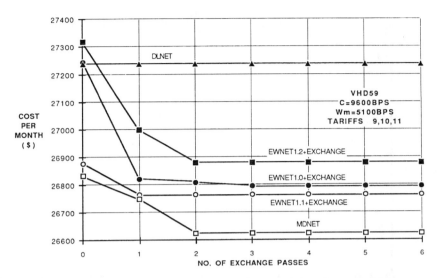

Figure 9.11 Costs per month versus the number of exchange passes for VHD59, Tariffs 9, 10, and 11, link capacity (C) = 9600bps, maximum allowable link rate (W_m) = 5100bps, and for directed path (DL), three variations of Esau-Williams (EW), and multidrop (MD) network topologies.

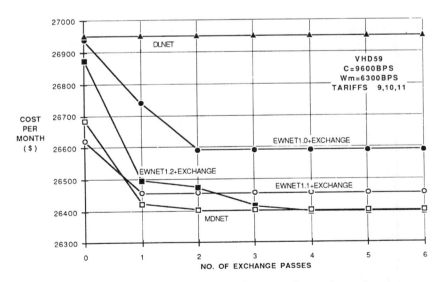

Figure 9.12 Costs per month versus the number of exchange passes for VHD59, Tariffs 9, 10, and 11, link capacity (C) = 9600bps, maximum allowable link rate (W_m) = 6300bps, and for directed path (DL), three variations of Esau-Williams (EW), and multidrop (MD) network topologies.

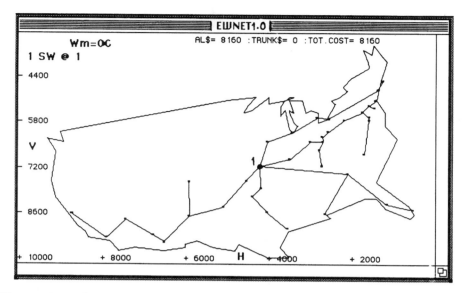

Figure 9.13 Esau-Williams (EW) network topology for VHD41, 1$ per mile tariff, and for the case W$_m$ is infinite.

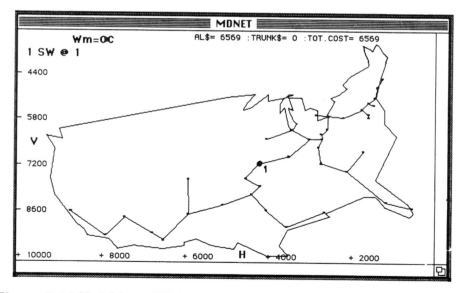

Figure 9.14 Multidrop (MD) network topology for VHD41, 1$ per mile tariff, and for the case W$_m$ is infinite.

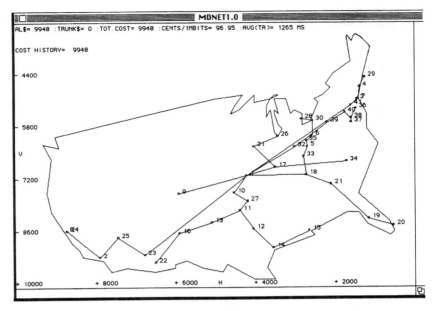

Figure 9.15 Unoptimized multidrop (MD) network topology for VHD41, 1$ per mile tariff, link capacity (C) - 9600bps, and maximum allowable link rate (W$_m$) = 5100bps.

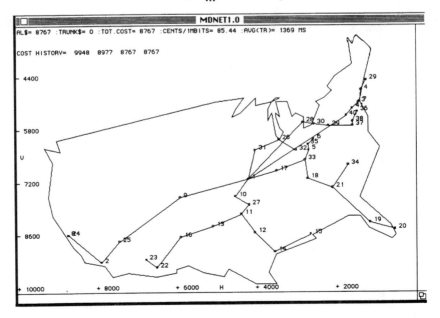

Figure 9.16 Optimized multidrop (MD) network topology for VHD41, 1$ per mile tariff, link capacity (C) - 9600bps, and maximum allowable link rate (W$_m$) = 5100bps.

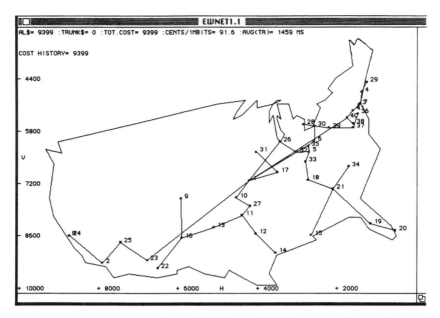

Figure 9.17 Unoptimized Esau-Williams (EW) network topology for VHD41, 1$ per mile tariff, link capacity (C) - 9600bps, and maximum allowable link rate (W_m) = 5100bps.

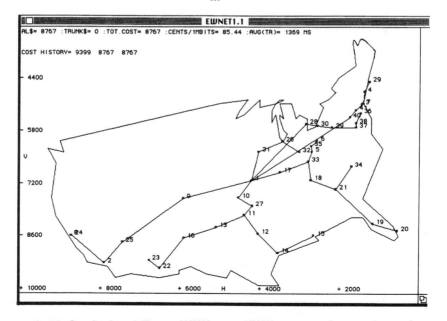

Figure 9.18 Optimized Esau-Williams (EW) network topology for VHD41, 1$ per mile tariff, link capacity (C) - 9600bps, and maximum allowable link rate (W_m) = 5100bps.

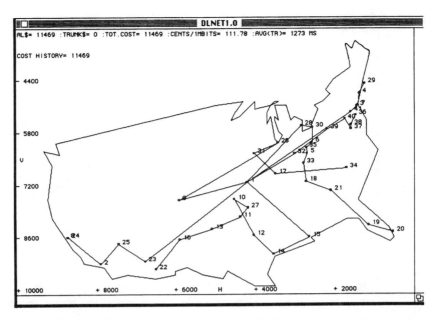

Figure 9.19 Unoptimized directed link (DL) network topology for VHD41, 1$ per mile tariff, link capacity (C) - 9600bps, and maximum allowable link rate (W$_m$) = 5100bps.

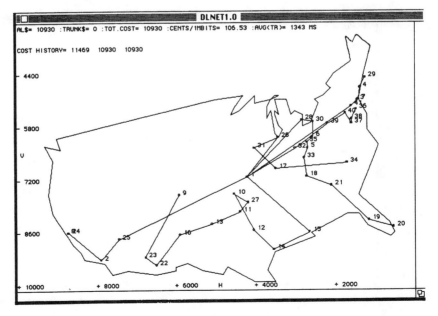

Figure 9.20 Optimized directed link (DL) network topology for VHD41, 1$ per mile tariff, link capacity (C) - 9600bps, and maximum allowable link rate (W$_m$) = 5100bps.

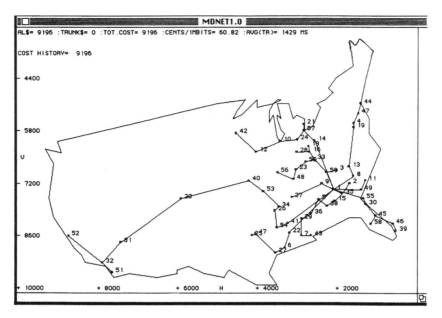

Figure 9.21 Unoptimized multidrop (MD) network topology for VHD59, 1$ per mile tariff, link capacity (C) - 9600bps, and maximum allowable link rate (W_m) = 5100bps.

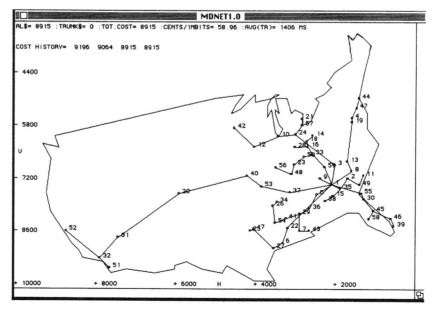

Figure 9.22 Optimized multidrop (MD) network topology for VHD59, 1$ per mile tariff, link capacity (C) - 9600bps, and maximum allowable link rate (W_m) = 5100bps.

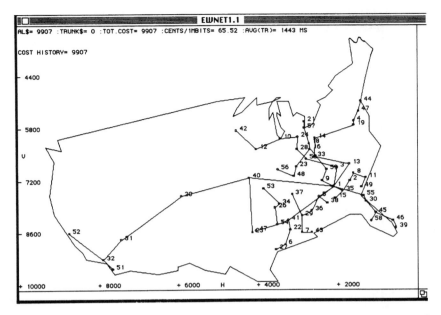

Figure 9.23 Unoptimized Esau-Williams (EW) network topology for VHD59, 1$ per mile tariff, link capacity (C) - 9600bps, and maximum allowable link rate (W$_m$) = 5100bps.

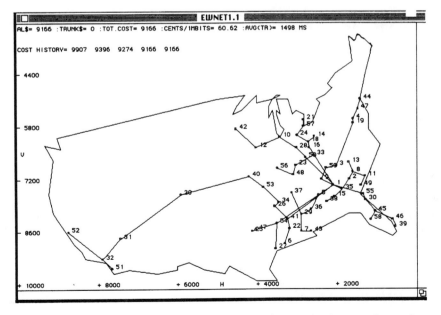

Figure 9.24 Optimized Esau-Williams (EW) network topology for VHD59, 1$ per mile tariff, link capacity (C) - 9600bps, and maximum allowable link rate (W$_m$) = 5100bps.

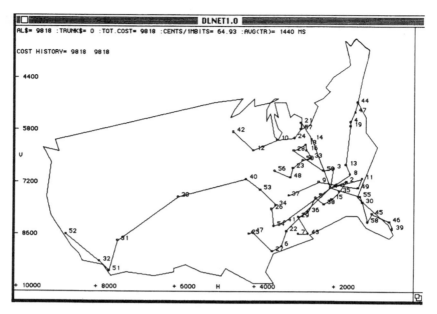

Figure 9.25 Optimized directed link (DL) network topology for VHD59, 1$ per mile tariff, link capacity (C) - 9600bps, and maximum allowable link rate (W$_m$) = 5100bps.

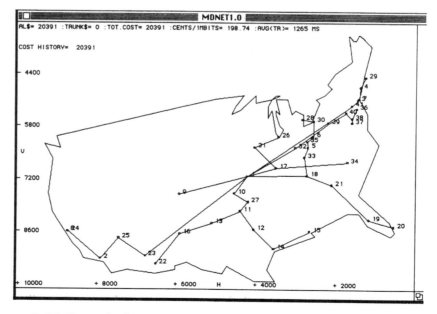

Figure 9.26 Unoptimized multidrop (MD) network topology for VHD41, Tariffs 9, 10, and 11, link capacity (C) - 9600bps, and maximum allowable link rate (W$_m$) = 5100bps.

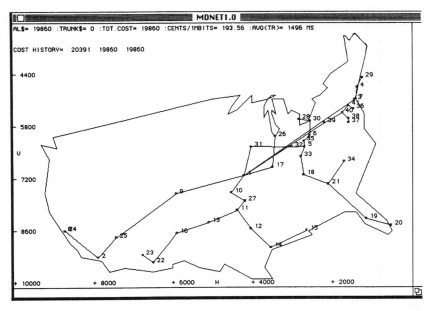

Figure 9.27 Optimized multidrop (MD) network topology for VHD41, Tariffs 9, 10, and 11, link capacity (C) - 9600bps, and maximum allowable link rate (W$_m$) = 5100bps.

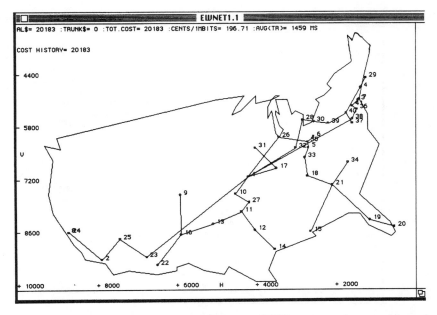

Figure 9.28 Unoptimized Esau-Williams (EW) network topology for VHD41, Tariffs 9, 10, and 11, link capacity (C) - 9600bps, and maximum allowable link rate (W$_m$) = 5100bps.

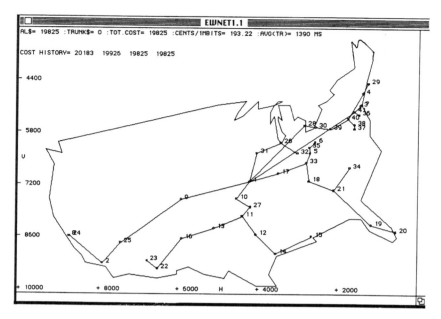

Figure 9.29 Optimized Esau-Williams (EW) network topology for VHD41, Tariffs 9, 10, and 11, link capacity (C) - 9600bps, and maximum allowable link rate (W$_m$) = 5100bps.

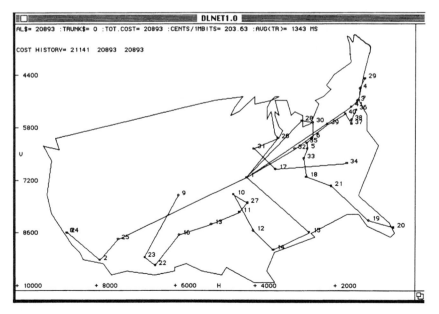

Figure 9.30 Optimized directed link (DL) network topology for VHD41, Tariffs 9, 10, and 11, link capacity (C) - 9600bps, and maximum allowable link rate (W$_m$) = 5100bps.

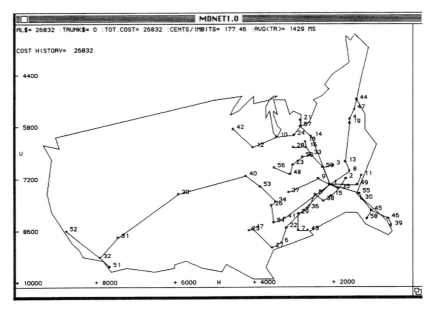

Figure 9.31 Unoptimized multidrop (MD) network topology for VHD41, Tariffs 9, 10, and 11, link capacity (C) - 9600bps, and maximum allowable link rate (W_m) = 5100bps.

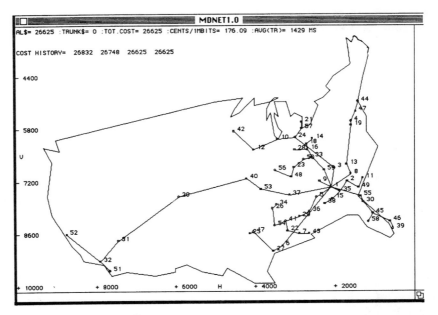

Figure 9.32 Optimized multidrop (MD) network topology for VHD41, Tariffs 9, 10, and 11, link capacity (C) - 9600bps, and maximum allowable link rate (W_m) = 5100bps.

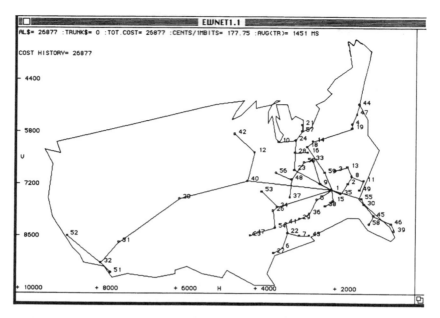

Figure 9.33 Unoptimized Esau-Williams (EW) network topology for VHD59, Tariffs 9, 10, and 11, link capacity (C) - 9600bps, and maximum allowable link rate (W$_m$) = 5100bps.

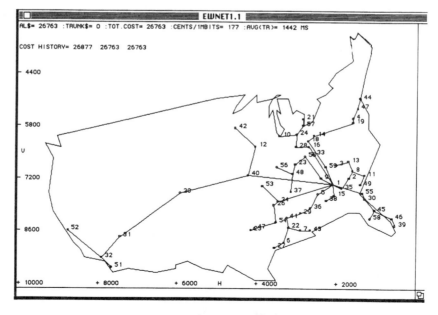

Figure 9.34 Optimized Esau-Williams (EW) network topology for VHD59, Tariffs 9, 10, and 11, link capacity (C) - 9600bps, and maximum allowable link rate (W$_m$) = 5100bps.

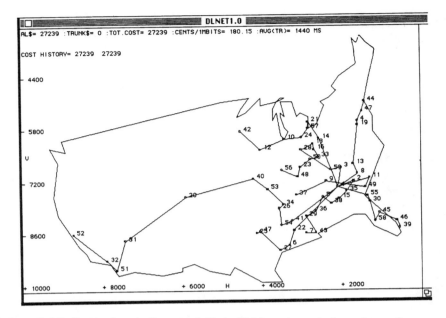

Figure 9.35 Optimized directed link (DL) network topology for VHD59, Tariffs 9, 10, and 11, link capacity (C) - 9600bps, and maximum allowable link rate (W$_m$) = 5100bps.

Tables 9.3 to 9.11 define some selected network topologies in terms of the nodes on each connection, the mileage and cost of each connection, and running totals of mileage and costs. The reader should ignore the slight discrepancy between the mileage and cost of a link for the 1$/mile tariff. It is caused by computational round-offs. The data of Tables 9.3 to 9.11 should be combined with the corresponding topology graphs to study the underlying differences between MDNet and EWNet algorithms. The EWNet1.1 is always sensitive to the order in which the various nodes are selected for integration into the network. For the MDNet or DLNet algorithm, only the farthest unused node to switch is considered at the beginning of a netlink design process. It then considers only the closest

Table 9.3 DATANET Link Analysis for VHD41 and DLNet with No Optimization

NO. OF SWITCHES=1　　　　　　　　AT　　　　　　1
MAXIMUM ALLOWED LINK RATE (Wm)= 5100 BPS

SW#	LK	#LKWT	AQT	ATAT	E(TR)		NODES									
1	1	5100	6	497	1063 ***	8	24	2	25	23	1					
1	2	4500	5	452	884 ***	29	4	7	3	41	1					
1	3	3600	3	872	1789 ***	20	19	21	18	33	5	35	6	30	28	1

Continued

Table 9.3 *Continued*

SW#	LK	#LKWT	AQT	ATAT	E(TR)	NODES
1	4	5100	6	513	1144 *** 36 38	37 40 39 32 1
1	5	3500	3	860	1713 *** 22 16	13 11 27 10 12 14 15 1
1	6	4500	5	536	1049 *** 34 17	31 26 9 1

MAXTR= 1789   AVG. TR=1273

OVERALL SYSTEM AVERAGE (TR)-MS = 1273

FROM	TO	MILES	COST$	TOT$	TMILES
8	24	7	8	8	7
24	2	345	346	354	353
2	25	227	227	581	581
25	23	256	257	838	838
23	1	1045	1045	1883	1883
29	4	92	92	92	92
4	7	98	98	190	191
7	3	9	9	199	200
3	41	68	68	267	268
41	1	1015	1015	1282	1284
20	19	203	204	204	203
19	21	413	413	617	617
21	18	213	213	830	830
18	33	153	153	983	984
33	5	89	89	1072	1073
5	35	48	48	1120	1121
35	6	65	65	1185	1186
6	30	125	126	1311	1312
30	28	90	91	1402	1403
28	1	641	641	2043	2044
36	38	89	90	90	89
38	37	35	35	125	125

Continued

FROM	TO	MILES	COST$	TOT$	TMILES
37	40	94	94	219	219
40	39	164	164	383	384
39	32	327	328	711	712
32	1	451	451	1162	1163
22	16	314	315	315	314
16	13	271	271	586	586
13	11	244	244	830	830
11	27	98	98	928	929
27	10	129	129	1057	1058
10	12	336	336	1393	1395
12	14	223	223	1616	1618
14	15	318	318	1934	1937
15	1	675	676	2610	2613
34	17	566	566	566	566
17	31	242	242	808	808
31	26	207	207	1015	1015
26	9	917	917	1932	1933
9	1	556	557	2489	2489

TOTAL AL COST= 11469
TOTAL AL, SL MILES=11469, 0
TOTAL AL AND SL COSTS= 11469, 0
TOTAL NUMBER OF SPECIAL LINKS= 0

Table 9.4 DATANET Link Analysis for VHD41 and DLNet with Optimization

NO. OF SWITCHES= 1   AT   1
MAXIMUM ALLOWED LINK RATE (Wm)= 5100 BPS

SW#	LK#	LKWT BPS	AQT MS	ATAT MS	E(TR) MS		NODES
1	1	4200	4	409	735 ***	8 24 2 25 1	
1	2	4500	5	452	884 ***	29 4 7 3 41 1	
1	3	3600	3	872	1789 ***	20 19 21 18 33 5 35 6 30 28 1	

Continued

Table 9.4 *Continued*

SW#	LK#	LKWT BPS	AQT MS	ATAT MS	E(TR) MS	NODES
1	4	5100	6	513	1144 ***	36 38 37 40 39 32 1
1	5	4800	5	1122	2807 ***	9 23 22 16 13 11 27 10 12 14 15 1
1	6	4200	4	392	703 ***	34 17 31 26 1

MAXTR= 2807   AVG. TR= 1343

OVERALL SYSTEM AVERAGE(TR)-MS = 1343

FROM	TO	MILES	COST$	TOT$	TMILES
8	24	7	8	8	7
24	2	345	346	354	353
2	25	227	227	581	581
25	1	1139	1139	1720	1720
29	4	92	92	92	92
4	7	98	98	190	191
7	3	9	9	199	200
3	41	68	68	267	268
41	1	1015	1015	1282	1284
20	19	203	204	204	203
19	21	413	413	617	617
21	18	213	213	830	830
18	33	153	153	983	984
33	5	89	89	1072	1073
5	35	48	48	1120	1121
35	6	65	65	1185	1186
6	30	125	126	1311	1312
30	28	90	91	1402	1403
28	1	641	641	2043	2044
36	38	89	90	90	89
38	37	35	35	125	125

Continued

FROM	TO	MILES	COST$	TOT$	TMILES
37	40	94	94	219	219
40	39	164	164	383	384
39	32	327	328	711	712
32	1	451	451	1162	1163
9	23	582	582	582	582
23	22	106	106	688	688
22	16	314	315	1003	1003
16	13	271	271	1274	1275
13	11	244	244	1518	1519
11	27	98	98	1616	1617
27	10	129	129	1745	1747
10	12	336	336	2081	2084
12	14	223	223	2304	2307
14	15	318	318	2622	2625
15	1	675	676	3298	3301
34	17	566	566	566	566
17	31	242	242	808	808
31	26	207	207	1015	1015
26	1	410	410	1425	1426

TOTAL AL COST = 10930
TOTAL AL,SL MILES = 10930, 0
TOTAL AL AND SL COSTS= 10930, 0 -
TOTAL NUMBER OF SPECIAL LINKS= 0

Table 9.5 NET Link Analysis for VHD41 and EWNet1.1 with No Optimization

NO. OF SWITCHES= 1   AT   1
MAX ALLOWED LINK RATE= 5100 BPS

SWAP	LK#	LKWT	AQT MS	ATAT MS	E(TR) MS		NODES
1	1	5100	6	497	1057	***	8 24 2 25 23 1
1	2	3600	3	897	1772	***	12 14 9 22 16 13 11 27 10 1

Continued

Table 9.5 *Continued*

SWAP	LK#	LKWT	AQT MS	ATAT MS	E(TR) MS	NODES
1	3	2400	2	213	283	*** 17 31 1
1	4	5100	6	1171	3048	*** 18 35 15 19 20 21 34 5 33 32 26 1
1	5	5100	6	524	1162	*** 36 37 39 30 28 6 1
1	6	5100	6	623	1432	*** 41 3 4 29 7 40 38 1

MAXTR= 3048   AVG. TR= 1459

OVERALL SYSTEM AVERAGE(TR)-MS = 1459

FROM	TO	MILES	COST$	TOT$	TMILES
8	24	7	8	8	7
2	25	227	227	235	235
23	25	256	257	492	492
1	23	1045	1045	1537	1537
2	24	345	346	1883	1883
12	11	188	188	188	188
14	12	223	223	411	411
9	16	331	331	742	743
22	16	314	315	1057	1057
13	11	244	244	1301	1302
27	11	98	98	1399	1400
10	27	129	129	1528	1530
1	10	177	177	1705	1707
16	13	271	271	1976	1978
17	1	238	238	238	238
31	17	242	242	480	480
18	33	153	153	153	153
35	5	48	48	201	201
15	21	424	424	625	626
19	20	203	204	829	830

Continued

FROM	TO	MILES	COST$	TOT$	TMILES
34	21	226	226	1055	1056
32	5	99	99	1154	1155
26	32	164	164	1318	1313
1	26	410	410	1728	1730
19	21	413	413	2141	2143
21	18	213	213	2354	2357
5	33	89	89	2443	2446
36	37	123	124	124	123
39	30	114	114	238	237
28	30	90	91	329	328
6	30	125	126	455	454
1	6	618	618	1073	1073
39	37	190	190	1263	1263
41	3	68	68	68	68
4	29	92	92	160	160
7	3	9	9	169	170
40	38	68	68	237	238
1	40	941	941	1178	1180
40	41	77	78	1256	1258
4	7	98	98	1354	1356

TOTAL AL COST = 9399
TOTAL AL & SL MILES = 9399,   0
TOTAL AL AND SL COSTS= 9399,   0
TOTAL NUMBER OF SPECIAL LINKS= 0

Table 9.6 NET Link Analysis for VHD41 and MDNet with No Optimization

NO OF SWITCHES= 1   AT   1
MAX LINK RATE= 5100 BPS

SW#	LK#	LKWT	AQT MS	ATAT MS	E(TR) MS	NODES
1	1	5100	6	497	1057***	8 24 2 25 23 1
1	2	4500	5	452	879 ***	29 4 7 3 41 1

Continued

Table 9.6 *Continued*

SW#	LK#	LKWT	AQT MS	ATAT MS	E(TR) MS	NODES
1	3	3600	3	872	1774***	20 19 21 18 33 5 35 6 30 28 1
1	4	5100	6	513	1137***	36 38 37 40 39 32 1
1	5	3600	3	860	1700***	22 16 13 11 27 10 12 14 15 1
1	6	4500	5	536	1043***	34 17 31 26 9 1

MAXTR= 1774   AVG.TR (FOR MAIN & SPEC. LINKS)= 1265
OVERALL SYSTEM AVERAGE(TR)-MS = 1265

FROM	TO	MILES	COST$	TOT$	TMILES
8	24	7	8	8	7
2	25	227	227	235	235
23	25	256	257	492	492
1	23	1045	1045	1537	1537
2	24	345	346	1883	1883
29	4	92	92	92	92
7	3	9	9	101	101
41	3	68	68	169	170
1	41	1015	1015	1184	1186
7	4	98	98	1282	1284
20	19	203	204	204	203
21	18	213	213	417	417
33	5	89	89	506	506
35	5	48	48	554	554
6	35	65	65	619	619
30	28	90	91	710	710
1	18	472	472	1182	1182
30	6	125	126	1308	1308
33	18	153	153	1461	1461
21	19	413	413	1874	1875
36	38	89	90	90	89

Continued

FROM	TO	MILES	COST$	TOT$	TMILES
37	38	35	35	125	125
40	38	68	68	193	193
39	40	164	164	357	358
32	39	327	328	685	685
1	32	451	451	1136	1137
22	16	314	315	315	314
13	11	244	244	559	558
27	11	98	98	657	657
10	27	129	129	786	787
12	11	188	188	974	975
14	12	223	223	1197	1198
15	14	318	318	1515	1517
1	10	177	177	1692	1694
13	16	271	271	1963	1965
34	17	566	566	566	566
31	26	207	207	773	773
9	1	556	557	1330	1330
1	17	238	238	1568	1568
31	17	242	242	1810	1811

TOTAL AL COST = 9948
TOTAL AL & SL MILES = 9948,   0
TOTAL AL & SL CASTS= 9948,   0
TOTAL NUMBER OF SPECIAL LINKS= 0

Table 9.7 NET Link Analysis for VHD41 and MDNet/EWNet1.1 with Optimization

NO. OF SWITCHES= 1   AT   1
MAX LINK RATE= 5100 BPS

SW#	LK#	LKWT	AQT MS	ATAT MS	E(TR) MS			NODES				
1	1	4500	5	484	941 ***	8	24	2	25	9	1	
1	2	4800	5	521	1102***	29	4	7	3	41	40	1

Continued

Table 9.7 *Continued*

SW#	LK#	LKWT	AQT MS	ATAT MS	E(TR) MS	NODES
1	3	5100	6	908	2296***	20 19 34 21 6 35 5 33 18 17 1
1	4	5100	6	513	1137***	36 38 37 39 30 28 1
1	5	4500	5	1022	2353***	23 22 16 15 14 13 12 11 27 10 1
1	6	2400	2	278	388 ***	32 26 31 1

MAXTR= 2353   AVG.TR (FOR MAIN & SPEC. LINKS)= 1369

OVERALL SYSTEM AVERAGE(TR)-MS = 1369

FROM	TO	MILES	COST$	TOT$	TMILES
8	24	7	8	8	7
2	25	227	227	235	235
9	1	556	557	792	792
9	25	603	603	1395	1395
2	24	345	346	1741	1741
29	4	92	92	92	92
7	3	9	9	101	101
41	3	68	68	169	170
40	41	77	78	247	248
1	40	941	941	1188	1189
7	4	98	98	1286	1288
20	19	203	204	204	203
34	21	226	226	430	430
6	35	65	65	495	495
5	35	48	48	543	543
33	5	89	89	632	632
18	33	153	153	785	785
17	1	238	238	1023	1024
17	33	241	241	1264	1265
18	21	213	213	1477	1478

Continued

FROM	TO	MILES	COST$	TOT$	TMILES
21	19	413	413	1890	1891
36	38	89	90	90	89
37	38	35	35	125	125
39	30	114	114	239	239
8	30	90	91	330	330
1	28	641	641	971	971
39	37	190	190	1161	1161
23	22	106	106	106	106
16	13	271	271	377	378
15	14	318	318	695	696
12	11	188	188	883	884
27	11	98	98	981	983
10	27	129	129	1110	1112
1	10	177	177	1287	1289
12	14	223	223	1510	1513
11	13	244	244	1754	1757
16	22	314	315	2069	2072
32	26	164	164	164	164
31	26	207	207	371	371
1	31	248	249	620	620

TOTAL AL COST = 8767
TOTAL AL & SL MILES= 8767, 0
TOTAL AL & SL COSTS= 8767, 0
TOTAL NUMBER OF SPECIAL LINKS= 0

unused node to the previously chosen node of a netlink. All of the algorithms considered here apply the capacity constraint at each decision point. Whereas the MDNet and EWNet algorithms apply the minimal spanning tree algorithm to the nodes of each netlink, the DLNet does not alter the order of netlink nodes. For further clarification, study the algorithms as defined in Chapter 5.

Table 9.8 DATANET Link Analysis for VHD59 and DLNet with No Optimization

NO. OF SWITCHES= 1   AT   1

MAXIMUM ALLOWABLE LINE RATE= 5100 BPS

SWAP	LKt	LKWT BPS	AQT MS	ATAT MS	E(TR) MS	NODES
1	1	5100	6	1080	2836	*** 52 32 51 31 20 40 53 34 26 54 41 1
1	2	5100	6	593	1373	*** 42 12 10 24 57 21 14 1
1	3	5100	6	433	927	*** 25 17 27 6 22 1
1	4	5100	6	485	1082	*** 44 47 4 19 13 8 1
1	5	5100	6	481	1073	*** 39 46 45 58 30 55 1
1	6	5100	6	778	1986	*** 56 48 23 50 33 16 18 28 59 3 1
1	7	5100	6	690	1708	*** 7 43 29 36 5 38 15 35 2 1
1	8	2700	2	357	538	*** 37 9 49 11 1

MAXTR= 2836   AVG. TR= 1440

OVERALL SYSTEM AVERAGE(TR)-MS = 1440

FROM	TO	MILES	COST$	TOT$	TMILES
52	32	350	350	350	350
32	51	112	112	462	463
51	31	263	263	725	726
31	20	603	603	1328	1329
20	40	556	557	1885	1886
40	53	146	146	2031	2032
53	34	178	178	2209	2210
34	26	46	46	2255	2257
26	54	141	142	2397	2399
54	41	95	96	2493	2495
41	1	455	455	2948	2951
42	12	220	220	220	220
12	10	207	207	427	427
10	24	141	141	568	568

Continued

FROM	TO	MILES	COST$	TOT$	TMILES
24	57	91	91	659	659
57	21	53	53	712	712
21	14	162	162	874	874
14	1	433	433	1307	1308
25	17	30	31	31	30
17	27	223	223	254	254
27	6	86	86	340	340
6	22	132	132	472	472
22	1	499	499	971	972
44	47	81	82	82	81
47	4	89	90	172	171
4	19	35	35	207	207
19	13	329	329	536	536
13	8	84	84	620	621
8	1	193	193	813	814
39	46	63	64	64	63
46	45	150	150	214	213
45	58	78	78	292	292
58	30	170	170	462	462
30	55	52	52	514	515
55	1	246	246	760	761
56	48	136	136	136	136
48	23	82	83	219	219
23	50	100	100	319	319
50	33	70	70	389	389
33	16	72	72	461	461
16	18	48	48	509	509
18	28	103	103	612	613

Continued

Table 9.8 *Continued*

FROM	TO	MILES	COST$	TOT$	TMILES
28	59	287	288	900	901
59	3	80	80	980	982
3	1	163	163	1143	1145
7	43	74	74	74	74
43	29	161	161	235	235
29	36	87	87	322	323
36	5	135	135	457	458
5	38	83	83	540	541
38	15	77	77	617	618
15	35	83	83	700	701
35	2	105	106	806	807
2	1	140	140	946	947
37	9	267	267	267	267
9	49	316	316	583	584
49	11	83	83	666	667
11	1	264	264	930	932

TOTAL AL COST = 9818

TOTAL AL.,SL MILES = 9818   0

TOTAL AL AND SL COSTS= 9818   0

TOTAL NUMBER OF SPECIAL LINKS=   0

*******SYSTEM DESIGN PARAMETERS*******

USER PORT SPEED(BPS)= 9600   HOST PORT SPEED(BPS)= 9600
INS MSG LENGTH(BYTES)= 8   OUT MSG LENGTH(BYTES)= 64
AVG # CUs/VMPTC= 8   AVG MSGS/SEC/VMPTC= 2.78E-03
HOST THINK TIME(SEC)= 1E-03   SHARED MEDIA SPEED(BPS)= 9600
NODAL PKT PROC TIME(MS)= 10   HALF MODEM DELAY(MS)= 4
PACKET SIZE(BYTES)= 64   INF BYTES/PACKET= 56
PROPAGATION CONSTANT(MS/MI)= .01   ANALYSIS TYPE= 3
MAXIMUM ALLOWABLE LINK RATE= 5100   TRAFFIC GROWTH FACTOR= 1

Table 9.9 NET Link Analysis for VHD59 and EWNet1.1 with No Optimization

NO. OF SWITCHES= 1   AT   1

MAX ALLOWED LINK RATE (Wm) = 5100 BPS

SW#	LK#	LKWT BPS	AQT MS	ATAT MS	E(TR) MS	NODES
1	1	5100	6	596	1371	*** 12 42 10 24 28 50 3 1
1	2	3000	3	467	780	*** 15 8 11 49 2 35 1
1	3	5100	6	824	1893	*** 25 51 32 52 31 20 40 1
1	4	5100	6	585	1344	*** 38 37 7 43 29 36 5 1
1	5	5100	6	485	1076	*** 46 39 45 58 30 55 1
1	6	5100	6	590	1357	*** 47 44 4 19 14 33 13 1
1	7	5100	6	768	1886	*** 53 34 26 17 54 6 27 22 41 1
1	8	4800	5	788	1844	*** 57 59 21 18 16 23 56 48 9 1

MAXTR= 1893   AVG. TR= 1443

OVERALL SYSTEM AVERAGE(TR)-MS = 1443

FROM	TO	MILES	COST$	TOT$	TMILES
12	10	207	207	207	207
42	12	220	220	427	427
24	28	104	104	531	531
50	28	106	107	638	638
3	1	163	163	801	802
3	50	253	253	1054	1055
24	10	141	141	1195	1196
15	35	83	83	83	83
8	2	65	65	148	148
11	49	83	83	231	231
1	35	76	76	307	308
11	8	104	104	411	412
2	35	105	106	517	518

Continued

Table 9.9 *Continued*

FROM	TO	MILES	COST$	TOT$	TMILES
25	40	459	460	460	459
51	32	112	112	572	572
52	32	350	350	922	922
31	32	227	227	1149	1150
20	40	556	557	1706	1707
1	40	674	674	2380	2381
31	20	603	603	2983	2985
38	5	83	83	83	83
37	29	194	194	277	277
7	43	74	74	351	352
36	29	87	87	438	440
1	5	140	141	579	580
7	29	139	139	718	720
36	5	135	135	853	855
46	39	63	64	64	63
45	58	78	78	142	142
30	55	52	52	194	194
1	55	246	246	440	441
30	45	123	123	563	564
45	46	150	150	713	714
47	44	81	82	82	81
4	19	35	35	117	117
14	33	154	154	271	271
13	1	226	226	497	497
13	33	280	280	777	777
14	19	326	326	1103	1104
4	47	89	90	1193	1194

Continued

FROM	TO	MILES	COST$	TOT$	TMILES
53	34	178	178	178	178
26	34	46	46	224	224
17	54	178	178	402	402
6	27	86	86	488	488
22	41	84	84	572	572
1	41	455	455	1027	1028
41	54	95	96	1123	1124
6	22	132	132	125S	1256
54	26	141	142	1397	1398
57	21	53	53	53	53
59	9	100	100	153	153
18	16	48	48	201	201
23	48	82	83	284	284
56	48	136	136	420	421
1	9	103	103	523	524
23	16	182	182	705	706
18	57	134	134	839	840
59	16	217	217	1056	1058

TOTAL AL COST = 9907
TOTAL AL & SL MILES = 9907,   0
TOTAL AL AND SL COSTS= 9907,   0
TOTAL NUMBER OF SPECIAL LINKS= 0

*******SYSTEM DESIGN PARAMETERS*******

USER PORT SPEED(BPS)= 9600
INP MSG LENGTH(BYTES)= 8
AVG # CUs/VMPTC= 8
NODAL PKT PROC TIME(MS)= 10
PACKET SIZE(BYTES)= 64
PROPAGATION CONSTANT(MS/MI)= .01
HOST THINK TIME(SEC)= 1E-03

HOST PORT SPEED(BPS)= 9600
OUT MSG LENGTH(BYTES)= 64
AVG # MSGS/SEC/VMPTC= 2.7E-03
HALF MODEM DELAY(MS)= 4
INF BYTES/PACKET= 56
ANALYSIS TYPE= 3
TRAFFIC GROWTH FACTOR= 1

Table 9.10 NET Link Analysis for VHD59 and MDNet with Optimization

NO. OF SWITCHES= 1   AT   1

MAX ALLOWED LINK RATE(Wm)= 5100 BPS

SW#	LK#	LKWT MS	AQT MS	ATAT MS	E(TR)	NODES
1	1	5100	6	791	1880 ***	52 32 51 31 20 40 53 37 1
1	2	5100	6	634	1458 ***	42 12 21 10 57 24 3 1
1	3	5100	6	433	922 ***	25 17 27 6 22 1
1	4	5100	6	485	1076 ***	44 47 4 19 13 8 1
1	5	5100	6	481	1066 ***	39 46 45 58 30 55 1
1	6	5100	6	824	2084 ***	56 14 28 18 16 23 48 50 33 59 1
1	7	4800	5	762	1783 ***	54 26 7 34 41 43 29 36 5 1
1	8	3000	3	571	985 ***	11 49 38 2 9 15 35 1

MAXTR= 2084   AVG TR (FOR MAIN & SPEC LINKS)= 1406

OVERALL SYSTEM AVERAGE(TR)-MS = 1406

FROM	TO	MILES	COST$	TOT$	TMILES
52	32	350	350	350	350
51	32	112	112	462	463
31	32	227	227	689	690
20	40	556	557	1246	1247
53	40	146	146	1392	1393
37	53	230	230	1622	1624
1	37	336	336	1958	1960
20	31	603	603	2561	2563
42	12	220	220	220	220
21	57	53	53	273	273
10	24	141	141	414	414
3	1	163	163	577	577
3	24	401	401	978	979
24	57	91	91	1069	1070

Continued

FROM	TO	MILES	COST$	TOT$	TMILES
10	12	207	207	1276	1277
25	17	30	31	31	30
27	6	86	86	117	116
22	6	132	132	249	249
1	22	499	499	748	748
27	17	223	223	971	972
44	47	81	82	82	81
4	19	35	35	117	117
13	8	84	84	201	201
1	8	193	193	394	395
13	19	329	329	723	724
4	47	89	90	813	814
39	46	63	64	64	63
45	58	78	78	142	142
30	55	52	52	194	194
1	55	246	246	440	441
30	45	123	123	563	564
45	46	150	150	713	714
56	48	136	136	136	136
14	18	65	65	201	201
28	16	99	99	300	300
23	48	82	83	383	383
50	33	70	70	453	453
59	33	146	146	599	599
1	59	154	154	753	753
33	16	72	72	825	826
16	18	48	48	873	874
50	23	100	100	973	974

Continued

Table 9.10 *Continued*

FROM	TO	MILES	COST$	TOT$	TMILES
54	41	95	96	96	95
26	34	46	46	142	142
7	43	74	74	216	217
29	36	87	87	303	304
5	36	135	135	438	439
1	5	140	141	579	580
29	41	112	112	691	693
7	29	139	139	830	832
26	54	141	142	972	974
11	49	83	83	83	83
38	15	77	77	160	160
2	35	105	106	266	266
9	1	103	103	369	369
1	35	76	76	445	445
35	15	83	83	528	528
2	49	108	108	636	637

TOTAL AL COST = 8915
TOTAL AL & SL MILES = 8915,   0
TOTAL AL & SL COSTS= 8915,   0
TOTAL NUMBER OF SPECIAL LINKS= 0

*******SYSTEM DESIGN PARAMETERS*******

USER PORT SPEED(BPS)= 9600
INP MSG LENGTH(BYTES)= 8
AVG # CUs/VMPTC= 8
NODAL PKT PROC TIME(MS)= 10
PACKET SIZE(BYTES)= 64
PROPAGATION CONSTANT(MS/MI)= .01
HOST THINK TIME(SEC)= 1E-03

HOST PORT SPEED(BPS)= 9600
OUT MSG LENGTH(BYTES)= 64
AVG # MSGS/SEC/VMPTC= 2.7E-03
HALF MODEM DELAY(MS)= 4
INF BYTES/PACKET= 56
ANALYSIS TYPE= 3 (SNA/SDLC)
TRAFFIC GROWTH FACTOR= 1

Table 9.11 NET Link Analysis for VHD59 and MDNet with Optimization (Tariffs 9, 10, 11)

NO. OF SWITCHES= 1   AT   1
MAX LINK RATE= 5100 BPS

SWI	LKI	LKWT MS	AQT MS	ATAT MS	E(TR)	NODES
1	1	5100	6	791	1880 ***	52 32 51 31 20 40 53 37 1
1	2	5100	6	634	1458 ***	42 12 21 10 57 24 3 1
1	3	4800	5	366	711 ***	25 17 27 6 1
1	4	5100	6	485	1076 ***	44 47 4 19 13 8 1
1	5	5100	6	481	1066 ***	39 46 45 58 30 55 1
1	6	5100	6	824	2084 ***	56 14 28 18 16 23 48 50 33 59 1
1	7	5100	6	859	2174 ***	54 22 26 7 34 41 43 29 36 5 1
1	8	3000	3	571	985 ***	11 49 38 2 9 15 35 1

MAXTR= 2174   AVG.TR (FOR MAIN & SPEC. LINKS)= 1429

OVERALL SYSTEM AVERAGE(TR)-MS = 1429

FROM	TO	MILES	COST$	TOT$	TMILES
52	32	350	592	592	350
51	32	112	440	1032	463
31	32	227	513	1545	690
20	40	556	706	2251	1247
53	40	146	462	2713	1393
37	53	230	515	3228	1624
1	37	336	583	3811	1960
20	31	603	721	4532	2563
42	12	220	509	509	220
21	57	53	382	891	273
10	24	141	458	1349	414
3	1	163	472	1821	577
3	24	401	625	2446	979

Continued

Table 9.11 *Continued*

FROM	TO	MILES	COST$	TOT$	TMILES
24	57	91	422	2868	1070
10	12	207	500	3368	1277
25	17	30	334	334	30
27	6	86	417	751	116
1	6	621	727	1478	738
27	17	223	511	1989	962
44	47	81	413	413	81
4	19	35	345	758	117
13	8	84	415	1173	221
1	8	193	491	1664	395
13	19	329	578	2242	724
4	47	89	421	2663	814
39	46	63	393	393	63
45	58	78	409	802	142
30	55	52	381	1183	194
1	55	246	526	1709	441
30	45	123	447	2156	564
45	46	150	464	2620	714
56	48	136	455	455	136
14	18	65	395	850	201
28	16	99	431	1281	300
23	48	82	414	1695	383
50	33	70	400	2095	453
59	33	146	461	2556	599
1	59	154	467	3023	753
33	16	72	402	3425	826
16	18	48	374	3799	874

Continued

FROM	TO	MILES	COST$	TOT$	TMILES
50	23	100	432	4231	974
54	41	95	427	427	95
22	41	84	415	842	179
26	34	46	371	1213	226
7	43	74	405	1618	301
29	36	87	418	2036	388
5	36	135	454	2490	523
1	5	140	458	2948	664
29	41	112	440	3388	777
7	22	95	427	3815	872
26	54	141	459	4274	1014
11	49	83	414	414	83
38	15	77	407	821	160
2	35	105	435	1256	266
9	1	103	434	1690	369
1	35	76	407	2097	445
35	15	83	414	2511	528
2	49	108	437	2948	637

TOTAL AL COST = 26625
TOTAL AL & SL MILES = 8959   0
TOTAL AL & SL COSTS= 26625   0
TOTAL NUMBER OF SPECIAL LINES= 0

*******SYSTEM DESIGN PARAMETERS*******

USER PORT SPEED(BPS)= 9600
INP MSG LENGTH(BYTES)= 8
AVG # CUs/VMPTC= 8
NODAL PKT PROC TIME(MS)= 10
PACKET SIZE(BYTES)= 64
PROPAGATION CONSTANT(MS/MI)= .01
HOST THINK TIME(SEC)= 1E-03

HOST PORT SPEED(BPS)= 9600
OUT MSG LENGTH(BYTES)= 64
AVG # MSGS/SEC/VMPTC= 2.7E-03
HALF MODEM DELAY(MS)= 4
INF BYTES/PACKET= 56
ANALYSIS TYPE= 3
TRAFFIC GROWTH FACTOR= 1

One can now increase the traffic growth factor (or TGF) to study the cost of transmitting 1Mbits. The computations were based on the assumptions that there are 5 peak hours per business day and there are 20 business days in a typical month. The results are shown in Figures 9.36 and 9.37 for TGF values of 1, 5, 10, and 20 for both network types and the three values of W_m. The costs per 1Mbits for VHD41 are somewhat higher than those for VHD59 since the nodes are scattered farther away from the switch location. Such unit network cost will be used to study other networks later.

The star topology is also frequently used to synthesize data networks based on AT&T's ISN/Datakit and CCITT's X.25 architectures. In particular the star topology is an ideal solution for interconnecting LANs distributed over a campus or a metropolitan environment. The network topologies obtained by using the STarNet item under the NETWORKING menu with properly modified input files of VHD41, VHD59, VHD100, and VHD200 will appear identical to those for the voice networks as described in Chapter 7. Each netlink will serve exactly two nodes—one at the switch location and the other at one of the remaining locations. If a non switch location node has a traffic load (W_i) greater than $W_m$bps, then exactly $[\text{int}(W_i/W_m)+1]$ special netlinks, each with a load of W_m and the last netlink with a load equal to the Remainder= W_i MOD W_m. The cost of a data network based on the star topology will be generally higher than that based on the multidrop (DL or MD) topology, especially for small values of TGF. Figure 9.38 shows a comparison of costs for the MDNet and STarNet as a function of TGF and for VHD41 input file and two values of W_m= 3900 and 6300bps. The results show that the cost difference disappears as the value of TGF becomes

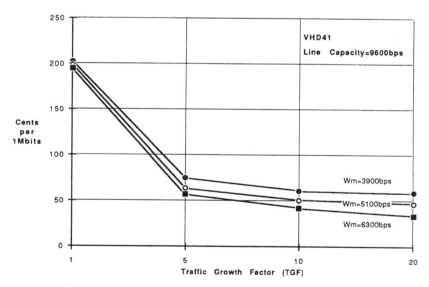

Figure 9.36 Cents per 1 Mbits versus TGF for VHD41, Tariffs 9, 10, and 11, link capacity (C) = 9600bps, and maximum allowable link rates (W_m) = 3900bps, 5100bps, and 6300bps.

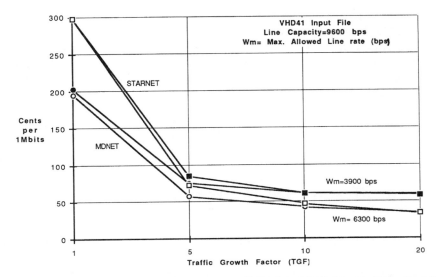

Figure 9.38 Cents per 1 Mbits versus TGF for VHD41, Tariffs 9, 10, and 11, link capacity (C) = 9600bps, and maximum allowable link rates (W_m) = 3900bps, 5100bps, and 6300bps, and MDNet/StarNet topologies.

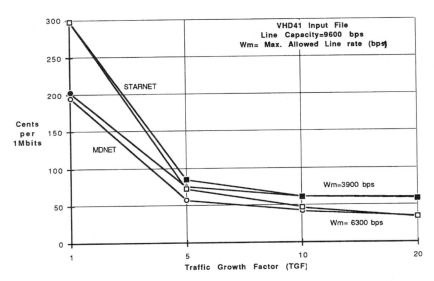

Figure 9.38 Cents per 1 Mbits versus TGF for VHD41, Tariffs 9, 10, and 11, link capacity (C) = 9600bps, and maximum allowable link rates (W_m) = 3900bps, 5100bps, and 6300bps, and MDNet/StarNet topologies.

larger than 5. This is due to the fact that the special links of an MDNet are identical to the netlinks of a STarNet as TGF increases.

The response times can be studied for DLNet, MDNet, and StarNet topologies using the ATP values of 1, 2, and 3, respectively. The ATP values of 1, 2, and 3 represent the ASYNC, Doelz's virtual multipoint and the SNA multipoint protocols, respectively. The results using the EWNet module should be similar to those for the MDNet module. The average bid-to-start (BST) delays or Avg. (T_r) are presented in Figures 9.39 through 9.41 as a function of TGF for the VHD41 input file. The value of BST is identical to the average access time as defined by Schwartz (Schwartz, 1987). Four interesting observations can be made.

1. The value of BST increases as W_m increases. According to the queuing theory as described in Chapter 4, the end-to-end queuing delays will tend to go to infinity as the value of W_m nears the value of circuit capacity, C. The average values of BST as plotted are valid for the actual number of controllers (N) for each synthesized multipoint netlink and $N_{cu} = 4$ for each virtual multipoint circuit on the ESR. The value of Ncu is assumed to equal N if N is less than 4. The Doelz network is less sensitive to heavy traffic load than a conventional MPT network; however, the Doelz network exhibits slightly higher access delays for low traffic loads as expected. See Chapter 4 for an explanation based on the virtual packet switching technique employed for resource sharing

2. The response times are significantly reduced with TGF greater than 5. This can be explained by the fact that the number of nodes (N or Ncu) per netlink are reduced as the nodal traffic loads are increased. This is especially true for the increase in the number of special links, each consisting of only two nodes as is the case with a StarNet.

3. The value of BST for the virtual point-to-point ASYNC operation is always much lower than the BST value for virtual multipoint or conventional multipoint operations. This can be attributed to the elimination of polling, which is employed only to achieve a centralized control through the mainframe or host. The ASYNC operation employs a distributed control. Its main disadvantage is related to the loss of stability at very high traffic loads. Since an ESR-based network achieves traffic flow control through the use of a virtual circuit concept, such instability is prevented. Since an ESR-based network also achieves relatively low queuing delays even at very heavy traffic loads, it becomes an ideal vehicle for ASYNC operation

4. The response times for a StarNet is always less than any other type of a multidrop network and for any given protocol. Only at high values of TGF, the response time of a StarNet becomes equivalent to those other network topologies for any given protocol.

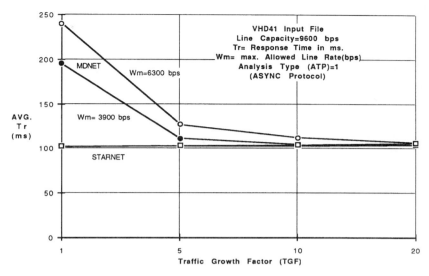

Figure 9.39 Average response time (T$_r$) versus TGF for VHD41, Tariffs 9, 10, and 11, link capacity (C) = 9600bps, maximum allowable link rates (W$_m$) - 3900bps, 5100bps, and 6300bps, and ASYNC traffic.

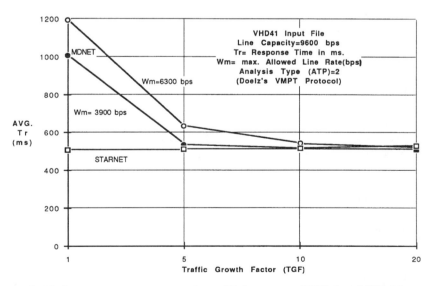

Figure 9.40 Average response time (T$_r$) versus TGF for VHD41, Tariffs 9, 10, and 11, link capacity (C) = 9600bps, maximum allowable link rates (W$_m$) - 3900bps, 5100bps, and 6300bps, and Doelz architecture.

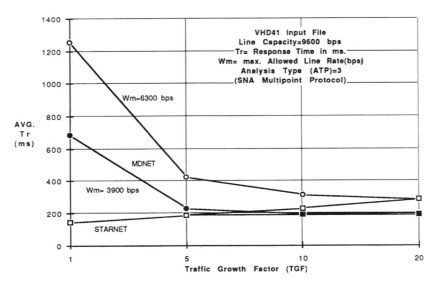

Figure 9.41 Average response time (T$_r$) versus TGF for VHD41, Tariffs 9, 10, and 11, link capacity (C) = 9600bps, maximum allowable link rates (W$_m$) - 3900bps, 5100bps, and 6300bps, and SNA architecture.

In the analysis presented earlier, we maintained a fixed line capacity of 9600bps. It should be quite instructive to study the network costs as a function of line capacities. See Figures 9.42 and 9.43 for the plots of costs versus TGF and three line capacities. The effects of bypass on the 56Kbps and 1.544Mbps facilities are also shown. The plots show that the use of a 9600bps line results in the cheapest network. The results also reveal that the use of a 56Kbps line would have been justified only for TGF > 20. A VHD41 network for TGF=1 would have consisted of only one 56Kbps netlink serving all nodes according to a true minimal spanning tree topology. A network designed for TGF=5 will consist of four or more netlinks. The plots suggest that the use of a 1.544Mbps line capacity will always result in the most expensive network for the range of TGF values considered. The use of a T1 line should result in the cheapest network for TGF values much greater than 40. The lesson of Figures 9.42 and 9.43 should now be quite obvious. One must always study the effect of line capacity on network costs before studying the network topologies in detail. This can be done by simply changing the tariff number for a given link type on the NLT input file. See Chapter 6 for directions.

9.2.3 A Study of Narrowband Data Network Topologies with Multiple Switches

In order to synthesize an optimum network topology, one must vary the number of optimally located switches in the network as shown in the previous chapter

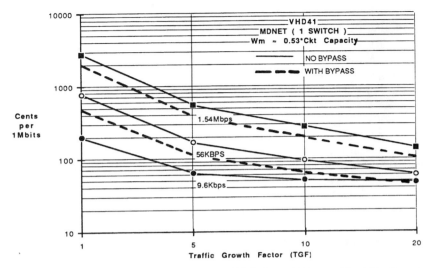

Figure 9.42 Cents per 1Mbits versus TGF for VHD41, Tariffs 9, 10, and 11, maximum allowable link rates (W$_m$) = 0.53 * C, and link capacities (C) = 9600bps, 5100Kbps, and 1.544Mbps.

dealing with voice applications. The COG approach was shown to be quite effective for selecting any desired number of switch locations.

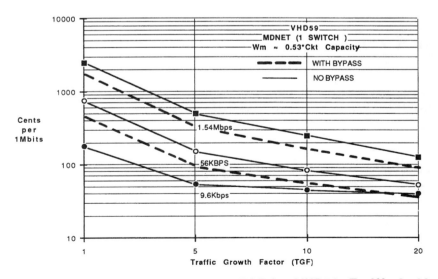

Figure 9.43 Cents per 1Mbits versus TGF for VHD59, Tariffs 9, 10, and 11, maximum allowable link rates (W$_m$) = 0.53 * C, and link capacities (C) = 9600bps, 5100Kbps, and 1.544Mbps.

The design process can be illustrated by selecting the VHD41 data file and increasing the number of switches from one to five in the network employing 9600bps circuits. Table 9.12 illustrates the trunking analysis summarizing the design assumptions, and the total link and trunk costs for two data switches located at nodes 16 and 39. The table shows that 9600bps links were used to implement FDX trunks and that maximum utilization of the trunk was 0.8. The table also shows that a backbone trunking factor (BBTF) of 2 was employed for determining the traffic flows on the backbone network as defined by the Original Trunk Traffic matrix and for TGF=1.

According to the 2x2 matrix, the intranodal traffic flows of 7446bps and 14346bps occur at switch nodes 1 and 2, respectively. The 2x2 matrix also shows that 3354bps flow on both directions of the trunk connecting nodes 1 and 2. Such a symmetric flow results from the nature of the simplified mathematical model as discussed in Chapter 2. Only the use of a From-To traffic file (see

Table 9.12 Trunking Analysis for a 2-Switch Data Communications Network

TRUNK LINK/NODE TYPE O (@9600bps)
FDX FACTOR = 1 : BBNTF=2
AX TRIML;PAD = .8 : TGF=1
TOTAL INTER/INTRA NODAL TRAFFIC= 28499.98
TARIFFS 9,10,11
NO. OF SWITCHES= 2 AT 16 39

ORIGINAL TRUNK TRAFFIC MATRIX AS FOLLOWS

7446.31	3353.68
3353.68	14346.31

MODIFIED TRUNK TRAFFIC MATRIX IS AS FOLLOWS

0	3353
0	0

MODIFIED TRUNK MATRIX IS AS FOLLOWS

0	1
0	0

TRUNK COST MATRIX IS AS FOLLOWS

FROM	TO	MILES	COST	TOTALS
16	39	1493	1006	1006

*****SUMMARY OF SYSTEM COSTS*****

ACCESS LINK COST= 18697
TOTAL TRUNK COST= 1006
TOTAL SYSTEM COST= 19703
CENTS PER 1MBITS= 192.03

Chapter 11 for an example) can provide actual traffic flows between the nodes of a backbone. The 2x2 Modified Trunk Traffic matrix shows only the effective traffic flow on the backbone network for computing the number of trunks needed. The 2x2 Modified Trunk matrix shows that only one trunk is needed to connect the two switch nodes. The final Trunk Cost matrix defines each trunk bundle, its mileage, unit trunk cost, and the total cost the trunk bundle.

The cost summaries of Table 9.12 and other similar tables were used to draw curves as plotted in Figure 9.44 for TGF values of 1, 5, 10, and 20. Although most of the curves are rather flat, it is possible to see that a network topology with 3 switches generally produces a least-cost network for higher values of TGF. Two sample network topologies are shown in Figures 9.45 and 9.46 for the 9600bps network with 2 and 3 optimally located switches. Similar results can be obtained for the VHD59 and other data files.

The results clearly show that unit network costs are affected significantly by increasing the value of TGF and/or selecting the appropriate line capacity (56Kbps versus 9600bps). The optimum choice of switches does not influence the network costs as much in comparison. This observation is similar to that for the voice networks where increasing the TGF and use of multiplexed lines also decreased the unit network costs significantly.

9.2.4 A Study of Data Network Topologies with Satellites

Recent technological developments have made data communications networks based on very small aperture terminals (VSATs) economically quite attractive.

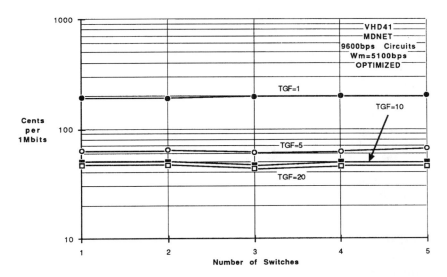

Figure 9.44 Cents per 1 Mbits versus the number of optimally located switches for VHD41, Tariffs 9, 10, and 11, link capacity (C) = 9600bps, and TGF = 1, 5, 10, and 20.

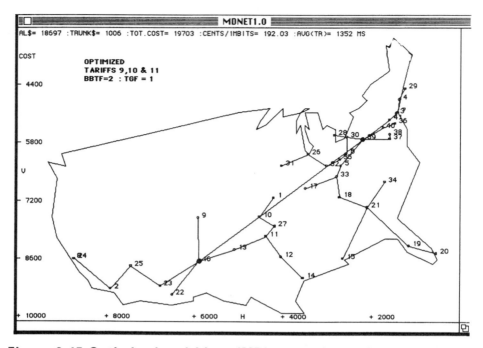

Figure 9.45 Optimized multidrop (MD) network topology for VHD41, Tariffs 9, 10, and 11, two optimally located switches, TGF = 1, BBTF = 2, link capacity (C) = 9600bps, and maximum allowable link rate (W_m) = 5100bps.

High-powered Ku-band satellites coupled with the advancements in the design of a master ground station (MGS) enable the use of a large number of VSATs on customer premises. The use of VSATs makes it possible to economically bypass both the LECs and IECs for low data rate communications applications, especially for the case when a large number of customer locations are widely scattered. Examples of such applications are IBM SNA/SDLC, CCITT's X.25/X.75, and DB management/sharing. Applications requiring very heavy traffic loads from every location will tend to justify a terrestrial network.

The major cost of a VSAT network deals with the hardware. The cost of a VSAT node will vary between $6,000 and $20,000 depending upon the number of ports served and performance requirements dealing with traffic, protocol, and reliability. The cost of an MGS will vary between $1 million and $2 million, depending upon the traffic, protocol, and reliability requirements. The monthly cost of the space segment (or the leased transponder bandwidth) per VSAT ranges from $30 to $150 per month depending upon the bandwidth and nodal intelligence. More intelligence incorporated into the VSAT design implies a lower cost of the space segment. In contrast, almost all of the cost of a terrestrial data communications network is associated with the leased private LEC and IEC lines. The cost of switching hardware on a monthly basis is generally negligible.

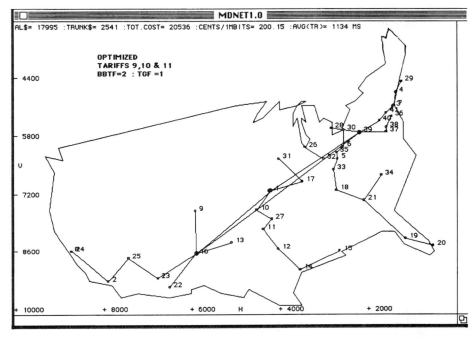

Figure 9.46 Optimized multidrop (MD) network topology for VHD41, Tariffs 9, 10, and 11, three optimally located switches, TGF = 1, BBTF = 2, link capacity (C) = 9600bps, and maximum allowable link rate (W_m) = 5100bps.

To study the applicability of a VSAT network, one can use the plots of Figure 9.47. The four networks consisting of 41, 59, 100, and 200 nodes were considered. Their VHD files have been defined in this and the previous chapters. The costs of the terrestrial networks were computed using the MDNet module with the correct number of switches. The costs of VSAT networks were computed using the following assumptions:

1. Each VSAT node costs $10,000 including the electronics and 1-meter dish

2. Each master ground station costs $1,500,000

3. A cost of $100 per VSAT per space segment

4. A five-year life cycle with 10 percent interest rate

5. A 35 percent corporate tax rate and no income tax credit

According to the return-on-investment (ROI) analysis of Chapter 2, the one-time hardware cost should be divided by a factor of 47.065 to compute the monthly cost due to hardware. The results show that a VSAT network appears very attractive for over 100 locations. Of course, other design considerations such as response times and security/privacy must not be ignored while the selection is made.

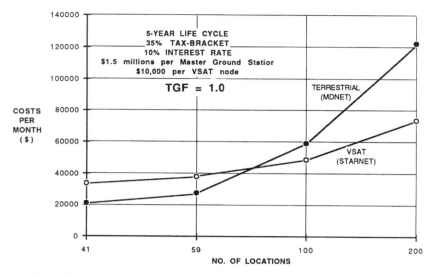

Figure 9.47 Costs per month versus the number of locations for a terrestrial network with MD topology and a satellite network with a single hub.

9.3 A Study of Broadband Data Network Topologies

9.3.1 Introductory Remarks

Previous chapters have focused on synthesizing a large variety of narrowband data network topologies for relatively low-speed facilities. Figures 9.36 through 9.43 have conclusively shown that as traffic grows, and values of W_m remain small and constant, the MDNet topologies become equivalent to star topologies due to the creation of a large number of special links(SLs). The performance of such networks also becomes equivalent to that for data networks based on the pure star topologies. We will now illustrate the design of multilevel StarDataNets and MDNet topologies for a 200-node network. Each node can be considered as a data LAN. It is assumed that 20 percent of the total traffic handled by a LAN is internetworked. We will use a VHD200 input file as simulated in Chapter 7 but its values are now multiplied by 10,000 to represent data traffic intensity in bits per second. In order to study two traffic intensities we will use two values of TGF: 1 and 10. Accordingly, the first node (or CPE) generates internodal BHR traffic intensities of 40,000bps and 400,000bps, respectively, for the two TGF values. Traffic intensities for other nodes never go below 10,000 and 100,000bps. If these values appear too low, one can either increase the TGF value in SDF file or increase the values in the 4ᵗʰ column of the VHD200 input file by a desired multiplier factor using the VHDTMultiplier item of the File menu.

In order to model a broadband data network, one should alter the design parameters UPR and HPR from the existing values of 9600bps to at least 56,000bps for achieving lower latency. In order to model a 3-level data network (AL-level 1, AL-level 2, and trunks), one needs to accomplish the following tasks:

1. Create a LINK200 file with correct values of link types entered at corresponding base nodes and at concentrator nodes. For simplicity, link type equal to 1 represents the lowest AL type for connecting the CPE nodes to the closest concentrating switching node. Link type equaling 2 defines the next level AL type for connecting the lowest level switches to the tandem switches. Therefore, the number "2" appears at all those nodes where concentrating switch nodes are to be located.

2. Create several NLT files with definitions of link type mixes at levels 1 and 2. Link type of 1 can represent a 9600bps or 56,000bps or a T1 line. Or link types 1 and 2 can both represent a 56,000bps line. Or link types 1 and 2 can both represent a T1 line. The designer may use only one NLT file and create many LINK files, each representing a different mix of link types as defined in a fixed NLT file. We chose to use only one LINK file and created two additional NLT files to model several combinations of link type/tariff combinations. Such an approach works fine since an NLT file is only a short file. Since the values of C and W_m are relatively high, one must avoid making wrong entries by missing one or two zeroes. Such mistakes can alter the output results drastically.

3. Set F_{lk} equal to 1 if a LINK file is to be used. Set F_{lk} equal to 0 if no LINK file is used. For that case, only a 2-level data network is synthesized. For that case, the SDF parameters ALT and TKLT determine the AL type and the trunk link type, respectively. For each run, choose the correct TKLT. The parameter ALT defines only the AL type only when F_{lk} is equal to 0. The parameter TKLT must be defined for both values of F_{lk}.

4. The designer can model an FPS or an ATM system by setting the SDF parameter ATP equal to 1 for the ASYNC operation as illustrated in Figure 4.1. We chose the values of BKL and ICPB as 64 and 56, respectively, since these reflect the exact values for several FPS systems considered here and since the values of 53 and 48 for an ATM system are pretty close to our chosen values. The design parameters T_{np} and T_{hm} also influence the response times. For the analysis data as tabulated and illustrated we selected very conservative values of T_{np}=10 ms and T_{hm}=4 ms. We will later compute the response times by using realistic values of T_{np} and T_{hm} and show a drastic reduction in response times.

The response times don't include the time required to set up the switched VC (or SVC). Such a time (generally a fixed value given by Equation 4.26) needs only to be added to the computed average response time (T_r) at the start of the message. To model a Frame Relay system, one must perform two special tasks: (1) set ATP=1, and (2) set BKL and ICPB equal to the average number of bytes per packet and the information bytes per packet, respectively. To model an X.25 PS system, one must perform two special tasks: (1) set ATP=4, and (2) set BKL and ICPB to actual number of bytes per packet and number of information bytes per packet, respectively. To model an SMDS-based system one must first make the values of BKL and ICPB identical to those for ATM system. Since SMDS provides a connectionless service and since each packet must be fully queued at each node (as in X.25), one should set the value of ATP=4. For all these analyses, one must select appropriate values of T_{np} and T_{hm}.

9.3.2 Discussion of Some Output Results

Several StarDataNet configurations were modeled for the 2-level and 3-level broadband data networks using a combination of 56,000bps and T1 circuits. Narrowband circuits were eliminated from analysis due to very high costs. We employed 15 concentrator switch nodes and the number of tandem switches was varied from 1 to 7 for each 3-level data network. The locations of the 15 concentrator switch nodes and the tandem switches were determined by using the FindCOGs item of the Networking menu. The COG results are tabulated in Table 9.13. The monthly costs and the response times are listed in Tables 9.14 and 9.15. The plots of monthly costs versus the number of optimally located switches are shown in Figures 9.48 and 9.49 for TGF=1 and 10, respectively. The legend of these charts (and corresponding tables) show the number of levels (e.g., L2 or L3) followed by the exact number of link types (1 implies a 9600bps circuit, 2

Table 9.13 COGs and Closest Nodes and Their V&H Coordinates

NAME OF VHD FILE=VHD200   NUMBER OF NODES=200

NO. OF DECOMPOSITIONS=7

INITIAL DECOMPOSITION FACTOR=0   (1 FOR HORIZ - 0 FOR VERT)

COGs and Closest Nodes and their V-H COORDs are as follows:

COG-V	COG-H	Node#	Node-V	Node-H
7224	4730	31	7348	4859
7455	6969	14	7331	6802
7010	2643	89	7208	2733
8568	7097	23	8853	6918
6476	6856	198	6541	6837
7892	2509	15	7753	2748
6083	2784	127	6245	2807
8423	8207	41	8553	8152

Continued

COG-V	COG-H	Node#	Node-V	Node-H
8706	6046	72	8751	5813
6539	7851	75	6394	7889
6409	5794	189	6179	5937
8061	3402	55	8113	3389
7713	1559	138	7811	1415
5971	3624	53	6031	3529
6189	1998	1	6375	1919

implies a 56,000bps circuit and a 3 implies a T1 link). The integers in the top horizontal bar of a table specify the number of switches (equal to S), and the last vertical bar also specifies the applicable TGF value for each network configuration. To illustrate, Figures 9.48 and 9.49 clearly emphasize the importance of selecting the correct link types for each level of the network. *Minimum* monthly cost penalties of $48,000 and $168,000 are incurred if optimum configurations are not selected for TGF =1 and 10, respectively. Monthly penalties greater than $1,000,000 are possible when no network planning is employed for TGF =10. The size of monthly penalties is bound to grow when integrated broadband networks are deployed using OC-N facilities.

The response times for corresponding 2- and 3-level broadband networks based on star topologies are plotted in Figures 4.50 and 4.51. The response times were observed to be in the range of 100ms to 300ms, depending upon the link

Table 9.14 Summary of Total Monthly Costs (in K$) for Various Broadband StarData Network Configurations

Config.	1	2	3	4	5	6	7	S
L2,2,2	509	464	456	441	434	449	442	TGF=1
L3,2,2,2	544	527	519	501	493	500	500	TGF=1
L3,1,2,2	761	752	737	725	710	709	710	TGF=1
L3,2,3,2	426	417	408	435	470	523	576	TGF=1
Config.	**1**	**2**	**3**	**4**	**5**	**6**	**7**	**S**
L2,3,3	2082	1850	1676	1625	1565	1600	1634	TGF=10
L3,3,3,3	1566	1536	1508	1523	1539	1569	1603	TGF=10
L3,2,2,3	3000	2900	2835	2790	2713	2698	2690	TGF=10
L3,2,3,3	2045	2015	1988	2003	2019	2050	2082	TGF=10

Table 9.15 Summary of Average Response Times (in ms.) for Various StarData Network Configurations

Config.	1	2	3	4	5	6	7	S
L2,2,2	107	133	128	129	128	128	128	TGF=1
L3,2,2,2	111	140	136	137	135	136	134	TGF=1
L3,1,2,2	273	310	302	303	305	305	306	TGF=1
L3,2,3,2	94	112	112	113	113	113	112	TGF=1
Config.	**1**	**2**	**3**	**4**	**5**	**6**	**7**	**S**
L2,3,3	75	85	87	86	85	84	83	TGF=10
L3,3,3,3	65	85	87	86	85	84	83	TGF=10
L3,2,2,3	136	150	156	154	155	154	153	TGF=10
L3,2,3,3	122	143	145	144	144	143	142	TGF=10

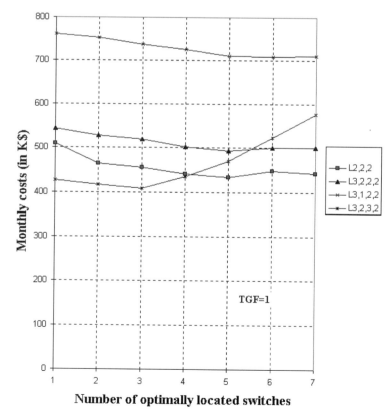

Figure 9.48 Monthly costs versus number of optimally located switches.

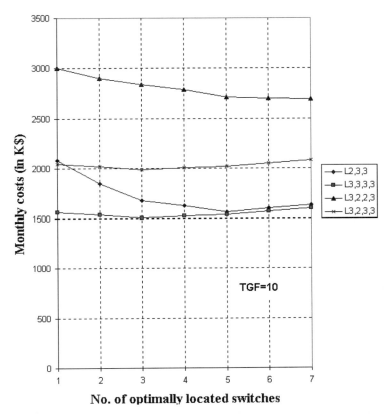

Figure 9.49 Monthly costs versus number of optimally located switches.

speeds at both levels. Reducing the values of modem time (T_{hm}) and nodal processing times (T_{np}) can further improve the systems responsiveness. In order to estimate the overall system response time in a multilevel network, one must first study the response times at each network level. The model assumes an identical subscriber and destination links for each level for computing the turnaround times. Such an assumption requires the subtraction of a major portion of the response time observed at the lowest network level from response times observed at higher levels, and then adding these terms together to estimate the overall system response times (see Table 9.16).

Two StarDataNet topologies, one for a 2-level network and one for a 3-level network, are shown in Figures 9.52 through 9.53 for the case of 5 tandem switches (or S=5) and TGF=10. Although these topologies reflect only a part of Table 9.14 data, these do show the components of the total monthly costs at all network levels. The results also show that 56000bps circuits are optimum for TGF=1 and T1 circuits are optimum for TGF=10. Furthermore, one can estimate the system response time for Figure 9.53 topology as equal to about 98 ms or [62.55+ (85.13-50)].

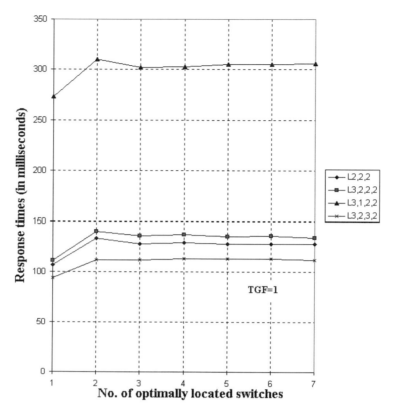

Figure 9.50 Response times versus number of optimally located switches.

Two MDNet topologies, one for a 2-level network and one for a 3-level network, are shown in Figures 9.54 through 9.55 for the case TGF=10. Although these topologies are not part of Table 9.14 data, these clearly show that significant savings can be achieved by using the MD topologies in place of the pure star topologies when the value of W_m is not too small compared to the nodal traffic intensities (W_i). To illustrate, monthly savings of about \$320,000 and \$137,000 can be achieved for 2-level and 3-level data networks respectively by choosing a MD topology. In order to achieve these savings, one needs the FPS/ATM nodes that provide the nodal multiplexing capability as illustrated by Figure 4.10. These savings clearly show the importance of deploying optimum FPS/ATM network topologies. The value of system response time for the 3-level MDNet topology of Figure 9.55 can be estimated as 159 ms or [131+(128-100)]ms. This shows that monthly savings will cost about 61 milliseconds of delay in the average system response times. By using the realistic values of T_{np} =1ms and T_{hm} = 1ms, the response times can be reduced to about 25 ms (for StarData) and 50 ms (for MD). That clearly shows the need for selecting correct values of design parameters.

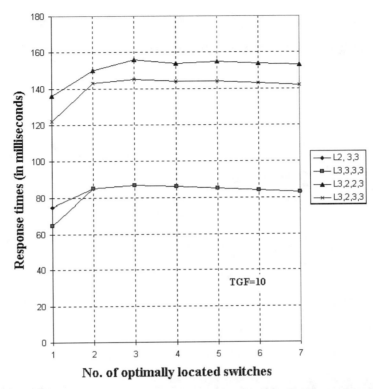

Figure 9.51 Response times versus number of optimally located switches.

Table 9.16 Summary of Average Response Times in Milliseconds versus Data Network Configuration

Config.	1	2	3	4	5	6	7	S
L3,3,3,3	1566	1536	1508	1523	1539	1569	1603	TotalCost
L3,2,2,3	3000	2900	2835	2790	2713	2698	2690	TotalCost
L3,2,3,3	2045	2015	1988	2003	2019	2050	2082	TotalCost
Config.	**1**	**2**	**3**	**4**	**5**	**6**	**7**	**S**
L2,2,2	107	133	128	129	128	128	128	Tr(ms)
L3,2,2,2	111	140	136	137	135	136	134	Tr(ms)
L3,1,2,2	273	310	302	303	305	305	306	Tr(ms)
L3,2,3,2	94	112	112	113	113	113	112	Tr(ms)

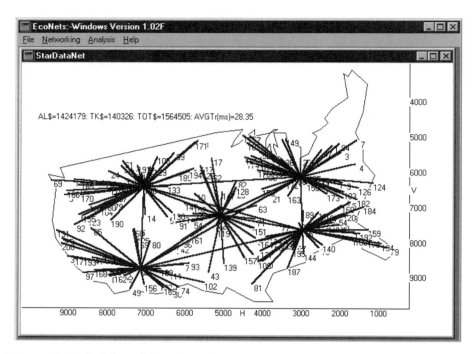

Figure 9.52 A 2-level StarDataNet topology (L2,3,3; S = 5; TGF = 10).

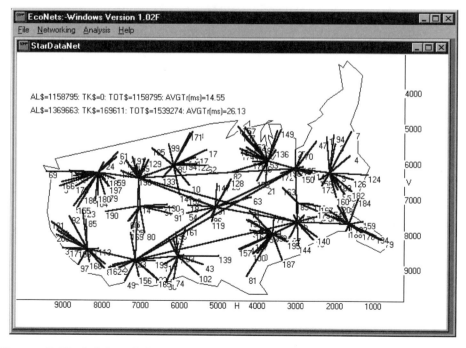

Figure 9.53 A 3-level StarDataNet topology (L3,3,3; S = 5; TGF = 10).

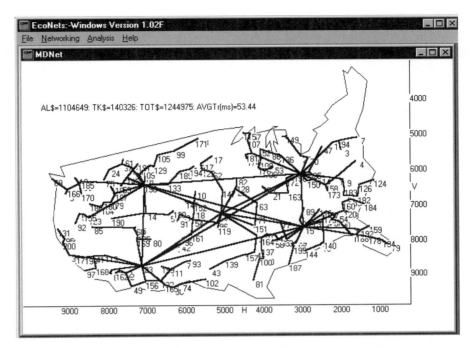

Figure 9.54 A 2-level MDNet topology (L2,3,3; S = 5; TGF = 10).

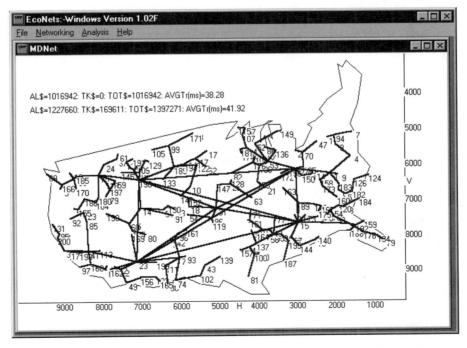

Figure 9.55 A 3-level MDNet topology (L3,3,3; S = 5; TGF = 10).

The tabulated data also show that a 3-level StarData network is cheaper than the 2-level data network for both TGF=1 and 10 for identical mix of facilities. Monthly savings of about $30,000 and $60,000 are possible for the two TGF values respectively. It is also interesting to observe that the optimum number of switches equals 3 for a 3-level data network and equals 5 switches for a 2-level data network. Furthermore, a 3-level MD network is not cheaper than the 2-level counterpart. A study of Figures 9.54 and 9.55 will convince the reader. Some other configurations may prove otherwise. It just shows that common sense does not always prevail. Combinatorics as related to networks will always surprise. Only a user-friendly tool can let one investigate most of the useful "what if" solutions joyfully.

9.4 Summary of Data Network Syntheses and Conclusions

Several algorithms were employed to synthesize capacitated minimal spanning trees (MSTs), directed link, and star topology networks. Two techniques, MDNet and three variations of the EWNet algorithms, were compared for the synthesis of capacitated MSTs. It was shown that the MDNet algorithm was more efficient than the EWNet algorithm for synthesizing cheaper networks. It was also shown that one could employ the exchange algorithm to further reduce the cost of any network based on the MDNet and EWNet technique. However, it was interesting to observe that MDNet-derived networks were usually the cheapest.

Another important observation deals with the fact that network topologies remain intact when only the tariffs are changed (e.g., when the 1$/mile tariff was changed to AT&T's Tariffs 9, 10, and 11 of Table 2.4). A slight change in a given tariff should have practically no effect on the network topology. These facts should allow one to accelerate the network design process significantly by using only simplified models of tariffs. The exact models of tariffs involving large data-bases should only be used for the final pricing effort.

Unit network costs were shown to be closely related to the value of TGF and the capacity of the circuits employed for synthesizing capacitated MSTs. The selection of an optimum number of switches did not reduce the network costs significantly. The use of concentrators can also be studied for networks with heavy traffic loads. In many cases, the unit cost of transporting 1Mbits can be reduced by increasing the number of network layers (or by using concentrators and switches). See Chapter 6 for directions related to the creation of special LINK files that specify the locations of concentrators at any level of the network hierarchy.

Although MDNet topologies were generally employed for synthesizing store-and-forward message switched (S/F MS) and later SNA type networks, their importance can not be ignored when considering systems based on fast packet switching (FPS). The broadband data networks considered in this chapter and the personal communication systems in Chapter 12 can employ the DLNet and MDNet topologies to reduce costs significantly.

Most of the data networks employing modern LANs (e.g., token rings and Ethernets) now employ star topologies, thanks to the invention of the intelligent hub (IH). Data internetworks employing bridges and routers do not employ structured topologies. Such internetworks are generally based on point-to-point connections. EcoNets can be used to design/analyze the resulting topologies through the use of its GivenNet item of the Networking menu. Enterprises have already discovered that significant topological simplifications and cost reductions can be achieved through the use of a private/public collapsed backbone in a data internetwork. One can consult any marketing brochure on public Frame Relay and SMDS networks to see the benefits.

Data communications networks using VSATs were also studied and their costs were compared with those obtained with terrestrial networks and multidrop topologies. It was shown that a VSAT network appears very attractive when the number of locations exceed 100 and when relatively low traffic volumes are involved. Other design considerations such as response times and privacy/security must also be considered while making the selection. A detailed analysis is specially recommended when traffic volumes are high.

We also modeled several 2-level and 3-level broadband data networks using a mix of 56,000bps and T1 facilities for several values of TGF. The analysis clearly shows the need for network planning, especially for broadband data application. In the absence of a copious "what if" type analysis, the enterprise can lose "big bucks. " The risk is bound to grow for broadband networks employing OC-N type facilities. The 3-level networks are cheaper than the 2-level data network. One can achieve additional savings by using the proper hardware and an MD topology for those cases when W_m is not too small when compared to nodal traffic intensities. Star topologies should be selected only when the need for faster response times and/or reliabilities is of highest importance to an enterprise.

There may be a situation where most of the data terminals are clustered in buildings scattered over a campus or a metropolitan area or a single LATA. A data network based on the star topology (resulting from the StarDataNet option under the Networking menu) is recommended when several such campuses or metropolitan areas are interconnected to form a corporate data network. For this case, most of the links are leased from the local exchange carrier for connecting buildings together in a star topology. No OCCs are involved. In order to compute the correct costs, one must define the LATA number for each location through the LATA file. In the analyses discussed earlier, all of the links were assumed to be of the IEC variety used for WANs. The analysis presented here clearly shows that data networks based on the star topology always provide superior response time performance and equivalent cost performance at high traffic loads (as exemplified by large values of TGF and nodal traffic intensities far exceeding W_m). This is quite a revealing fact favoring the star topology especially in view of the recent increases in the cost of leasing multidrop line segments.

Exercises

9.1 Using VHD41 and the 1996 MCI tariffs, synthesize an optimum MDNet.

9.2 Using VHD59 and the 1996 MCI tariffs, synthesize an optimum MDNet.

9.3 Using VHD41 and the 1996 MCI tariffs, synthesize an optimum StarDataNet for TGF=10 and compare its performance with that for an optimum MDNet.

9.4 Using VHD59 and the 1996 MCI tariffs, synthesize an optimum StarDataNet for TGF=10 and compare its performance with that for an optimum MDNet.

9.5 Repeat the network modeling tasks of Section 9.3 for a 100-node network as defined by the VHD100 file of Chapter 6.

9.6 Repeat the network modeling tasks of Section 9.3 for a 200-node network as defined by the VHD200 using the MCI tariffs as defined in Table 2.5.

9.7 Using the VHD17, LATA17,and NAME17 files created using the design data of Table E7.1, and the enclosed DTP8 file, model the optimum data networks for both MDNet and StarDatNet topologies.

Bibliography

Bay Networks. 1995. "The World of Computer Networking: A Primer." Http://www.spark.nstu.su/BayNetworks/Products/Papers/wp/-primer.html

Callon, R., J. Halpern, J. Drake, H. Sandick, J. Jeffords. 1996. "Routing in a Multiprotocol over ATM Environment." *CONNEXIONS*, v. 10, no. 3, March 1996, pp. 34–48.

Chandy, K., and T. Lo. 1973. "The Capacitated Minimum Spanning Tree." *Networks*, vol. 3, no. 2, pp. 173–182.

Esau L., and K. Williams. 1966. "On Teleprocessing System Design." *IBM Systems Journal*, vol. 5, no. 3, pp. 142–147.

Klessig, Bob. 1995. "Integrating ATM Across the Enterprise Data Network." *3TECH, The 3COM Technical Journal*, vol.6, no. 2, April 1995, pp. 4–11.

Nolle, T. 1997. "Switching's Dark Side." *Network World*, February 10, 1997, p. 35.

Roberts, L. 1997. "CIF: Affordable ATM, At Last." *Data Communications*, April 1997, p. 96.

Sharma, R. 1983. "Design of an Economical Multidrop Network Topology with Capacity Constraints, IEEE Trans." on Comm., vol. COM-31, no. 4, pp. 590–591.

Sharma, R. L., P. T. De Sousa, and A. D. Ingle. 1982. *Network Systems*. New York: Van Nostrand Reinhold.

Sharma, R., and M. El-Bardai. 1972. "Sub-Optimal Communications Network Synthesis." Proc. 1970 International Conference on Communications, San Francisco. Also see U.S. Patent No. 3,703,006 Algorithm for Synthesizing Economical Data Communications Network. Issued to R. Sharma on November 14, 1972.

10

Design Process for an Integrated Broadband Network

10.1 Introductory Remarks

Techniques for synthesizing network topologies for either voice or data applications have been described in preceding chapters. It should be stressed that such techniques are effective for designing not only WANs but also LANs. This chapter will discuss several design considerations for implementing integrated broadband networks for both WAN and LAN applications. Before proceeding further, we must define two words:

1. Integrated

2. Broadband

The word *integrated* will represent any network system implementation that handles many dissimilar applications, for example, voice, data, image, and video, while sharing transmission facilities or switching vehicles or both. In the discussions that follow, we will not include network systems that handle an aggregation of similar applications (e.g., 20 different data applications). It is generally much easier to integrate similar applications when compared to the task of integrating diverse applications on the same network.

The word *broadband* implies any network system that does not employ narrowband (with a capacity < 36Kbps) transmission facilities. In fact, one may go as far as stating that all modern broadband systems employ digital facilities with data rates exceeding 56Kbps.

Let us first consider those network systems that allow sharing of only transmission facilities among many applications. It is interesting to note that integration was first achieved through the use of shared T1 facilities during the late '70s. Later, many vendors began offering the so-called 56Kbps multiplexers at very reasonable prices. Such devices provided a capability for multiplexing 8 or 10 voice conversations on a single 56Kbps. These devices could also multiplex streams of regular data and digitally encoded voice traffic. It is interesting to note that whereas such systems employed broadband transmission facilities, these systems employed a mixture of switching nodes. A generalized topology of such a network is shown in Figure 10.1(a). It shows a backbone network consisting of T1/T3 trunk bundles controlled by dynamic bandwidth allocator (DBA) nodes (commonly called the intelligent T1/T3 multiplexing nodes). Such DBA nodes provided the correct channelized bandwidth (BW) as demanded by separate voice, data, and video switches. A DBA allocated these BWs not on a per-call switched basis but only on a prearranged basis. Consequently, a true integration of transmission and switching facilities was never achieved for a mix of diverse applications. Such a fact doesn't discourage many modern enterprises from using such an approach since they can still achieve significant savings per month. Such savings are mainly achieved through economies of scale associated with the broadband facilities needed to handle a large mix of dissimilar traffic types. We will model such a typical environment.

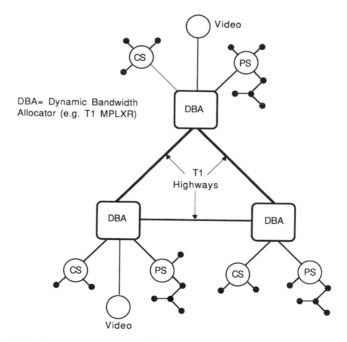

Figure 10.1(a) Topology of an IVDN employing dynamic bandwidth allocators.

The goal of per-call allocation of BW can now be finally achieved through the use of an ATM switch providing switched virtual circuits (SVCs) as opposed to only permanent virtual circuits (PVCs). This represents a new era of true integration of both transmission and switching facilities. Figure 10.1 (b) shows a new network topology involving several hierarchies of ATM switches. Many enterprises had started to use virtual services (e.g., SDN, VNet, or VPN) to handle their voice traffic. Their private networks handled only data traffic generated by their legacy LANs. But their private networks have become too complex due to the inherent dissimilarities between LANs and WANs. The time is now ripe to bring some sanity into their private networks through true integration. This may have many side benefits. The first one is an end to the soaring costs related to network management and control (NMC) resulting from network complexity. The second one is the possibility of bringing voice traffic back to the private network, resulting in additional savings through greater economies of scale.

Chapters 7 and 9 have already described methods to design voice and data networks using broadband facilities and several topologies. However, enterprises have been witnessing a rapid growth in multimedia traffic from every site. This new situation requires new solutions. The purpose of this chapter is to provide methodologies for designing integrated broadband networks using both the older and the new emerging technologies. A new ATM switching/multiplexing node that can handle a mix of applications with BW scalability and little latency can now replace the DBA node.

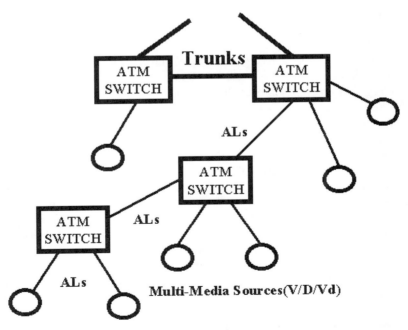

Figure 10.1(b) Topology of an integrated broadband network using ATM switches at access and backbone levels.

10.2 Design Considerations for Integrated Networks

Before we can discuss the actual design process, we must introduce some key design considerations related to switching nodes, available communication facilities, and premise wiring systems (PWSs)—the building blocks of an integrated broadband system.

10.2.1 Switching Nodes

A successful transition to a new enterprise network architecture generally demands the retention of existing CPEs in the form of dumb terminals (e.g., telephones, WSs), voice LANs (e.g., PABXs) and data LANs (e.g., Ethernet, token ring). Each enterprise has invested a great deal of money in such equipment. The new network architecture is supposed to bring about a seamless transition resulting in great recurring savings and simplification in NMC activity.

The most common type of a voice-switching node used in an enterprise is a digital PABX. Each major site is generally served by a single PABX. These PABXs are connected together to create an enterprise voice network via a proper number of PABXs with tandem overlays. For a small enterprise, voice grade lines are leased to create AL and trunk bundles. For a large enterprise, standalone digital tandem switches and T1/T3 multiplexers may be used to synthesize an economical backbone network. T1 multiplexers may be required to synthesize some large AL bundles to reduce the monthly cost. Some enterprises have chosen to employ only the virtual services such as SDN/Vnet/VPN to interconnect their sites. Some other enterprises have outsourced their entire voice internetworking needs. The desire to reduce the recurring NMC costs is generally cited while ignoring the maximum savings obtainable through network design. Although the voice requirements of a typical enterprise are growing at a rate of only 3 to 4 percent annually, they still determine a large portion of the monthly network costs. By aggregating packetized voice traffic (in the form of ATM cells) and employing statistical multiplexing with data traffic, one can derive big savings.

State-of-the-art digital PABXs and tandem switches can be purchased from a large number of vendors. A large variety of features and internetworking capabilities can be selected. The best part is that an enterprise can easily interconnect PABXs from different vendors to create a wide area network. However, each conversation still requires an unshared end-to-end circuit switched path. This is about to change through the use of ATM switches. The PABXs and tandem switches are generally judged by their throughput (number of BHR call attempts and BHR calls handled, and number of BHR erlangs handled) and number of terminations allowed for all types of ALs and trunks.

Most of the enterprises employ data LANs in the form of Ethernets and token rings for interchanging high-speed data between hosts, servers, workstations, and costly peripherals and data terminals. The LAN internetworks provide communications between LANs distributed over metropolitan and wide areas. The internetworks consist of bridges, routers, and leased transmission facilities. Due to the bursty nature of data traffic and their appetite for high bandwidths, costly broadband facilities are employed for both MAN and WAN configurations. Despite the introduction of star-topology-based intelligent hubs to replace the original physical bus/ring topologies, the original CSMA/CD and Token Ring protocols have survived. This has created an unnatural boundary between LANs (as based on connectionless PS and multicasting) and connection-oriented MANs and WANs. This in turn has required the development of new standards to achieve enterprise data networks, but resulting in a very rapid rise in monthly NMC costs. This is especially scary since the data traffic is increasing at the annual rate of about 20 percent.

Whereas some MANs and WANs employ fully owned PS nodes and leased broadband facilities, most of the data MANs and WANs employ the frame relay virtual service leased from a public data network employing FPS nodes based on the ITU standards Q.922. Some also employ virtual data services based on the connectionless Switched Multi-Megabit Data Service (SMDS) standard. A close study of all these existing data networks will show that point-to-point connections are installed on a piecemeal basis without performing a detailed network design. Bridges and routers are being installed at a rapid rate as internetworking devices on a node-by-node basis. The field of data networking is about to go out of control when multimedia traffic hits the enterprises.

The early version of a network node for integrating voice and data applications on shared transmission facilities came in the form of a simple T1 multiplexer that allocated 56Kbps streams (or DS0 interfaces) to either data or digitized voice on a backbone network route consisting of one or more T1 circuits. A fixed network configuration was created or updated manually on a network console unit each time the demand profile changed. An ideal broadband network node can be called a dynamic bandwidth allocator (DBA) that provides not only the required bandwidth on demand but also a fully integrated network management and control capabilities. This includes features like subrate multiplexing of low-bit data channels, multiple DS0 channel multiplexing, multiple T1 channel multiplexing, SONET-based optical facilities, network security, network-wide diagnostics, fault recovery, and automatic alternate path routing. See Figure 10.1(a) for an illustration of a backbone network based on the use of DBAs. As shown, each DBA serves many network nodes like PABXs, X.25 packet switches, data circuit switches (DCSs), video compression units (VCUs), hosts, PCs, consoles, and data terminals. Each of these devices has requirements for varying amounts of bandwidth for wide area networking. Since the use of T1 circuits results in great savings in monthly transmission costs when compared to the costs of individual voice and 56Kbps data networks, DBA has become a very cost-effective vehicle for wide area enterprise networking. The market for intelligent multiplexers (or DBAs) is still going strong.

Enterprise networks based on the presently available DBAs resemble the older ISDN technology. Each ISDN node is a composite of many individual switches and each user location is served by either the primary rate (23B+D) or the basic rate (2B+D) interface. A flat (nonhierarchical) backbone network consisting of T1, T2, and T3 circuits interconnects ISDN nodes. The circuit establishments and network control are provided by the tandem CS on a call-by-call basis. Eventually, the DBAs will evolve to provide such interfaces first on the user side and eventually on the trunk side. The marketplace will demand such an evolution. The high-speed packet switches based on the Frame Relay and ATM (or B-ISDN) technologies are also maturing to act as DBAs to handle multimedia (voice, data, image, and video) applications. In particular, the ATM technology appears to be the most attractive to provide a true integration both at the switching node and transmission facility levels. Although today's ATM technology allows only constant bit rate (CBR) voice encoding, future innovations will allow the implementation of variable bit rate (VBR) capability for packetized voice. Many vendors are already marketing ATM hardware to achieve significantly greater economies of scale achievable through VBR encoding of voice. The use of ATM switches at the LAN, MAN, and WAN hierarchies will also eliminate the artificial complexity created by the differences in the LAN and MAN/WAN protocols. The elimination of existing internetworking complexities will help bring down the NMC related costs to the pre-divestiture levels.

10.2.2 Premise Wiring Systems

The premise wiring systems (PWSs) constitute a major portion of the onetime costs for integrated broadband corporate network systems. The 1984 divestiture of AT&T forced all enterprises to own their own PWSs and switching nodes. Furthermore, each user must also worry about the costs of moves, changes, and equipment reconfigurations each time employees change offices and major company reorganizations occur. The user must also consider the need for an integrated voice, data transport to prepare for the emerging multimedia environments. As a result, each user is now obligated to give much attention to PWS planning for LAN, MAN, and WAN applications. Due the nature of interconnections at each customer premise, LAN considerations far outweigh those for the MAN and WAN. The broadband communication facilities that interconnect LANs are usually leased from a common carrier.

A modern PWS should start with the requirement of a standard information outlet (SIO) that will provide user-configurable standard jacks with six or eight pin configurations for a multivendor environment. In other words, up to two 4-pairs of twisted cables can be terminated in the 2-jack (popularly known as the RJ11 jack) outlet for handling all future multimedia requirements. Most popular voice and data terminals can now be easily connected to the premise LAN's hubs via UTPs as shown in Figure 10.2. Many other vendors sell products to replace moderately expensive cables (e.g., Apple Talk) with the unshielded twisted pairs. Use of expensive cables can thus be eliminated through the use of unshielded twisted pairs (UTPs). A recent paper by Townsend et al. (Townsend, Werner, and Nguyen, 1995) discusses a new modem technology that enables the transmission

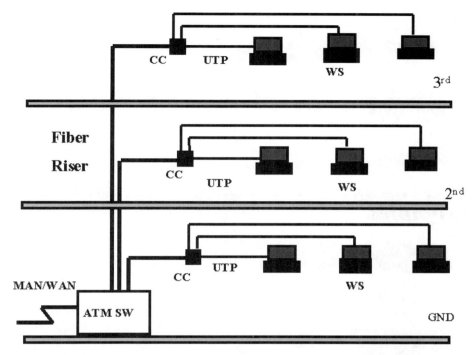

Figure 10.2 A premise wiring system using fiber riser cables and unshielded twisted pair, category 5 (UTP-5) cables to workstations (WSs) using a star topology.

of data rates as high as 51.84Mbps on UTPs over distances of up to 100 meters. This can bring ATM capability right to the desktop. It can be easily shown that a multimedia capability requires a bandwidth of 50 MHz or less if we employ the available data compression techniques. Of course, there are some WSs that create graphics at a rate higher than 100 MHz. For such cases, one will need horizontal fiber cables for connecting such WSs to the LAN hubs.

The other components of a PWS are the copper cross connect (CCC) and/or fiber cross connect (FCC) that reside in a satellite closet and are used to interface the horizontal cables to the riser fiber cables.

The third component deals with the cross connect between cables that are used in the building and those that are used to interconnect a MAN/WAN. See Figure 10.2 for an illustration of a modern PWS for a multifloor building.

Many vendors now offer proprietary wiring systems. Nortel offers the Integrated Building Distribution System (IBDS). AT&T offers the Premise Distribution System (PDS). Both of these systems are similar to the one described previously. Each employs the popular UTP-5 wiring with cable sizes varying from a 100 to several thousand twisted pairs. This technique is usually called the space division multiplexing (SDM). The fiber cable employs wave division multiplexing (WDM) to concentrate several streams of data into one. In a large building, the use of fiber cables can yield significant savings when compared to the copper twisted wire cables. Specially covered Plenum horizontal cables can be used for preventing smoke in case of fire.

Other well-known PWSs are the IBM Cabling System, DEConnect System from Digital Equipment Corporation, and the Universal Information Transport Plan (UITP) from Bell of Pennsylvania. There are still no international standards. One must be careful in selecting the vendor. The user must get at least three bids from the PWS vendors for their integrated PWS. The user should also insist on actual performance tests.

10.3 Integrated Broadband Network Design Examples

The topology of a broadband network can be synthesized through two methods:

1. Superimposing separate AL subnetworks onto a *common* backbone network. In other words, each network can be separately designed using the methodology presented in the previous chapters. Later, the backbone networks of each separate network can be superimposed on top of each other resulting in a so-called T1 or a T3 network. The voice/video ALs will obey the star topology and the data ALs can obey either the star or MD topologies depending on the GOS requirements.

2. Computing the total traffic in bps at each multimedia site (for multimedia environs) and synthesizing a single broadband network based on shared ALs/TKs and a desired topology.

The Networking menu items of the EcoNets package can be employed to synthesize an IVD network topology employing both methods.

Method 1 (Using Topology Superimposition). One will require two VHD files that define the V&H coordinates for sites that generate voice traffic in milliErlangs and data traffic in bps. In case a location has multiple sources of traffic, one should represent it as two locations, one handling only voice and the other handling only data. One will also need a LINK file in case one needs to connect the CPE sites with desired link types. Network topologies and their monthly costs were computed for VHD100 file for a total of 100 sites (about 25 voice-only sites and 75 data-only sites), backbone switches varying from 1 to 8, and for several combinations of design parameters. See Figure 10.3 for an illustration of a 100-node network topology obtained for 4 backbone switches, 7 concentrators, TGF=20, and three types of transmission facility designs. See Figure 10.4 for the plots of monthly costs for three types of IVDNs as a function of the number of backbone switches. The first type employed only the voice-grade type of circuits for ALs and trunks. The second type of IVDN employed T1 circuits for only trunks. The third type of the IVDN employed T1s for trunks and only those ALs where economics justified their use. It should be stressed that all of the IVDNs were designed to handle a traffic growth factor (TGF) of 20 when considering

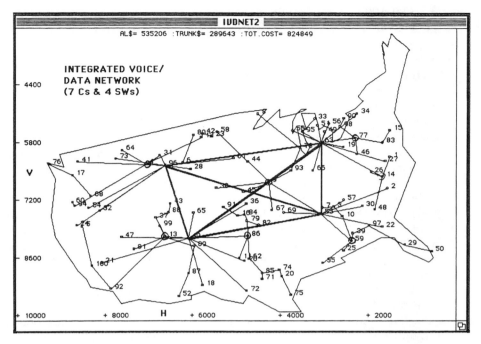

Figure 10.3 A 100-node IVDN topology with four switches and seven data concentrators.

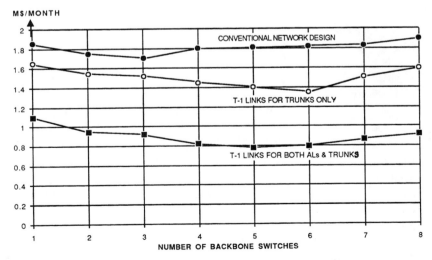

Figure 10.4 Monthly costs of transmission versus the number of optimally located switches for some useful design parameters for the 100-node IVDN.

the VHD100 file. The results show that monthly savings of up to $400,000 and $1,000,000 can be achieved by using the second and third types of networks, respectively. One can now appreciate the reason behind the popularity of T1-based private networks. For much higher traffic loads, T3 facilities could yield even greater savings than those observed for T1 facilities.

Method 2 (Using ATM). The second method is easy to apply and produce additional savings due to the use of higher-speed lines. Such a method will be an ideal solution for cases where all sites produce multimedia traffic. A multimedia environment is also ideal for the deployment of ATM switches. Each type of traffic can be converted into traffic intensity measured in bits per seconds considering the data encryption/compression employed. See Appendix B for the list of well-known data compression standards for speech, audio, images, and video. An ATM switch can then provide true integration on all AL and TK bundles based on the true Dynamic Bandwidth Allocation (DBA) technology. This in turn will yield the greatest economies of scale possible resulting from the use of higher-speed lines. One should work out Exercise 10.1 to learn that such a method actually helps eliminate standalone networks and yields significant savings even for a small network.

We will illustrate Method 2 with the VHD100 file database of Table 7.9 and a TGF of 20. We create a VHD100b file whose 4th column shows traffic intensity (TI) values in milliErlangs for voice or in bps for data. The TI values are obtained by multiplying the corresponding TI values of VHD100 file of Table 7.9 by a factor of 20,000. Next we run a voice network using voice-grade lines to determine the number of voice conversations (equal to N_{vc}) that must be supported by the AL bundle. To integrate the sample voice and data traffic, one must use the following relationship to compute the total number of bits handled by each AL bundle:

$$R_{tbr} = \sum_{i=1}^{N} R_c * R_{vci} + R_{dri} \qquad (10.1)$$

where

N = number of sites (or CPEs)

R_c = bit rate required per conversation using the desired voice *compression* technology

R_{vci} = total number of BHR voice conversations at each ith site

R_{dri} = data rate for regular data application at the ith site

R_{tbr} = total bit rate for both voice and data applications in our example

Equation (10.1) can be extended to other traffic sources like video, high-quality sound, and image. We employed a value of R_c equal to 7000bps (for CBR) and equal to 3500bps (for VBR). The variable bit rate (VBR) capability for ATM assumes a time asynchronous speech interpolation (TASI) technique for eliminating silent periods on either side of the speech path. Since we had employed a random number simulation for generating the original VHD100 file, approximate traffic intensity values for R_{tbr} were derived by multiplying the T1 values of VHD100b by a value of 5.4. The results are shown in VH100c, which is a part of the database (DB) shown in Table 10.1. It also computes a total BHR traffic of 29,160,000. This was pretty close to the expected value of 27,000,000.

The 8 percent increase should partly offset the degradation expected from the ATM overhead (53 bytes versus 48 bytes of information).

Table 10.1 IVDN Database (DB) for Using VBR/ATM Technology

***************NODALDEFINITIONDATA***************

N#	-V-	-H-	LOAD (BPS/MEs)	LATA	LINK	NAME
1	6053	1503	108000	0	0	
2	6807	1551	432000	0	0	
3	7262	2762	216000	0	0	
4	6230	7031	108000	0	0	
5	7731	8494	432000	0	0	
6	6189	6192	108000	0	0	
7	7273	3024	216000	0	0	
8	5026	4496	324000	0	0	
9	6660	4249	432000	0	0	
10	7488	2566	216000	0	0	
11	8509	4905	108000	0	0	
12	5587	5771	432000	0	0	
13	7966	6606	432000	0	0	
14	6525	1653	324000	0	0	
15	5414	1499	216000	0	0	
16	7469	4943	432000	0	0	
17	6498	8705	432000	0	0	
18	9159	5733	432000	0	0	
19	5816	2549	216000	0	0	
20	8940	3945	216000	0	0	
21	8617	8046	108000	0	0	
22	7741	1659	216000	0	0	
23	5554	5528	324000	0	0	
24	7705	8603	216000	0	0	

Continued

Table 10.1 *Continued*

N#	-V-	-H-	LOAD (BPS/MEs)	LATA	LINK	NAME
25	8320	2532	432000	0	0	
26	6418	1915	108000	0	0	
27	6153	1601	216000	0	0	
28	6343	6014	324000	0	0	
29	8173	1147	432000	0	0	
30	7236	2104	324000	0	0	
31	6017	6721	324000	0	0	
32	7345	8087	324000	0	0	
33	5143	3203	108000	0	0	
34	5013	2230	432000	0	0	
35	8618	3012	324000	0	0	
36	7186	4742	108000	0	0	
37	7526	6805	216000	0	0	
38	6801	5433	432000	0	0	
39	7894	2331	324000	0	0	
40	8018	6063	216000	0	0	
41	6167	8590	324000	0	0	
42	5493	5736	216000	0	0	
43	7187	6475	108000	0	0	
44	6161	4713	324000	0	0	
45	6900	4811	432000	0	0	
46	5991	2238	324000	0	0	
47	7987	7595	432000	0	0	
48	7317	1829	432000	0	0	
49	5454	2975	324000	0	0	
50	8337	538	108000	0	0	
51	5306	3144	432000	0	0	

Continued

N#	-V-	-H-	LOAD (BPS/MEs)	LATA	LINK	NAME
52	9420	6266	216000	0	0	
53	5419	3706	324000	0	0	
54	7281	8340	108000	0	0	
55	7421	3060	216000	0	0	
56	5260	2880	108000	0	0	
57	7096	2534	432000	0	0	
58	5462	5410	324000	0	0	
59	8075	2360	324000	0	0	
60	7154	8706	216000	0	0	
61	6085	5037	432000	0	0	
62	8521	4680	108000	0	0	
63	5710	3050	108000	0	0	
64	5900	7580	324000	0	0	
65	7406	5950	324000	0	0	
66	6385	3240	432000	0	0	
67	7340	4173	216000	0	0	
68	7002	8292	324000	0	0	
69	7396	3910	324000	0	0	
70	8574	4784	216000	0	0	
71	9011	4392	324000	0	0	
72	9297	4734	108000	0	0	
73	6095	7726	432000	0	0	
74	8782	4000	108000	0	0	
75	9393	3742	108000	0	0	
76	6226	9256	108000	0	0	
77	5597	2282	324000	0	0	
78	5837	3537	324000	0	0	
79	7597	4733	324000	0	0	

Continued

Table 10.1 *Continued*

N#	-V-	-H-	LOAD (BPS/MEs)	LATA	LINK	NAME
80	5533	5945	432000	0	0	
81	8278	7304	324000	0	0	
82	7695	4471	216000	0	0	
83	5705	1659	432000	0	0	
84	7450	4789	108000	0	0	
85	8866	4387	432000	0	0	
86	7962	4721	108000	0	0	
87	8869	6056	324000	0	0	
88	7396	6516	432000	0	0	
89	8210	5927	324000	0	0	
90	5131	2540	432000	0	0	
91	7267	5263	216000	0	0	
92	9240	7833	432000	0	0	
93	6383	3739	432000	0	0	
94	7237	8655	216000	0	0	
95	5453	3478	216000	0	0	
96	6257	6593	432000	0	0	
97	7701	1956	432000	0	0	
98	5323	2628	324000	0	0	
99	7718	6720	432000	0	0	
100	8692	8268	432000	0	0	

Tot.BHRTraffic = 29160000

**************Node(N)Link(L)Type(T) [NLT] FILE PRINTOUT**************

*****LEGEND****

{ C=Link Cap.: MaxR=Max. Allwd. Rate(Wm): MF=VMpxg.Fact.: FPF=Priv.Fac. Fact.}

LType	LinkC	MaxLinkR	MF	Tariff#	FPF
1	9600	6300	1	1	1.0
2	56000	48000	8	2	1.0

Continued

LType	LinkC	MaxLinkR	MF	Tariff#	FPF
3	1544000	1440000	24	3	1.0
4	45000000	40000000	672	4	1.0

*************** TARIFF DATA PRINTOUT ***************

TARIFF #=1 AVG. LOCAL LOOPS CHARGES ($)=294

MILEAGE BANDS:

50	100	500	1000	10000

FIXED COSTS ($):

72.98	149.28	229.28	324.24	324.24

COST PER MILE ($):

2.84	1.31	0.51	0.32	0.32

TARIFF #=2 AVG. LOCAL LOOPS CHARGES ($)=492

MILEAGE BANDS:

50	100	500	1000	10000

FIXED COSTS ($):

232	435	571	1081	1081

COST PER MILE ($):

7.74	3.68	2.32	1.3	1.3

TARIFF #=3 AVG. LOCAL LOOPS CHARGES ($)=2800

MILEAGE BANDS:

50	100	10000	10000	10000

FIXED COSTS ($):

1770	1808	2008	2500	2500

COST PER MILE ($):

10	9.25	7.25	7.25	7.25

TARIFF #=4 AVG. LOCAL LOOPS CHARGES ($)=8000

MILEAGE BANDS:

10000	10000	10000	10000	10000

FIXED COSTS ($):

16600	16600	16600	16600	16600

COST PER MILE ($):

47	47	47	47	47

Continued

Table 10.1 *Continued*

*************** SYSTEM DESIGN PARAMETERS ****************

ATP/D=1	UPR/D=56000	HPR/D=56000	IML/D=28	RML/D=300
Ncu/D=4	Rmph/D=100	HTT/D=0.001	Fopt/D=0	Tnp/D=1
Thm/D=0	Kpg/D=0.01	BKL/D=64	ICPB/D=56	TGF/C=1
Flk/C=0	Fnn/C=0	Flt/C=0	Fftd/C=0	NA=0
ALT/V/D=3	NA=1	Bal/V/A=0.01	ECC/V=13.33	ECD/V/A=300
DREQ/A=60	PEXD/A=0.15	Clbr/A=23	Frst/A=1	ACDT/A=1
TKLT/V/D=3	NA=1	Btk/V=0.01	Ffdx/D=1	MTKU/D=0.8
BBTF/C=2	Vmin/C=3000	Vmax/C=10000	Hmin/C=0	Hmax/C=10000
Fvc0/C=0	Fvc1/C=0	Fvc2/C=0	Fvc3/C=0	Fvc4/C=0
Fvc5/C=0	Fvc6/C=30	Fvc7/C=0	Fsh/D=0	Fnp/C=1
DPM/A=30	Fdis/C=1	NA=0	TFXC/A=1	NDEC/C=7
DECT/C=1	Legend:	/A=ACD	/C=Common	/D=Data

*************** NAMES OF INPUT FILES ****************

VHD100c * LINK200 * MAPusa * NLT * TARIFF * SDF2 * NAME17 * FTF1 * LATA17 * FILES.TXT * CSABDS * UTBL * WUTBL * MUTBL * RSTBL * DTP8 * Swf4 *

*************** DAILY TRAFFIC PROFILE ****************

Hour Numbers & Corresponding Fractions of Daily Traffic are as follows:

1	0	2	0	3	0	4	0	5	0	6	0
7	0.05	8	0.1	9	0.1	10	0.1	11	0.1	12	0.1
13	0.1	14	0.1	15	0.1	16	0.1	17	0.05	18	0
19	0	20	0	21	0	22	0	23	0	24	0

******* Switch File Definition *******

Number of Switches =4 @ 96, 13, 63, 55.

Networking Menu Item No. Employed = 2

We will also use AT&T's 1991/92 simplified tariffs as defined in Table 2.5. We will first compute the monthly costs for several separate voice and data networks topologies using B_{al} = 0.01, B_{tk} = 0.01, ECC= 13.33 for voice, UPR=56000, HPR= 56000, T_{np} = 1, T_{hm} = 0 for data applications. The results are tabulated in Table 10.2 and plotted in Figure 10.5.

Table 10.2 Total Monthly Costs (K$) versus Application and Network Topology

Application	S=1	S=2	S=3	S=4	S=5	S=7
Voice(56KbpsCBR)	2240	2451	2435	2466	2401	2453
Voice(T1,CBR)	3715	3830	3770	3820	3723	3820
Data(56Kbps)	392	387	376	367	362	367
Data(T1)	1007	850	826	811	814	910
Voice(56Kbps) (VBR/ATM)	1180	1280	1270	1285	1256	1284
IVD(T1) (New VHD100c, VBR/ATM)	1007	876	857	811	814	910

NOTES:

New VHD100c reflects the total bit rate for voice and normal data applications (VHD100b for each).

MF=16 was used to model the voice application for 56Kbps-VBR capability.

The above modeling employed the StarDataNet item of the Networking menu.

S defines the number of optimally located switches

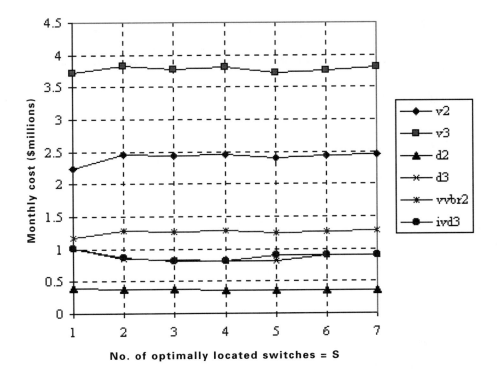

No. of optimally located switches = S

Figure 10.5 Monthly costs versus network topology.

The results show that the optimum standalone data network requires 56Kbps links for both ALs and trunks, and should employ four or five optimally located ATM switches. The optimum standalone voice network topology based on constant bit rate (CBR) ATM technology will also require 56Kbps digital facilities for both ALs and trunks with a multiplexing factor (MF) of 8, and four or five optimally located switches.

One can also model a standalone voice network using the variable bit rate (VBR) ATM technology. It removes all the pauses in a conversation. This technology was originally developed as time asynchronous speech interpolator (TASI) to enhance the utilization of expensive ocean submerged cables. A conservative model of VBR/ATM should require MF to equal 16 (instead of 8 for the constant bit rate capability). A standalone voice network shows a monthly reduction of about $1.17 million for 56Kbps links.

But the real payoff comes when an ATM-based network handles both voice and data in an integrated manner. The aggregation of all bit rates will require higher speed circuits. The newly derived VHD100c file shown as part of Table 10.1 DB is used for designing an integrated broadband network. In our case, the optimum IVD network with a star topology requires four switches and T1 circuits. The topology is illustrated in Figure 10.6. The average response time is about 26 milliseconds. It assumes a turnaround from a host and 1 millisecond of cell processing delay in every ATM switch. These values reflect a very conservative system

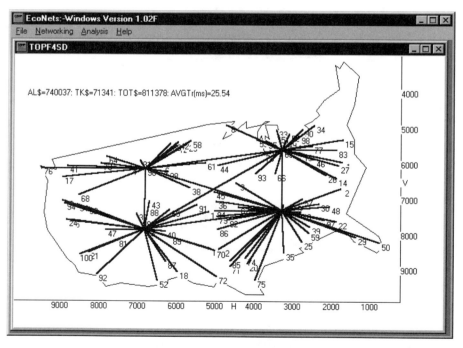

Figure 10.6 An IVDN based on star topology and ATM technology.

implementation. For example, a one-way connection should cut down the response time by about half. By halving the value of nodal delay, one can reduce the response time by additional 3 milliseconds.

By comparing the sum of monthly costs of the best standalone voice and data networks with that of an integrated broadband network, we find a monthly savings of $[(1.285 + 0.367) - 0.8114]$ million or simply \$741,000 per month (or a monthly saving of about \$2.4M if one considers the CBR-based voice solution). These are sizable savings even for the small size of the network and low traffic volume. A network with much higher traffic loads should employ T3 or OC-3 or even OC-N circuits and consequently should result in even bigger savings. Both Method 1 and Method 2 result in significant savings, but Method 2 is ideal for the emerging multimedia environments. The bigger the aggregation of bit rates from all types of information sources, the bigger the savings.

If cost reduction is the main criterion, other topologies must not be ignored. Many vendors like StrataCom and Ascom Timeplex offer hardware solutions for DL/MD topologies. Additional monthly savings of \$158,000 can be realized by employing the MDNet topology of Figure 10.7; however, one has to sacrifice some reliability and responsiveness in the network. For example, the average response time is about 34 milliseconds (which is greater than the average response time of 26 milliseconds observed for the StarDataNet topology). To reduce the response time,

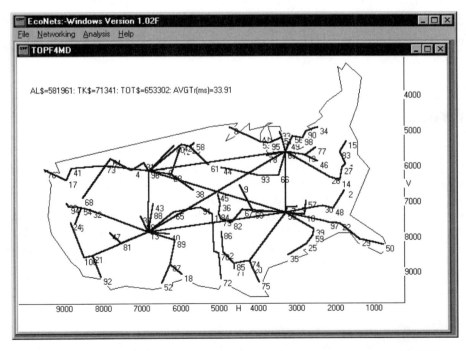

Figure 10.7 An IVDN based on the multidrop topology and ATM technology.

one could make W_m=1,110,000bps to design an MDNet. That will reduce the average response time to about 30 milliseconds but increase the total monthly cost by about $11,000. The star topology is almost an ideal one for most situations when highly reliable and responsive integrated broadband networks are desired.

10.4 Concluding Remarks

A methodology has been presented for designing integrated broadband networks. Selection criteria for switching nodes and premise wiring systems were introduced before describing two methods for designing integrated broadband networks. The first method is ideal where each site generates only one type of traffic. The first method superimposes separate AL subnetworks onto a common backbone network to design a broadband network. The second method is ideal for the case where all sites generate multimedia traffic (converted into bits per second for all applications). It first computes the total bit rate created by a mix of applications before designing a truly integrated network topology using EcoNets. It was shown that these methodologies could yield significant monthly savings. However greater economies of scale can result through ATM-based solutions as traffic increases and the need for higher capacity links becomes paramount. The variable bit rate (VBR) capability of an ATM network is one of the main reasons for such savings and better performance. Since the ATM technology provides seamless and scalable solutions, NMC operations are bound to get simplified. This in turn results in additional sizable savings.

Exercises

10.1 Using the database of Table E7.1 (associated with VHD17 file), design an optimum voice network for B_{al} and B_{tk}=0.1 and ECC=13.33, and an optimum StarDataNet. Tabulate the costs of individual networks. Using the output files, create an IVDNet feasibility table that shows (1) the total bit rate required for voice using the applicable MF value, (2) total bit rate required for data, (3) total bit rate required by both voice and data applications, (4) total available raw capacity in bps, and (5) available excess capacity in bps from each site. Create a new VHD17a file with the total bits per second rate required for both voice and data and model an optimum data network topology using ATM switches. Is it a star or an MD topology? What is the net monthly savings?

10.2 Using the VHD100 file database of Table 7.9, repeat the steps of Exercise 10.1 for voice and data applications using a TGF of 50. Compare the savings with and without the VBR capability.

Bibliography

Townsend, R., J. Werner, and M. Nguyen. 1995. "Using Technology to Bring ATM to the Desktop." *AT&T Technical Journal*, July/August 1995, pp. 25–37.

11

Design Process for a Backbone Network (BBNet)

11.1 Introductory Remarks

A backbone network provides important traffic concentration onto shared trunk facilities, thus providing a significant economy of scale when compared to a network with CPEs or LANs interconnected by point-to-point transmission facilities. As the size of an enterprise WAN increases, the need for an optimized backbone network becomes very urgent. This author knows of many enterprises that still connect their CPEs with hundreds of T1 and T3 facilities on a point-to-point basis. Consequently, these enterprises could achieve significant savings by employing a shared (also called a collapsed) backbone network.

The public broadband networks as administered by BOCs and IECs are generally characterized by large backbones consisting of hundreds of switches. Some backbones have a ring topology spanning a distance of many thousands of miles and dozens of switches. Pure ring topologies provide very high reliabilities and are capable of switching to an alternate path within 50 milliseconds. Meshed ring topologies are somewhat cheaper in transmission costs (savings achieved through sharing of transmission paths) but are slower in self-healing. EcoNets always provides fully connected backbone networks as a first step. Then, using iterations and "divide and conquer" approach, any backbone can be optimized for cost reduction and/or reliability enhancement.

The cost optimization of a backbone network can be carried out using either the available algorithms as described by Kershenbaum (Kershenbaum, 1993) or the iterative approach using EcoNets. Since reliability enhancement is the major force behind BBNet topology optimization effort, use of complex mathematical algorithms is no longer attractive for designing ring or meshed ring topologies.

11.2 The Basic Optimization Process for a Backbone Net

EcoNets always designs a fully connected backbone network for a brand-new network using a methodology as described in Chapter 2. The exact model of a backbone network for an existing network with known CPE to CPE traffic flows can be obtained by using an FTD file. Both of these backbone networks can be optimized further to achieve either additional cost savings or higher reliabilities through an improved topology (e.g., a ring topology for SONET). Generally, cost savings are possible only when excess capacities are available on most trunk bundles. But that is not the case when an economical, fully connected BBNet is designed for a brand-new system. Higher BBNet availabilities can be achieved by employing either the ring or meshed ring topologies. *The correct BBNet optimization procedure must always start with the initial BBNet created by EcoNets.* Otherwise, the original traffic flow information will be lost. It is almost meaningless to start a BBNet optimization process without any knowledge of traffic flows, exact or approximate.

A correct methodology should be to first design an enterprise WAN with AL subnetworks and a common backbone network using EcoNets. If the value of SDF parameter Fvc5 is maintained as "0" during the WAN design process, one will obtain an FTD file named FTF0 that defines all the meaningful traffic flows (with units expressed in either erlangs for voice or bps for data) in the corresponding BBNet. The BBNet can then be optimized iteratively performing the following steps:

1. Create a new VHD file valid for only the backbone switch nodes (with IDs as 1, 2, 3, and 4).

2. Create all the necessary NLT, TARIFF, LATA, and SWF (for S switches employed by BBNet, numbered as 1,2,3,...S) files for modeling the BBNet.

3. Name these input files (VHD4, FTF0, etc.) in the FILES. To read FTF0, SDF parameter F_{ftd} must be set to 1. In order to prevent the creation of unwanted new FTD files, the SDF parameter Fvc5 must also be set to 1.

4. RUN the appropriate application (voice or data application that generated FTF0). This will result in a backbone network with no AL subnetworks.

5. Compare the cost of the BBNet with that obtained with the original run of the complete enterprise network. If it is the same, then proceed. Otherwise, correct the error.

6. Now use the BBNet Optimization item of the Analysis menu. The optimization process is based on removing one link at a time until the cost reduction is no longer achieved or the desired topology is obtained. Cost reduction is generally achieved by removing the link with the lowest traffic, one at a time, and routing this traffic on an alternate route involving two adjacent trunk bundles. This actually increases the total traffic handled by the BBNet. For some applications, it may be quite undesirable. On the other hand, this may reduce the cost since a trunk bundle is completely removed and the addition of traffic may not need an increase of circuits in the trunk bundles of an alternate path. Consequently, one requires a tremendous amount of excess capacity on some trunk bundles to achieve any cost reduction. The optimization process is identical for achieving a desired final topology. One trunk bundle is removed, one at a time, until the correct topology is obtained. Traffic on the removed trunk bundle must always be routed over the trunk bundles of an alternate path. For complex BBNets (with too many switch nodes), *one must always start with a link with nodes that are farthest apart.*

7. When the BBNet Optimization item of the Analysis menu is selected, one will see the familiar spreadsheet, inviting the designer to enter the nodal IDs of two nodes of a link to be removed. The program also invites the designer to define the ID of the en route node (associated with an alternate route). Now click on the Compute button. At this point, another FTF0i is created for each ith step. The designer will now see the modified cost of the BBNet and the previous cost. If the previous cost is correct, then proceed.

8. Continue this process until the cost of the BBNet stops decreasing or the correct topology for high reliability is achieved. One can model (for cost and topology) each intermediate BBNet by using the corresponding FTF0i file. Experience shows that BBNets are generally optimized for additional reliability. A properly designed fully connected BBNet generally does not possess excess bandwidth; therefore, cost reduction is not generally possible on such networks.

11.3 Discussion of Example BBNet Optimizations

One can illustrate the preceding process using a special VHD100c file as created in Chapter 10 for handling an IVD network. See Table 10.1 for the entire database of V&H information, LINK and LATA numbers used for creating an IVD WAN with a 4-switch backbone network. Also shown in Table 10.1 are the NLT data for four link types, and the list of all valid SDF parameters. Table 11.1 shows the original trunking analysis for the example WAN.

Table 11.1 IVDN Trunking Model and Design Data for a Fully Connected Backbone Circuit

Trunk LINK/NODE Type = 3 :   BackBone Trunking Factor =2
MAX ALLOWED Data TK LOAD =0.8:   FDX FACTOR/Data =1
TK Multiplexing FACTOR =0: VoiceTKBlocking =0.01 Fftd=0
Max. Allowed Data Rate(bps)/AL=1440000:   Traffic Growth Factor =1

TOTAL INTER/INTRA NODAL TRAFFIC=29160000

NO. OF SWITCHES=4   AT   13   96   55   63

ORIGINAL TRUNK TRAFFIC MATRIX AS FOLLOWS

4924800	777600	1152000	921600
777600	3499200	864000	691200
1152000	864000	5600000	1024000
921600	691200	1024000	4275200

MODIFIED TRUNK TRAFFIC MATRIX IS AS FOLLOWS

0	777600	1152000	921600
0	0	864000	691200
0	0	0	1024000
0	0	0	0

MODIFIED TRUNK MATRIX IS AS FOLLOWS

0	1	1	1
0	0	1	1
0	0	0	1
0	0	0	0

TRUNK COST MATRIX IS AS FOLLOWS

From	To	Miles	$/TK	TOT.$
13	96	541	8730	8730
13	55	1135	13037	13037
13	63	1332	14465	14465

Continued

From	To	Miles	$/TK	TOT.$
96	55	1177	13341	13341
96	63	1134	13030	13030
55	63	542	8738	8738

*****SUMMARY OF SYSTEM COSTS*****

ACCESS LINK COST=740037
TOTAL TRUNK COST=71341
TOTAL SYSTEM COST=811378
Cents per 1 MBITS=2.57

See Table 11.2 for a database of design data employed for modeling a fully connected 4-switch BBNet with the help of an FTD file. It shows the data of a 4-node VHD4 file and all other useful files used to optimize a BBNet. Since a From-To-File (FTF0) was used to represent the traffic flows in a backbone network, the traffic intensity values as defined by the VHD file are ignored. The FTF0 is also defined in Table 11.2. Such a table was created because the SDF parameter Fvc5 was set to 0 during the WAN design process. Since we must suppress the creation of new FTF0 files during the BBNet optimization process, the SDF design parameter Fvc5 must now be set to 1 as shown in the database of Table 11.2.

Table 11.2 Backbone Network Design Database (DB)

*************** NODAL DEFINITION DATA ***************

N#	-V-	-H-	LOAD (BPS/MEs)	LATA	LINK	NAME
1	7966	6606	1425600	0	0	
2	6257	6597	1166400	0	0	
3	7421	3060	1520000	0	0	
4	5710	3050	1318400	0	0	

Tot. BHR Traffic =   5430400

************* Node(N)Link(L)Type(T) [NLT] FILE PRINTOUT *************

***** LEGEND*****

{ C=Link Cap.: MaxR=Max. Allwd. Rate(Wm): MF=VMpxg.Fact.: FPF=Priv.Fac. Fact.}

LType	LinkC	MaxLinkR	MF	Tariff#	FPF
1	9600	6300	1	1	1
2	56000	48000	8	2	1

Continued

Table 11.2 *Continued*

LType	LinkC	MaxLinkR	MF	Tariff#	FPF
3	1544000	1440000	24	3	1
4	45000000	40000000	672	4	1

*************** TARIFF DATA PRINTOUT ***************

TARIFF #=1 AVG. LOCAL LOOPS CHARGES ($)=294
MILEAGE BANDS:

50	100	500	1000	10000

FIXED COSTS ($):

72.98	149.28	229.28	324.24	324.24

COST PER MILE ($):

2.84	1.31	0.51	0.32	0.32

TARIFF #=2 AVG. LOCAL LOOPS CHARGES ($)=492
MILEAGE BANDS:

50	100	500	1000	10000

FIXED COSTS ($):

232	435	571	1081	1081

COST PER MILE ($):

7.74	3.68	2.32	1.3	1.3

TARIFF #=3 AVG. LOCAL LOOPS CHARGES ($)=2800
MILEAGE BANDS:

50	100	10000	10000	10000

FIXED COSTS ($):

1770	1808	2008	2500	2500

COST PER MILE ($):

10	9.25	7.25	7.25	7.25

TARIFF #=4 AVG. LOCAL LOOPS CHARGES ($)=8000
MILEAGE BANDS:

10000	10000	10000	10000	10000

FIXED COSTS ($):

16600	16600	16600	16600	16600

Continued

COST PER MILE ($):

| 47 | 47 | 47 | 47 | 47 |

*************** SYSTEM DESIGN PARAMETERS ***************

ATP/D=1	UPR/D=56000	HPR/D=56000	IML/D=28	RML/D=300
Ncu/D=4	Rmph/D=100	HTT/D=0.001	Fopt/D=0	Tnp/D=1
Thm/D=0	Kpg/D=0.01	BKL/D=64	ICPB/D=56	TGF/C=1
Flk/C=0	Fnn/C=0	Flt/C=0	Fftd/C=1	NA=0
ALT/V/D=4	NA=0	Bal/V/A=0.01	ECC/V=13.33	ECD/V/A=300
DREQ/A=60	PEXD/A=0.15	Clbr/A=23	Frst/A=1	ACDT/A=2
TKLT/V/D=4	NA=0	Btk/V=0.01	Ffdx/D=1	MTKU/D=0.8
BBTF/C=2	Vmin/C=3000	Vmax/C=10000	Hmin/C=0	Hmax/C=10000
Fvc0/A=0	Fvc1/A=0	Fvc2/A=0	Fvc3/A=0	Fvc4/A=0
Fvc5/A=1	Fvc6/A=30	Fvc7/A=0	Fsh/D=0	Fnp/C=1
DPM/A=30	Fdis/C=1	NA=0	TFXC/A=0	NDEC/C=
DECT/C=1	Legend:	/A=ACD/C=Common		/D=Data

*************** NAMES OF INPUT FILES ***************

VHD4*   LINK200*   MAPusa*   NLT*   TARIFF*   SDF*   NAME17*   FTF0*

LATA17* FILES.TXT* CSABDS* UTBL* WUTBL* MUTBL* RSTBL* DTP8* SWF4*

*************** DAILY TRAFFIC PROFILE ***************

Hour Numbers & Corresponding Fractions of Daily Traffic are as follows:

1	0	2	0	3	0	4	0	5	0	6	0
7	0.05	8	0.1	9	0.1	10	0.1	11	0.1	12	0.1
13	0.1	14	0.1	15	0.1	16	0.1	17	0.05	18	0
19	0	20	0	21	0	22	0	23	0	24	0

******* Switch File Definition *******

Number of Switches =4 @ 1, 2, 3, 4.

****** FROM-TO FILE (FTF) DATA IS AS FOLLOWS ******

From	To	TI
1	2	777600

Continued

Table 11.2 *Continued*

From	To	TI
1	3	1152000
1	4	921600
2	3	864000
2	4	691200
3	4	1024000

Total From-To TI = 5430400

Networking Menu Item No. Employed= 2

See Table 11.3 for the initial Trunking Model/Design. At the beginning, it shows some SDF design parameters implying a trunk link type of 3 (for T1 lines) and a full duplex trunk link. The FDX assumption implies (1) data travels in both directions, and (2) direction with a higher data rate defines the number of T1 trunks to be leased. An HDX assumption would have required the addition of bit rates in both directions before computing the number of T1 trunks. This choice may be useful for tariffs that allow the leasing of both HDX and FDX T1 circuits. The SDF parameter MTKU implies that the DBA node (equivalent to a T1 multiplexer) allows a maximum utilization of a T1 trunk equaling 0.9. In other words, each T1 trunk is constrained to carry a maximum of 1540000*0.9bps only.

Table 11.3 Model of a Fully Connected Backbone Network with Four Switches

Trunk LINK/NODE Type = 3 :   BackBone Trunking Factor =2
MAX ALLOWED Data TK LOAD =0.8:   FDX FACTOR/Data =1
TK Multiplexing FACTOR =0: VoiceTKBlocking =0.01 Fftd=1
Max. Allowed Data Rate(bps)/AL=1440000:   Traffic Growth Factor =1
TOTAL INTER/INTRA NODAL TRAFFIC=5430400

NO.OF SWITCHES=4   AT   1 2 3 4

ORIGINAL TRUNK TRAFFIC MATRIX AS FOLLOWS

0	777600	1152000	921600
0	0	864000	691200
0	0	0	1024000
0	0	0	0

MODIFIED TRUNK TRAFFIC MATRIX IS AS FOLLOWS

0	777600	1152000	921600
0	0	864000	691200

Continued

MODIFIED TRUNK TRAFFIC MATRIX IS AS FOLLOWS

0	0	0	1024000
0	0	0	0

MODIFIED TRUNK MATRIX IS AS FOLLOWS

0	1	1	1
0	0	1	1
0	0	0	1
0	0	0	0

TRUNK COST MATRIX IS AS FOLLOWS

From	To	Miles	$/TK	TOT.$
1	2	541	8730	8730
1	3	1135	13037	13037
1	4	1332	14465	14465
2	3	1178	13349	13349
2	4	1135	13037	13037
3	4	542	8738	8738

*****SUMMARY OF SYSTEM COSTS*****

ACCESS LINK COST=0
TOTAL TRUNK COST=71356
TOTAL SYSTEM COST=71356
Cents per 1 MBITS=1.21

Table 11.3 also shows the Original Trunk Traffic, Modified Trunk Traffic, and the Modified T1-Trunk matrices. Each of these matrices is 4x4 in size. The Original Trunk Traffic matrix represents the peak period traffic in bps flowing in both directions between any pair of switches. The Modified Trunk Traffic matrix represents the effective peak period traffic in bps flowing between the two switches irrespective of direction. These values are obtained after considering the FDX or HDX nature of trunks. The Modified Trunk matrix defines the number of T1 trunks connecting ith node to jth node where both i and j vary from 1 to 4 for this matrix. It also defines each T1 trunk by the nodes it serves, its mileage, the cost of each T1 trunk/bundle, and the total cost of the entire backbone network. It should be emphasized that the total cost of each trunk bundle is obtained by multiplying the unit cost of each trunk by the number of trunks from the Modified Trunk matrix. One should note that the AL costs are reduced to zero since no ALs are used. Figure 11.1 illustrates the topology of the

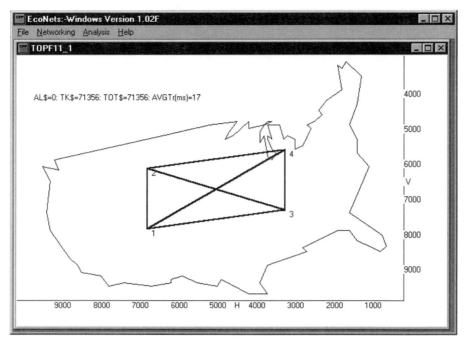

Figure 11.1 A fully connected backbone network with four switch nodes.

fully connected 4-node backbone network along with some pertinent cost summaries. Since the cost of a BBNet agrees with that of Table 11.1, we can proceed to the next step in the optimization process.

Table 11.4 defines the traffic flows between the various nodes for the case of a partially connected backbone network with backbone link (1,4) removed and its associated traffic flowing on the alternate path of (1,2,4). The direct traffic flow between Node 1 and Node 4 is now eliminated. Since there was no excess capacity in the trunk bundles of the alternate path, the cost of the BBNet actually increases. One could have shown a cost reduction if one had set TKLT=4 (which implies the use of T3 trunks). The corresponding BBNet topology is shown in Figure 11.2.

Similarly, the original traffic flow from Node 2 to Node 3 as defined in Table 11.4 is now routed over the tandem route of (2,4,3) and the direct traffic flow between Node 2 and Node 3 is eliminated. Table 11.5 shows the corresponding trunking analysis. Again the cost of the BBNet increased since the T1 links do not possess any excess capacity. Figure 11.3 shows the corresponding ring topology of a 4-node backbone network. The trunk bundles (1, 4) and (2, 3) are now missing.

Table 11.4 A Model of a Partially Connected Backbone Network with Four Switches

Trunk LINK/NODE Type = 3 :   BackBone Trunking Factor =2
MAX ALLOWED Data TK LOAD =0.8:   FDX FACTOR/Data =1
TK Multiplexing FACTOR =0: VoiceTKBlocking =0.01 Fftd=1
Max. Allowed Data Rate(bps)/AL=1440000:   Traffic Growth Factor =1
TOTAL INTER/INTRA NODAL TRAFFIC=6352000

NO. OF SWITCHES=4   AT   1 2 3 4

ORIGINAL TRUNK TRAFFIC MATRIX AS FOLLOWS:

0	1699200	1152000	0
0	0	864000	1612800
0	0	0	1024000
0	0	0	0

MODIFIED TRUNK TRAFFIC MATRIX IS AS FOLLOWS

0	1699200	1152000	0
0	0	864000	1612800
0	0	0	1024000
0	0	0	0

MODIFIED TRUNK MATRIX IS AS FOLLOWS:

0	2	1	0
0	0	1	2
0	0	0	1
0	0	0	0

TRUNK COST MATRIX IS AS FOLLOWS:

From	To	Miles	$/TK	TOT.$
1	2	541	8730	17460
1	3	1135	13037	13037
1	4	1332	14465	0
2	3	1178	13349	13349
2	4	1135	13037	26074
3	4	542	8738	8738

Continued

Table 11.4 *Continued*

*****SUMMARY OF SYSTEM COSTS*****

ACCESS LINK COST=0
TOTAL TRUNK COST=78658
TOTAL SYSTEM COST=78658
Cents per 1 MBITS=1.14

The reader should note the gradual increase in the average response time as the various trunk bundles are removed. The delays inside the backbone net are a significant part of the total end-to-end delays experienced by a user. One should bear in mind that not all calls travel over trunks. Each ATM switch also handles a sizable amount of traffic on an intranodal basis (i.e., call enters on one AL bundle and travels on another AL bundle served by a common ATM switch).

The preceding approach can be repeated by assuming T3 facilities (i.e., TKLT=4). One can now simulate the actual cost reductions in an iterative fashion since T3 trunks have a tremendous amount of excess capacity on all six trunk bundles. Although such a network starts with an unwanted excess capacity and a high price tag, it does show artificial savings after each trunk-bundle removal and/or an increase in reliability through the use of either a ring or a meshed-ring topology.

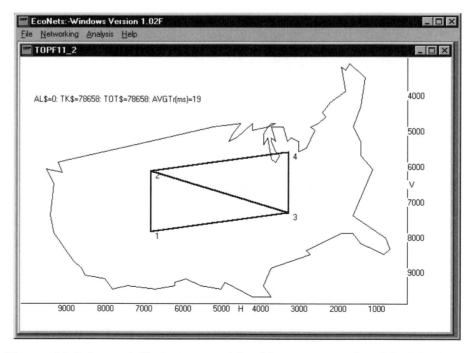

Figure 11.2 A partially connected backbone network with 4 Switch nodes.

Table 11.5 A Model of a Backbone Network with a Ring Topology with Four Switch Nodes

Trunk LINK/NODE Type = 3 :
MAX ALLOWED Data TK LOAD =0.8:
TK Multiplexing FACTOR =0: VoiceTKBlocking =0.01
Max. Allowed Data Rate(bps)/AL=1440000:
TOTAL INTER/INTRA NODAL TRAFFIC=7216000

BackBone Trunking Factor =2
FDX FACTOR/Data =1
Fftd=1
Traffic Growth Factor =1

NO. OF SWITCHES=4   AT   1 2 3 4

ORIGINAL TRUNK TRAFFIC MATRIX AS FOLLOWS:

0	1699200	1152000	0
0	0	0	2476800
0	0	0	1888000
0	0	0	0

MODIFIED TRUNK TRAFFIC MATRIX IS AS FOLLOWS:

0	1699200	1152000	0
0	0	0	2476800
0	0	0	1888000
0	0	0	0

MODIFIED TRUNK MATRIX IS AS FOLLOWS:

0	2	1	0
0	0	0	3
0	0	0	2
0	0	0	0

TRUNK COST MATRIX IS AS FOLLOWS:

From	To	Miles	$/TK	TOT.$
1	2	541	8730	17460
1	3	1135	13037	13037
1	4	1332	14465	0
2	3	1178	13349	0
2	4	1135	13037	39111
3	4	542	8738	17476

Continued

Table 11.5 *Continued*

*****SUMMARY OF SYSTEM COSTS*****
ACCESS LINK COST=0
TOTAL TRUNK COST=87084
TOTAL SYSTEM COST=87084
Cents per 1 MBITS=1.11

11.4 Concluding Remarks

A basic design process for an optimum BBNet has been described. Such a process can be employed to synthesize either a cost-effective BBNet or a BBNet with a desired topology (e.g., ring topology for SONET). The process was illustrated using a 4-switch BBNet for both cases. The starting BBNet was a subnetwork of the IVD network as developed in Chapter 10.

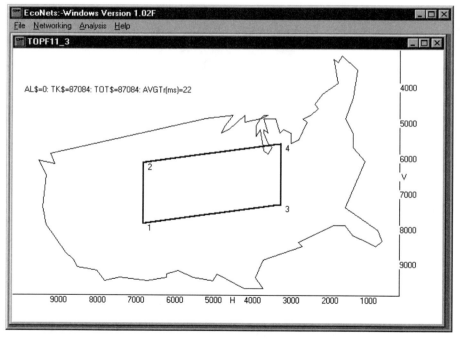

Figure 11.3 A ring topology with four switch nodes.

Exercises

11.1 Considering the data of Table 10.1 (and VHD100c), design a fully connected backbone network consisting of five switches. Design a ring consisting of five switches. Provide all the trunking design tables and the corresponding BBNet topology plots.

11.2 Considering the design data of Table 7.10 (and VHD200), create an optimum voice network for a value of TGF equal to 20. Using its trunking design and the resulting FTF0, model all the BBNets until a ring topology is obtained. Provide all the trunking designs and plot all the BBNet topologies.

Bibliography

Kershenbaum, A. 1993. *Telecommunications Network Design Algorithms*. New York: McGraw-Hill.

12

Modeling Process of a Given Network Topology

12.1 Introductory Remarks

The designer should always model the topology (which node is connected to what other nodes and by what link types) of an existing network before proceeding to design a cost-effective network. Such a task helps the designer quantify the improvements. Unfortunately, this task is never accomplished due to the lack of a user-friendly tool based on an iterative methodology. It is the purpose of this chapter to show how a given network topology can be easily modeled using EcoNets. Using the FindRoutes item of the Analysis menu, one can also create a Routing Table for any given network topology.

12.2 The Basic Modeling Process for a Given Network

The need to model a given network will require the creation of the following input files:

1. Create a valid VHD file for all the nodes of a given network topology. One may have trouble getting the V&H coordinates for nodes in a foreign country. For that case, it may be proper to apply the simple method as discussed in Chapter 6. One could get a map of the area from a good

atlas and then allocate V&H coordinates by overlaying a transparent grid. Be careful to change the design parameters Vmin, Vmax, Hmin, and Hmax, and Fdis of the SDF file.

2. Create all the necessary MAP, NLT, TARIFF, NAME, and LATA files. The most important file for modeling a given network topology is a specially designed FTD (or FTF) file. Each vector's first column represents the From-Node ID, the second element represents the To-Node ID, and the third element defines the number of circuits connecting the two nodes. The type of the link emanating from the From-Node is defined by the LINK file. If a LINK file is not read (when Flk=0), then the SDF design parameter ALT defines the link type for all connecting links.

3. Name these input files in the FILES. The specialized ACD-related files are ignored. The DTP and SWF files are also ignored.

4. RUN the Networking menu item GivenNet to model the given network as specified by the input files of FILES. The SDF parameter F_{ftd} is always assumed to equal 1 for this networking item. However, by making F_{ftd} equal to 1, one can force the printing of the FTF file as part of the database. This also forces EcoNets to recompute the new values of TI to be entered in the database. These recomputed values of TI may appear meaningless for this task.

5. The cost of the network is summarized in the network map. A detailed cost summary is in the output file named ALFGN. The network topology is saved in a file called TOPF.TXT. Both of these output files must be saved under a different name to prevent overwrites.

6. While the network map is on the screen, one can compute all the desirable routes between any two specified nodes of the Given Network by invoking the Analysis menu item called FindRoutes. Once this Analysis item is selected, the user is presented with a spreadsheet interface that invites the designer to enter the nodal IDs of the two nodes. Once that is done, the designer should then click on the Compute button. The list of desirable routes will be saved in an output file name RTBLi where i represent the ith click on the Compute button for each nodal pair IDs. If this process is repeated for all useful nodal pairs, one can create a routing table for the design of a switching node.

12.3 Discussion of Outputs for an Example Given Network

One can illustrate the preceding process using an example 41-node network as defined by the VHD41 file and the FTD file named as FTF1. The entire design data is shown in Table 12.1. Several observations are in order. The 4th column of a VHD41 file has been recomputed because F_{ftd} was equal to 1. If

F_{ftd} had been equal to 0, the normal TI values of VHD41 would have been listed. We are now using the actual values of the LATA file (as opposed to the null LATA values used in example networks of Chapters 10 and 11). We are now able to use the correct names of each site. There are no surprises in the NLT and Tariff files. The SDF parameters F_{lk}, F_{nn}, F_{lt}, and F_{ftd} should be carefully set to correct values. The SDF parameters ALT and TKLT are particularly important when F_{lk} is set to 0. The other SDF parameters V_{min}, V_{max}, H_{min}, H_{max}, and F_{dis} are important if the topologies span other countries or states. The only other parameter of interest is F_{np}. Its value is generally 1, implying that nodal IDs are printed next to each nodal point. The FTF1 file is printed as part of the database since F_{ftd} was set to 1. The 3rd column represents the correct number of circuits connecting the From-node and the To-node.

We will illustrate the modeling process by ignoring the LINK file and using the SDF parameter ALT equal to 2 (implying all 56Kbps circuits). The network topology is illustrated in Figure 12.1 with a cost summary. This cost represents the monthly cost for this network with 56Kbps lines only. The computed design data is shown in Table 12.2. The names of the most useful input files as named in FILES are listed. The values of ALT, the related Tariff number, TGF, and F_{lk} are listed before the access line connection data is presented.

Table 12.1 Design Database (DB) of a Given Network

******************* NODAL DEFINITION DATA *******************

N#	-V-	-H-	LOAD (BPS/MEs)	LATA	LINK	NAME
1	7027	4203	6	524	0	KANMO
2	9213	7878	4	730	0	LAXCA
3	5015	1430	0	224	0	NWKNJ
4	4687	1373	17	920	0	HFDCT
5	6263	2679	2	922	0	CINOH
6	5972	2555	4	324	0	COLOH
7	4997	1406	0	132	0	NYCNY
8	8492	8719	2	722	0	SFOCA
9	7501	5899	6	656	0	DENCO
10	7489	4520	3	532	0	WICKA
11	7947	4373	3	536	0	OKLOK
12	8436	4034	0	552	0	DALTX
13	8266	5076	0	546	0	AMLTX

Continued

Table 12.1 *Continued*

N#	-V-	-H-	LOAD (BPS/MEs)	LATA	LINK	NAME
14	8938	3536	12	560	0	HOUTX
15	8483	2638	4	490	0	NOLLA
16	8549	5887	9	664	0	ALBNM
17	6807	3482	4	520	0	STLMO
18	7010	2710	7	470	0	NASTN
19	8173	1147	6	952	0	TMPFL
20	8351	527	5	460	0	MIAFL
21	7260	2083	3	438	0	ATLGA
22	9345	6485	4	668	0	TCSAR
23	9135	6748	0	666	0	PHXAR
24	8486	8695	4	722	0	OAKCA
25	8665	7411	5	721	0	LASNE
26	5986	3426	10	358	0	CHIIL
27	7707	4173	4	538	0	TULOK
28	5536	2828	0	340	0	DETMI
29	4422	1249	2	128	0	BOSMA
30	5574	2543	2	320	0	CLEOH
31	6261	4021	2	635	0	CDRIA
32	6272	2992	6	336	0	INDIN
33	6529	2772	3	462	0	LOUKY
34	6657	1698	10	422	0	CHTNC
35	6113	2705	4	328	0	DATOH
36	5251	1458	0	228	0	PHLPA
37	5622	1583	0	236	0	WASDC
38	5510	1575	0	238	0	BALMA
39	5621	2185	2	234	0	PITPA

N#	-V-	-H-	LOAD (BPS/MEs)	LATA	LINK	NAME
40	5363	1733	0	226	0	HARPA
41	5166	1585	1	228	0	ALNPA

Total BHR Traffic = 156

**************Node(N)Link(L)Type(T) [NLT] FILE PRINTOUT **************
***** LEGEND *****

{ C=Link Cap.: MaxR=Max. Allwd. Rate(Wm): MF=VMpxg.Fact.: FPF=Priv.Fac. Fact.}

Ltype	LinkC	MaxLinkR	MF	Tariff#	FPF
1	9600	6300	1	1	1.0
2	56000	48000	8	2	1.0
3	1544000	1440000	24	3	1.0
4	45000000	40000000	672	4	1.0

************** TARIFF DATA PRINTOUT **************

TARIFF #=1 AVG. LOCAL LOOPS CHARGES ($)=294
MILEAGE BANDS:

50	100	500	1000	10000

FIXED COSTS ($):

72.98	149.28	229.28	324.24	324.24

COST PER MILE ($):

2.84	1.31	0.51	0.32	0.32

TARIFF #=2 AVG. LOCAL LOOPS CHARGES ($)=492
MILEAGE BANDS:

50	100	500	1000	10000

FIXED COSTS ($):

232	435	571	1081	1081

COST PER MILE ($):

7.74	3.68	2.32	1.3	1.3

TARIFF #=3 AVG. LOCAL LOOPS CHARGES ($)=2800
MILEAGE BANDS:

50	100	10000	10000	10000

FIXED   COSTS   ($):

1770	1808	2008	2500	2500

Continued

Table 12.1 *Continued*

COST PER MILE ($):

| 10 | 9.25 | 7.25 | 7.25 | 7.25 |

TARIFF #=4   AVG. LOCAL LOOPS CHARGES ($)=8000
MILEAGE BANDS:

| 10000 | 10000 | 10000 | 10000 | 10000 |

FIXED COSTS ($):

| 16600 | 16600 | 16600 | 16600 | 16600 |

COST PER MILE ($):

| 47 | 47 | 47 | 47 | 47 |

*************** SYSTEM DESIGN PARAMETERS ***************

=0

ATP/D=3	UPR/D=9600	HPR/D=9600	IML/D=28	RML/D=300
Ncu/D=4	Rmph/D=100	HTT/D=0.001	Fopt/D=0	Tnp/D=10
Thm/D=4	Kpg/D=0.01	BKL/D=64	ICPB/D=56	TGF/C=1
Flk/C=0	Fnn/C=1	Flt/C=1	Fftd/C=1	NA=0
ALT/V/D=2	NA=0	Bal/V/A=0.1	ECC/V=13.33	ECD/V/A=300
DREQ/A=60	PEXD/A=0.15	Clbr/A=23	Frst/A=1	ACDT/A=2
TKLT/V/D=2	NA=0	Btk/V=0.1	Ffdx/D=1	MTKU/D=0.8
BBTF/C=2	Vmin/C=3000	Vmax/C=10000	Hmin/C=0	Hmax/C=10000
Fvc0/C=0	Fvc1/C=0	Fvc2/C=0	Fvc3/C=0	Fvc4/C=0
Fvc5/C=1	Fvc6/C=30	Fvc7/C=0	Fsh/D=0	Fnp/C=1
DPM/A=30	Fdis/C=1	NA=0	TFXC/A=1	NDEC/C=3
DECT/C=1	Legend:	/A=ACD	/C=Common	/D=Data

*************** NAMES OF INPUT FILES ***************

VHD41 * LINK41 * MAPusa * NLT * TARIFF * SDF * NAME41 * FTF1*
LATA41*FILES.TXT*CSABDS*UTBL*WUTBL*MUTBL*RSTBL*DTP8*Swf4*

*************** DAILY TRAFFIC PROFILE ***************

Hour Numbers & Corresponding Fractions of Daily Traffic are as follows:

| 1 | 0 | 2 | 0 | 3 | 0 | 4 | 0 | 5 | 0 | 6 | 0 |
| 7 | 0.05 | 8 | 0.1 | 9 | 0.1 | 10 | 0.1 | 11 | 0.1 | 12 | 0.1 |

Continued

Table 12.1 *Continued*

13	0.1	14	0.1	15	0.1	16	0.1	17	0.05	18	0
19	0	20	0	21	0	22	0	23	0	24	0

******* Switch File Definition *******

Number of Switches =4 @ 96, 13, 63, 55.

******FROM-TO FILE (FTF) DATA IS AS FOLLOWS******

From	To	TIor#Ckts
2	25	2
2	16	3
2	9	2
2	22	2
16	14	6
16	1	8
16	21	3
14	15	5
14	4	10
14	39	5
25	33	2
25	11	3
22	34	4
22	10	3
9	26	5
9	18	6
6	4	4
6	30	4
34	4	7
34	20	7
33	4	4
26	4	4
1	26	4

Continued

Table 12.1 *Continued*

From	To	TIor#Ckts
1	21	1
21	34	1
10	18	4
18	34	4
15	19	4
19	20	4
26	32	4
32	35	4
35	5	4
19	21	4
11	27	4
27	17	4
17	32	4
8	24	4
24	25	4
31	26	5
29	4	4
41	4	2

Tot. TI or Termns. =168
Networking Menu Item No. Employed = 8

Each link's attributes (From-Node ID, To-Node ID, number of ALs connecting the nodal pair, computed length of the AL circuit in miles, cost of each AL, and the total cost of the AL-Bundle) are listed for each vector of FTF0 file. Finally, the total number of AL terminations and the total cost of ALs are also listed.

While the network topology of the given network is still on the computer screen, one can now select the FindRoutes item of the Analysis Menu. A spreadsheet appears, inviting the designer to enter the IDs of two nodes and then click on the Compute button to record the available routes between the two nodes. For each ith click on the Compute button, the results are recorded in an output file RTBLi. Three sets of nodal pairs were selected and three output files were created. These were merged together to create a single route (see Table 12.3). EcoNets discovered six routes between nodes 2 and 4, two routes between nodes 2 and 20, and only one route between nodes 2 and 15. It is interesting to note

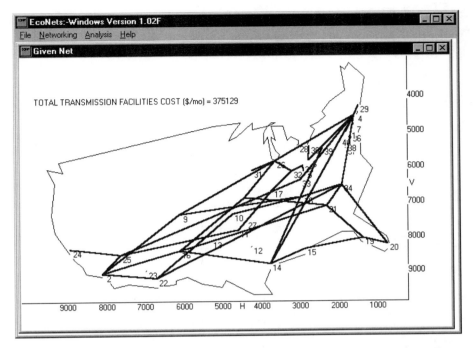

Figure 12.1 Topology of a 41-node given network using 56Kbps lines.

Table 12.2 Design Data of a 41-Node Given Network

********** APPLICABLE DESIGN PARAMETERS **********

FTD File = FTF1:   VHD File = VHD41:   LINK File = LINK41
ALT / Flk = 2 / 0:   Tariff No. = 2:   TGF = 1

********** ACCESS LINE CONNECTION DATA **********

From	To	ALs	MIs	$/AL	$/ALB
2	25	2	227	1591	3183
2	16	3	663	2436	7308
2	9	2	827	2649	5298
2	22	2	442	2090	4181
16	14	6	753	2553	15319
16	1	8	717	2506	20051
16	21	3	1270	3225	9675
14	15	5	318	1803	9015
14	4	10	1508	3534	35346

Continued

Table 12.2 *Continued*

From	To	ALs	MIs	$/AL	$/ALB
14	39	5	1132	3045	15229
25	33	2	1615	3673	7347
25	11	3	987	2857	8572
22	34	4	1736	3831	15324
22	10	3	854	2684	8053
9	26	5	917	2766	13831
9	18	6	1020	2900	17401
6	4	4	552	2291	9167
6	30	4	125	1355	5421
34	4	7	631	2394	16762
34	20	7	651	2420	16944
33	4	4	731	2524	10098
26	4	4	768	2572	10290
1	26	4	410	2016	8065
1	21	1	674	2450	2450
21	34	1	226	1589	1589
10	18	4	592	2343	9375
18	34	4	338	1849	7397
15	19	4	481	2181	8724
19	20	4	203	1536	6145
26	32	4	164	1445	5783
32	35	4	103	1304	5217
35	5	4	48	1103	4412
19	21	4	413	2023	8093
11	27	4	98	1291	5165
27	17	4	358	1895	7583
17	32	4	229	1596	6386
8	24	4	7	293	1175

Continued

From	To	ALs	MIs	$/AL	$/ALB
24	25	4	409	2014	8056
31	26	5	207	1545	7727
29	4	4	92	1269	5076
41	4	2	165	1448	2896
# Terminations = 168:		Tot. AL$ =375129			

that the listed routes have noncontiguous path numbers. EcoNets throws away some paths due to either repetitive occurrences or too many loops.

12.4 Concluding Remarks

A basic methodology for modeling a given network topology has been presented. An example network was modeled for 56Kbps lines. Modeling of a given network is so easy that no designer should have any excuse to ignore such an important exercise just before developing better solutions. It provides a basis for future improvements. A methodology is also discussed for finding the routes between any two nodes. This capability can then be used to develop a routing table useful in the design of a switch.

Table 12.3 A List of Routes between Three Sets of Nodel Pairs

Following is a partial list of routes between nodes #2 and #4 (RTBL1)

2	25	33	4		Path No. 1
2	16	14	4		Path No. 2
2	926	4			Path No. 3
2	22	34	4		Path No. 4
2	16	1	26	4	Path No. 7
2	9	18	34	4	Path No. 9

End of Output File

Following is a partial list of routes between nodes #2 and #20 (RTBL2)

2	22	34	20		Path No. 4
2	9	18	34	20	Path No. 7

End of Output File

Following is a partial list of routes between nodes #2 and #15 (RTBL3)

2	16	14	15	Path No. 2

End of Output File

Exercises

12.1 Using VHD41 and the FTF1 files, compute the costs of the given network with VG, 56000, T1, and T3 circuits. Tabulate the results. Create two LINK files for two mixes of links.

12.2 Using VHD41 and the FTF1 files, compute the costs of the given network with VG, 56000, T1, and T3 with two samples of LINK files that employ all four types of links. Tabulate the results.

12.3 Using the VHD41 and the FTF41 files, compute a Routing table for at least 10 pairs of nodes. Tabulate the results.

12.4 Repeat Exercise 12.1 using the MCI tariffs of 1996.

12.5 Repeat Exercise 12.2 using the MCI tariffs of 1996.

12.6 Repeat Exercise 12.3 using the MCI tariffs of 1996.

13

Design Process for a Personal Communication System (PCS)

13.1 Introductory Remarks

Digital Personal Communication Systems (PCS) employ the wireless technology for providing interpersonal communications between persons who are moving on foot or in vehicles. Digital PCSs comprises four types of accesses:

1. GSM (also known as the GSM/PCS 1900 standard) Access

2. TDMA (also known as the IS-136 standard, previously called the IS-54)

3. CDMA (also known as the IS-95 standard)

4. Fixed Wireless Access

Although the fourth type of access began with fixed-location residential/business applications, these systems now allow user mobility. Both GSM and TDMA employ the time division multiaccess. The GSM obeys the strict ITU standard and is widely deployed in Europe and other countries. The CDMA employs the code division multiaccess. Services based on these standards make use of the radio spectrum in the 1.8 GHz to 1.9 GHz range—a spectrum freed up by the FCC recently for mobile personal communications services. Such services include both voice and data communications. The reader should consult the recent papers (Mathias and Rysavy, 1994) and (Wexler, 1995) to get a deeper understanding of the PCS field. A recent *Business Week* article (Arnst, 1994) provides a keen insight into the buying of PCS frequency band licenses.

Both GSM and TDMA boast a threefold to sixfold increase in capacity over the analog version of the cellular technology. CDMA claims at least a tenfold increase in capacity over the analog cellular technology. It also boasts a superior interchannel interference and privacy. Whereas GSM and TDMA are mature technologies, CDMA is a relatively new offering. Market pressures are forcing the development of a universal handset for seamless interoperability for these major technologies.

A metropolitan area PCS network (hereon called the PCS/MAN) consists of towers supporting radios that accept analog voice signals from mobile users and transmit the digitized signals toward network nodes. The radio equipment is called the Base Station Transceiver (BST). Each BST serves a geographical area called the cell. Since each cell is designed to handle a certain amount of traffic intensity (TI), the cells may differ from one another in geographical shape. The second component of a PCS network is a base station controller (BSC) that serves a certain number of BSTs. The function of a BSC is to concentrate traffic flowing toward a Mobile Switching Center (MSC) where ultimate switching of traffic takes place. A BSC also handles call handoffs occurring as a result of a caller moving from one cell to another. A call handoff may or may not require the call handling functions to be transferred from one BSC to another. Such handoffs may cause some slight perturbations in the traffic flow patterns. There may be one or more switching centers (i.e., MSCs) in a PCS network. A MSC handles not only the call switching functions but also manages the inter-PCS/MAN traffic and specialized databases for providing service to roaming mobile users.

13.2 The Basic Design Process for a PCS

The design of cells and their structures requires an in-depth analysis of terrain as a function of spectrum limitations. Such an effort clearly falls out of the scope of this book. The basic design process for a PCS/MAN must strive to obtain an optimum network topology consisting of the correct number and location of BSCc and MSCs, and the correct mix of transmission facilities that connect BSTs to BSCs and BSCs to MSCs. See Figures 13.1, 13.2, 13.3, and 13.4 for applicable PCS topologies for connecting either BSTs to a BSC or connecting several BSCs to an MSC. Some vendors combine the functions of a BSC with those of an MSC thus enabling the designer to treat both BSC and MSC as a switch. Therefore, several switches can be connected together to form a fully connected mesh. On the other end of the spectrum, some MSCs are so complex, reliable, and feature rich that one can afford only a single MSC in the PCS/MAN. For that case, Figure 13.3 is quite appropriate.

Most of the design challenges are related to a PCS/MAN. To create a nationwide PCS/WAN, one can interconnect all of the Gateway MSCs (from each PCS/MAN) via another trunk hierarchy. Such a network will function as a

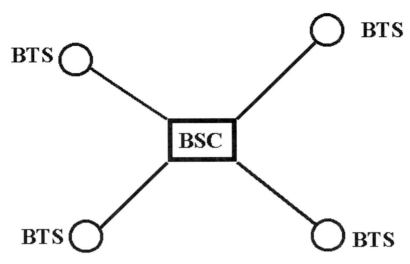

Figure 13.1 A star topology connecting BTSs to a BCS.

national backbone network. It can be modeled and optimized using the methodology already discussed in Chapter 11. These Gateway MSCs can employ a CCS network to provide fast connections and a variety of value-added features to users. The design process for a CCS network is described in Chapter 14. The PCS network design will require the successful execution of the following tasks:

1. Create a valid VHD file for BSTs. One may have trouble getting the V&H coordinates for each BST due to the peculiar cellular structures. It may be proper to apply the simple method as discussed in Chapter 6. One could get an architect's map of the area to be covered by PCS and then allocate V&H coordinates by overlaying a transparent grid. Be careful to change the design parameters V_{min}, V_{max}, H_{min}, and H_{max}, and F_{dis} of an SDF file.

2. Create all the necessary NLT, TARIFF, LATA, and SWF (with the number of switches, and the switch node IDs). A LINK file may be useful if one needs to connect each BSC to an MSC with a different link type. For that case, use Fvc0=1. However, the normal procedure will be to use the LINK file for defining the number of BSCs (used here as concentrators) and the link type needed to connect the BSCs to the MSCs. This is accomplished by making Fvc0=0 and Flk=1.

3. Name these input files in the FILES.

4. RUN the appropriate application (StarData orDLNet or MDNet).

5. Compute the cost of the 2-level networks (single level ALs and a BBNet) or 3-level networks (lower-level ALs, higher-level ALs, and a BBNet) by varying the mix of topologies (e.g., MD and Star) and link types (e.g., Leased T1s or fully owned digital microwave links). Plot these results to learn about any trends.

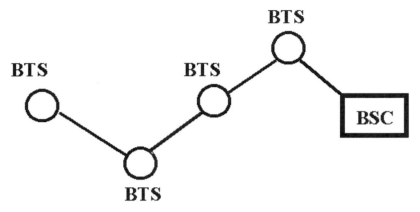

Figure 13.2 A DL topology connecting BTSs to a BCS.

13.3 Discussion of Example PCS Network Topologies

One can illustrate the preceding process using an example 250-node PCS network for the Dallas–Fort Worth area (DFW) by running the desired Networking modules of the EcoNets software. Table 13.1 defines the corresponding database of V&H information, traffic intensities, LATA numbers, and LINK types employed. The VHD file is only partially displayed in Table 13.1 since the V&H coordinates and TI values were simulated using random numbers. The reader should note the range of values (between 10,000 to 900,000) and a total of 164,614,020bps during a busy hour. Assuming 19.2Kbps per voice conversation, it represents a TI of about 857 erlangs for TGF=1. The value of TI becomes 8570

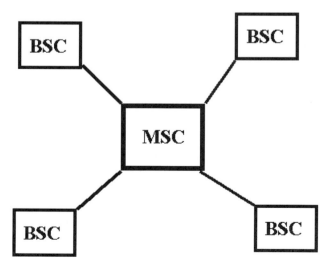

Figure 13.3 A star topology connecting BTSs to an MSC.

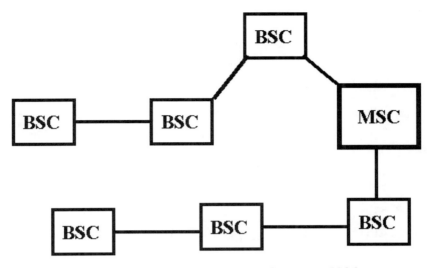

Figure 13.4 A DL topology connecting BTSs to an MSC.

and 17,140 erlangs for TGF equaling 10 and 20, respectively. Considering the DTP8 file employed (about 10 equivalent busy hours a day), the smaller TI value represents about 154 million call-minutes per month. Depending upon one's viewpoint, it may represent a small or a giant operation. See Table 13.2 for the definition of 15 COGs that can be used to select 1, 2, 3,…15 locations for BSCs (also called local traffic controllers or concentrators) and/or switches.

Table 13.1 PCS Network Design Database

************** NODAL DEFINITION DATA ***************

N#	-V-	-H-	LOAD (BPS/MEs)	LATA	LINK	NAME
1	8433	4189	380520	552	1	
2	8441	4242	754080	552	1	
3	8485	3934	251580	552	1	
4	8545	4157	535100	552	1	
5	8456	4213	577900	552	2	
6	8433	4148	821200	552	1	
7	8548	4207	347660	552	1	
8	8482	3920	1030000	552	1	
9	8428	4158	745640	552	1	
10	8582	4231	442140	552	2	**Continued**

Table 13.1 *Continued*

N#	-V-	-H-	LOAD (BPS/MEs)	LATA	LINK	NAME
240	8407	4141	359940	552	1	
241	8523	4299	453040	552	1	
242	8422	4055	1568000	552	1	
243	8464	4087	666400	552	1	
244	8457	4247	201120	552	1	
245	8412	3908	240020	552	1	
246	8514	3954	292260	552	1	
247	8407	4112	748660	552	1	
248	8417	3998	329160	552	1	
249	8571	3980	657780	552	1	
250	8443	4086	243560	552	1	

Tot. BHR TI (bps) = 164614020

************** Node(N)Link(L)Type(T) [NLT] FILE PRINTOUT **************
***** LEGEND *****
{ C=Link Cap.: MaxR=Max. Allwd. Rate(Wm): MF=VMpxg.Fact.: FPF=Priv.Fac. Fact.}

LType	LinkC	MaxLinkR	MF	Tariff#	FPF
1	1544000	1444000	24	2	1
2	1544000	1444000	24	2	1
3	6174000	5444000	24	3	1

************** TARIFF DATA PRINTOUT **************

TARIFF #=1   AVG. LOCAL LOOPS CHARGES ($)=0
MILEAGE BANDS:

4	8	25	50	10000

FIXED COSTS ($):

122.44	122.44	178.52	205.8	205.8

COST PER MILE ($):

51.5	51.5	44.98	43.39	43.39

Continued

TARIFF #=2 AVG. LOCAL LOOPS CHARGES ($)=0
MILEAGE BANDS:

| 10 | 20 | 40 | 60 | 10000 |

FIXED COSTS ($):

| 550 | 630 | 810 | 890 | 900 |

COST PER MILE ($):

| 1 | 1 | 1 | 2 | 2 |

TARIFF #=3 AVG. LOCAL LOOPS CHARGES ($)=0
MILEAGE BANDS:

| 10000 | 10000 | 10000 | 10000 | 10000 |

FIXED   COSTS ($):

| 579 | 0 | 0 | 0 | 0 |

COST PER MILE ($):

| 243 | 1 | 1 | 1 | 1 |

*************** SYSTEM DESIGN PARAMETERS ***************

ATP/D=3	UPR/D=56000	HPR/D=56000	IML/D=28	RML/D=300
Ncu/D=4	Rmph/D=100	HTT/D=0.001	Fopt/D=0	Tnp/D=1
Thm/D=0	Kpg/D=0.01	BKL/D=64	ICPB/D=56	TGF/C=20
Flk/C=1	Fnn/C=0	Flt/C=1	Fftd/C=0	NA=0
ALT/V/D=1	NA=1	Bal/V/A=0.1	ECC/V=13.33	ECD/V/A=300
DREQ/A=60	PEXD/A=0.15	Clbr/A=23	Frst/A=1	ACDT/A=1
TKLT/V/D=1	NA=1	Btk/V=0.1	Ffdx/D=1	MTKU/D=0.8
BBTF/C=2	Vmin/C=8325	Vmax/C=8625	Hmin/C=3875	Hmax/C=4325
Fvc0/C=0	Fvc1/C=0	Fvc2/C=0	Fvc3/C=0	Fvc4/C=0
Fvc5/C=0	Fvc6/C=30	Fvc7/C=0	Fsh/D=0	Fnp/C=1
DPM/A=30	Fdis/C=1	NA=0	TFXC/A=1	NDEC/C=3
DECT/C=1	Legend:	/A=ACD	C=Common	/D=Data

*************** NAMES OF INPUT FILES ***************

VHD250* LINK250* MAPdfw* NLT1* TARIFF1* SDFpcs* NAME17* FTF1*
LATA250* FILES.TXT* CSABDS* UTBL* WUTBL* MUTBL*RSTBL* DTP8* SWF5*

Continued

Table 13.1 *Continued*

*************** DAILY TRAFFIC PROFILE ***************

Hour Numbers & Corresponding Fractions of Daily Traffic are as follows:

1	0	2	0	3	0	4	0	5	0	6	0
7	0.05	8	0.1	9	0.1	10	0.1	11	0.1	12	0.1
13	0.1	14	0.1	15	0.1	16	0.1	17	0.05	18	0
19	0	20	0	21	0	22	0	23	0	24	0

******* Switch File Definition *******
Number of Switches =5 @ 84, 147, 1, 75, 110.
Networking Menu Item No. Employed = 2

Table 13.2 COGs and Closest Nodes and Their V&H Coordinates

NAME OF VHD FILE=VHD250   NUMBER OF NODES=250
NO. OF DECOMPOSITIONS=7
INITIAL DECOMPOSITION FACTOR=1   (1 FOR HORIZ - 0 FOR VERT)

COGs, the Closest Nodes and their V-H Coordinates are as follows:

COG-V	COG-H	Node#	Node-V	Node-H
8496	4089	84	8500	4090
8491	4186	133	8501	4186
8500	3998	203	8500	4005
8537	4183	147	8539	4182
8442	4189	1	8433	4189
8549	4007	11	8549	4008
8453	3990	110	8456	3980
8548	4234	136	8550	4235
8527	4132	180	8515	4136
8445	4238	2	8441	4242
8437	4132	232	8441	4139
8550	4047	233	8559	4040
8547	3956	115	8525	3956
8455	4041	129	8463	4053
8451	3940	107	8453	3940

-END OF FILE

The designer may be required to create several LINK files to represent the number/locations of the BSCs. The designer must also design several SWF files to represent the number and locations of the MSCs.

Table 13.1 also shows the data of an NLT file with three link types and a Tariff file with three tariffs. The first link or tariff type defines the intra-LATA leased T1. One should notice the mileage bands of 4, 8, 25, 50, and larger than 50 for the Southwestern Bell Company tariffs. The second link type defines a modeled digital microwave channel. The total cost of DMW dish and electronics was amortized over seven years to get a monthly rate for the shortest distance of 10 miles. For each next mileage range, the cost of a larger dish and/or electronics was taken into account and the new monthly charges were computed. The third link type represents a DMW-T2 link (with a capacity of about 6Mbps) but it was not used since it represented an overkill solution for the problem under consideration.

The SDF parameters as shown in Table 13.1 assume switches based on FPS/Virtual Cell technology. Therefore, we fixed ATP=1, UPR=56Kbps, HPR=56Kbps, BLKL=64 bytes, ICPB=56, T_{np}=1 ms, T_{hm}=1ms (considering delays in DSU/RSU for T1), F_{lk}=0 or 1 depending whether a LINK file was used to create a 3-level PCS network. The reader is advised to study other design parameters that influence the final results only indirectly.

The contents of the DTP8 file and the SWF are also included in Table 13.1.

A large number of runs were made using EcoNets. See Figures 13.5 through 13.8 for charts relating monthly transmission costs versus network topology (i.e., the number of switches, type of connectivity [star or DL], and the link types employed). The first two sets of charts deal with 2-level PCS networks. The last two sets of charts deal with 3-level PCS networks.

The first set of charts assumes a value 10 for TGF. The curves show that a DL topology is almost insensitive to the number of switches or link types employed. For those who prefer the star topology for reliability, responsiveness, and easy reconfigurability, only DMW facilities are recommended for only a slight cost penalty. The second set of charts for TGF=20 show the same conclusions with one observation: the advantage of DMW facilities over leased T1 lines is increased. The charts for a 3-level PCS network reinforce the previous conclusions with one additional observation: the 3-level PCS networks seem to be much cheaper than the 2-level counterpart with star topology with one or two switches.

Figures 13.9 and 13.10 illustrate two sample 3-level networks synthesized with five switches using star and DL topologies.

These charts and topologies show that the use of DMW facilities is the least risky. In case a star topology is preferred, five switches provide a better reliability through rich connectivity. If the cost of an MSC is too high, and the thought of a distributed-switched PCS/MAN network is threatening to some, one should select the DL topology, which provides additional cost incentives.

Once a PCS/MAN is designed optimally, it is a straightforward process to design a PCS/WAN. One switch in every PCS/MAN can be assigned to function as a gateway switch. Such gateway switches from metropolitan areas can then be interconnected to create a higher-level BBNet. Consult Chapter 11 for a methodology to model and optimize a nationwide PCS/WAN.

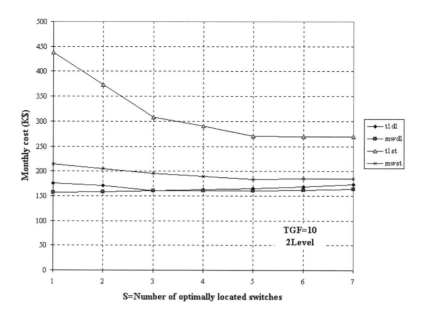

Figure 13.5 Monthly costs versus topology.

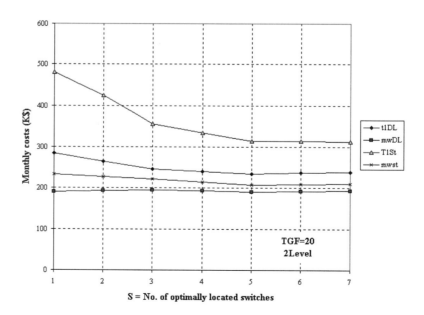

Figure 13.6 Monthly costs versus topology.

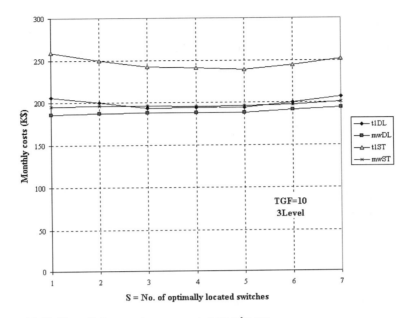

Figure 13.7 Monthly costs versus topology.

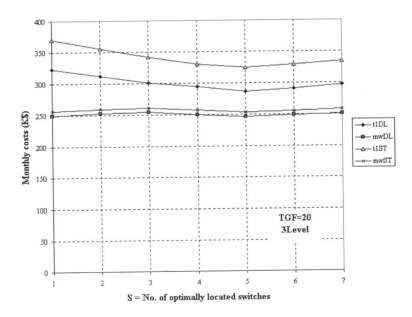

Figure 13.8 Monthly costs versus topology.

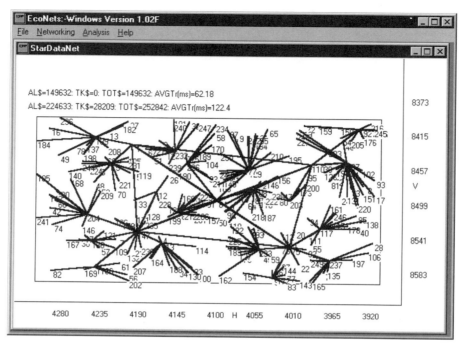

Figure 13.9 A 3-level PCS network based on star topology.

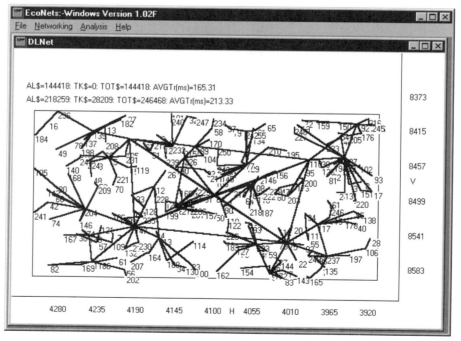

Figure 13.10 A 3-level PCS network based on DL topology.

13.4 Concluding Remarks

A basic design process for a digital PCS has been described. Such a process can be employed to synthesize either a cost-effective PCS/MAN or a PCS/WAN. The process was illustrated using a 250-node PCS for the Dallas–Fort Worth region. The results show that an optimum PCS network topology consists of five switches and digital microwave facilities for high traffic intensities.

Exercises

13.1   Using the data of Table 13.1, model other combinations of concentrators and switches for a 3-level PCS/MAN and plot the results to draw some conclusions.

Bibliography

Arnst, C. 1994. "The New Wireless Looks a Bit Pricey." *Business Week*, December 5, 1994.

Mathias, C., and P. Rysavy. 1994. "The ABCs of PCS." *Network World*, November 7, 1994, pp. 53–54.

Wexler, J. 1995. "Getting to Know PCS." *Network World*, May 22, 1995, p. 28.

14

Design Process for a Common Channel Signaling System (CCSS)

14.1 Introductory Remarks

The EcoNets tool can be employed to plan and design an Intelligent Network (IN) or an Advanced Intelligent Network (AIN) using the building blocks of a common channel signaling system (CCSS). A CCSS is based on transmitting signaling information on separate, shared facilities instead of on traffic carrying trunks. A CCSS includes networks based on either the SS6 or SS7 signaling standard. The older dumb PTNs used the technique of in-band signaling. Since a route was established, one link at a time, according to a fixed table; the trunk utilization suffered a great deal. In 1975, AT&T employed the common channel interoffice signaling system (CCIS which was equivalent to ITU's SS6 standard) to separate signaling from the traffic carrying trunks. However, its protocols were too rigid (unlike that of the multilayered OSI data processing standard), its messages were too short to be useful, and the link speeds were too slow (2400–4800bps) to be responsive. The SS7 system eliminated all those problems, at least for the time being.

A CCSS consists of three types of switching points: service switching points (SSPs), a proper number of optimally located signaling transfer points (STP), and service control point (SCP) nodes. A subset of the CCSS network can be con-

436 Network Design Using EcoNets

sidered as a shared backbone network that provides intelligence to a large number of service switching points (SSPs) that are analogous to either Class 5 COs or Class 4 Regional Offices requiring signaling service. All North American CCSS are designed to satisfy the ANSI Message Transfer Point/Signaling Connection Control Part (MTP/SCCP) signaling standard. ITU has equivalent standards for other countries. Modern CCSSs are also called CCS7 or SS7 systems.

The SSPs are equivalent to CPEs in the network design process. According to the network design methodology presented in previous chapters, the STP nodes should be optimally located at a subset of CPEs. This fact has motivated some vendors to deliver integrated (i.e., SSP+STP) nodes for providing economies of scale. STPs are equivalent to packet switches in a distributed network. SCPs are equivalent to database management systems (DBMSs). But all of these signaling points possess PS capabilities.

A fully generalized CCS7 network is shown in Figure 14.1. All of its links have a minimum capacity of 56Kbps (with T1-capability on the horizon) and are designated with letters A to F described as follows:

1. A-Links (also called access links) allow SSPs and SCPs to access the STPs.

2. B-Links (also called bridge links) join mated STP pairs.

3. C-Links (also called cross-links) interconnect primary and secondary STPs to create an STP pair.

4. D-Links (also called diagonal links) interconnect primary and secondary STP pairs, serving as an alternate route for linking pairs of *unequal hierarchy*.

5. E-Links (also called extended links) connect SSPs and SCPs to remote STP pairs for transaction service only.

6. F-Links (also called fully associated links) connect SSPs or SCPs to one another directly, without STP as an enroute node. The F-links become associated with the signaling messages for controlling the regular traffic trunks between two SSPs.

From the previous discussion, it should be clear that a CCSS network requires very high availability. Transmission facilities, STPs, and SCPs are provided in redundant pairs, with availability to handle the BHR traffic even when half of the network is dead. The STP-pair nodes must be in different cities to enforce drastically different link routes for survivability. There are three basic CCSS architectures (as subsets of the generalized architecture of Figure 14.1) described as follows:

1, *Fully Associated Architecture of Figure 14.2.* It has a small number of SSPs connected to one another in the form of a fully connected mesh via the F-Links (in addition to the regular traffic carrying trunks). No STPs are employed in the network. Such a network is suitable for small customers with few SSPs. As the number of SSPs increases, the meshed network becomes too expensive.

2. *Hybrid Network Architecture of Figure 14.3.* It employs F-Links and A-Links to connect all the SSPs and STPs. Only the mated STPs are connected via C-Links. This architecture has been widely used in some foreign countries to economize on links.

3. *Quasi-Associated Architecture of Figure 14.4.* It employs A-Links, B-Links, and C-Links to connect all STPs, physically paired through C-Links. But the two SSPs are not connected to one another by F-Links, thereby resulting in the category of *quasi-associated.* This architecture provides high availability and a balanced load on network resources. Consequently, some call it the *balanced* architecture. It is also more modular and easier to operate than the fully associated and the hybrid architectures.

It should be pointed out that Figures 14.2, 14.3, and 14.4 don't show the SCP nodes. One or more SCPs can be connected to any subset of SSP/STP nodes. No IN or AIN can exist without the SCP nodes. The reader should consult a book by Kessler (Kessler, 1990) for a deeper insight into the SS7 standard and protocols.

An IN or an AIN can be realized with an addition of Intelligent Peripherals (IPs) to the generalized CCSS network of Figure 14.1. These IPs are high-powered WSs that reside at the service provider's premises and these can demand agreed upon services by sending meaningful messages to the connected SSPs of the public network. The class of services can also be altered quickly.

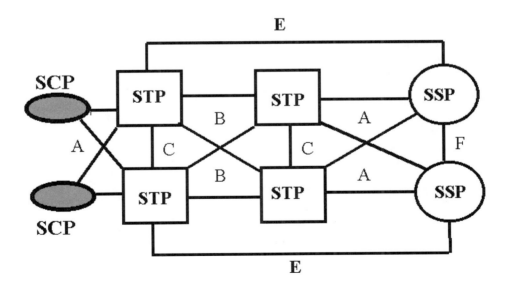

Figure 14.1 A generalized CCSS architecture.

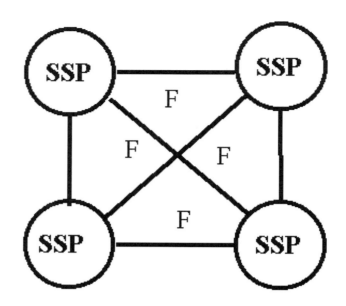

Figure 14.2 A fully associated CCSS architecture.

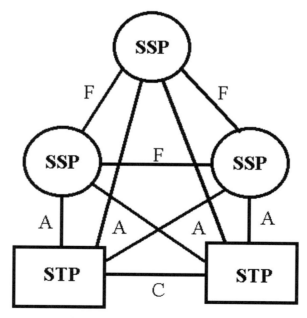

Figure 14.3 A hybrid CCSS architecture.

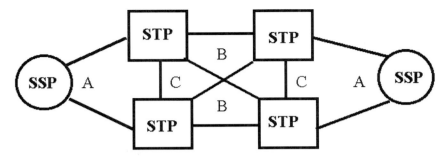

Figure 14.4 A quasi-associated CCSS architecture.

14.2 The Basic Design Process for a CCSS Network

The need to model a given network will require the creation of the following input files:

1. Create a valid VHD file for all the SSP nodes of an existing network. One may have trouble getting the V&H coordinates for nodes in a foreign country. For that case, it may be proper to apply the simple method as discussed in Chapter 6. One could get a map of the area from a good atlas and then allocate V&H coordinates by overlaying a transparent grid. This effort creates a proper MAP file.

2. One should carefully check the SDF design parameters ATP, UPR, HPR, T_{np}, T_{hm} (for type of FPS nodes employed), and V_{min}, V_{max}, H_{min}, and H_{max}, and F_{dis} (for types of maps employed). The design parameters ALT and TKLT define the link types in the AL and TK subnetworks since $F_{lk}=0$.

3. Create the necessary NLT, TARIFF, NAME, and LATA files. A LINK file is not generally required for designing CCSS networks. Therefore, the SDF design parameters ALT and TKLT define the link type for all ALs and trunks.

4. Name these input files in the FILES.

5. RUN the Networking menu item StarDataNet (essential for modeling reliable data networks) to model a 2-level data network consisting of (1) the A-Links connecting SSPs and SCPs to STPs, and (2) the trunks connecting STPs in the form of B-Links and C-Links. In this methodology, the SCP nodes are equivalent to Hosts directly attached to a subset of STPs. The CCSS traffic is in the form of messages with average length of 25 bytes.

6. In order to create a double connectivity between an SSP and two STPs of a regional STP pair, one can make *several* passes at creating the AL subnetworks, one for each subset of STPs. The cost of these AL subnetworks must be added.

7. The trunk subnetwork is created by considering the SSPs as CPEs, and STPs as switches. The total AL traffic should be equally divided among the SCPs employed.

8. One can use slightly modified versions of (1) the automatically created FTF0 file that defines the trunk bundles for all the STP switches, and (2) the automatically created FTFGN files that specify all the AL connections (From-Node, To-Node, No. of Circuits) for both StarDataNet runs to model the final CCSA network using the GivenNet item of Networking Menu and verify the total monthly cost.

9. The network topologies are saved in associated output files called ALSDF1s, ALFGN, TKFs, and TOPFs that must be saved under a different name to prevent overwrites.

14.3 Discussion of Outputs for an Example SS7 Network

One can illustrate the preceding process using example VHD100c file for designing a large but robust CCSS network. See Table 14.1 for a database for the CCSS network design. Since VHD100c was obtained through simulation, only a part of the file is illustrated. See Chapter 10 to learn how its BHR traffic intensities are related to VHD100b file. The DB shows that a BHR total TI is equal to about 29Mbps which implies a BHR throughput of about 100,000 SS7messages per second assuming an average of 25 characters per message. The NLT file will allow us to use link types 2 and 3 for 56Kbps and T1 circuits respectively. No attempt will be made to use the 9600bps and T3 circuits. The corresponding Tariffs 2 and 3 are also defined in Table 14.1. The SDF design parameters assume an FPS environment. If one wants to use the X.25 PS technology, one can make the necessary changes in SDF design parameters in a manner as discussed in Chapter 10. We will illustrate the modeling process by ignoring the LINK file and using SDF parameter ALT equal to 2 (which also implies a Tariff #2) for 56Kbps lines, and TKLT equal to 3 implying the use of T1 circuits for trunks. A simple run showed that 56Kbps circuits are too costly for trunks.

Table 14.1 CCSS Network Design Database (DB)

*************** NODAL DEFINITION DATA ***************

N#	-V-	-H-	LOAD (BPS/MEs)	LATA	LINK	NAME
1	6053	1503	108000	0	0	
2	6807	1551	432000	0	0	
3	7262	2762	216000	0	0	
4	6230	7031	108000	0	0	**Continued**

N#	-V-	-H-	LOAD (BPS/MEs)	LATA	LINK	NAME
5	7731	8494	432000	0	0	
6	6189	6192	108000	0	0	
7	7273	3024	216000	0	0	
8	5026	4496	324000	0	0	
9	6660	4249	432000	0	0	
10	7488	2566	216000	0	0	
95	5453	3478	216000	0	0	
96	6257	6593	432000	0	0	
97	7701	1956	432000	0	0	
98	5323	2628	324000	0	0	
99	7718	6720	432000	0	0	
100	8692	8268	432000	0	0	

Tot. BHR Traffic = 29160000

********* Node(N)Link(L)Type(T) [NLT] FILE PRINTOUT *********
***** LEGEND*****

{ C=Link Cap.: MaxR=Max. Allwd. Rate(Wm): MF=VMpxg.Fact.:
FPF=Priv.Fac. Fact.}

LType	LinkC	MaxLinkR	MF	Tariff#	FPF
1	9600	6300	1	1	1
2	56000	48000	8	2	1
3	1544000	1440000	24	3	1
4	45000000	40000000	672	4	1

*************** TARIFF DATA PRINTOUT ***************

TARIFF #=1   AVG. LOCAL LOOPS CHARGES ($)=294
MILEAGE BANDS:

50	100	500	1000	10000

FIXED COSTS ($):

72.98	149.28	229.28	324.24	324.24

COST PER MILE ($):

2.84	1.31	0.51	0.32	0.32

Continued

Table 14.1 *Continued*

TARIFF #=2   AVG. LOCAL LOOPS CHARGES ($)=492
MILEAGE BANDS:

50	100	500	1000	10000

FIXED COSTS   ($):

232	435	571	1081	1081

COST PER MILE ($):

7.74	3.68	2.32	1.3	1.3

TARIFF #=3 AVG. LOCAL LOOPS CHARGES ($)=2800
MILEAGE BANDS:

50	100	10000	10000	10000

FIXED COSTS ($):

1770	1808	2008	2500	2500

COST PER MILE ($):

10	9.25	7.25	7.25	7.25

TARIFF #=4   AVG. LOCAL LOOPS CHARGES ($)=8000
MILEAGE BANDS:

10000	10000	10000	10000	10000

FIXEDCOSTS($):

16600	16600	16600	16600	16600

COST PER MILE ($):

47	47	47	47	47

*************** SYSTEM DESIGN PARAMETERS ***************

=0

ATP/D=1	UPR/D=56000	HPR/D=56000	IML/D=28	RML/D=30
Ncu/D=4	Rmph/D=100	HTT/D=0.001	Fopt/D=0	Tnp/D=1
Thm/D=1	Kpg/D=0.01	BKL/D=64	ICPB/D=56	TGF/C=1
Flk/C=0	Fnn/C=0	Flt/C=0	Fftd/C=0	NA=0
ALT/V/D=2	NA=0	Bal/V/A=0.1	ECC/V=13.33	ECD/V/A=300
DREQ/A=60	PEXD/A=0.15	Clbr/A=23	Frst/A=1	ACDT/A=2
TKLT/V/D=3	NA=0	Btk/V=0.1	Ffdx/D=1	MTKU/D=0.8

Continued

*************** SYSTEM DESIGN PARAMETERS ***************

BBTF/C=2	Vmin/C=3000	Vmax/C=10000	Hmin/C=0	Hmax/C=10000
Fvc0/C=0	Fvc1/C=0	Fvc2/C=0	Fvc3/C=0	Fvc4/C=0
Fvc5/C=0	Fvc6/C=30	Fvc7/C=0	Fsh/D=0	Fnp/C=1
DPM/A=30	Fdis/C=1	NA=0	TFXC/A=1	NDEC/C=3
DECT/C=1	Legend:	/A=ACD	/C=Common	/D=Data

*************** NAMES OF INPUT FILES ***************

VHD100c * LINK200 * MAPusa * NLT * ARIFF * SDF * NAME41* FTF1* LATA41 *
FILES.TXT * CSABDS * UTBL * WUTBL * MUTBL * RSTBL * DTP8 * Swf4b *

*************** DAILY TRAFFIC PROFILE ***************

Hour Numbers & Corresponding Fractions of Daily Traffic are as follows:

1	0	2	0	3	0	4	0	5	0	6	0
7	0.05	8	0.1	9	0.1	10	0.1	11	0.1	12	0.1
13	0.1	14	0.1	15	0.1	16	0.1	17	0.05	18	0
19	0	20	0	21	0	22	0	23	0	24	0

******* Switch File Definition *******
Number of Switches =4 @ 96, 13, 63, 55.
Networking Menu Item No. Employed = 2

The published vendor data showed that an STP can handle at least 255 terminations, each rated at 56Kbps, with plans to handle 512 terminations, and a lesser number of high-speed lines (HSLs) each rated at 1.54Mbps. A simple analysis shows that 255 lines can handle about 57,000 messages per second assuming 25 bytes per message and 80 percent utilization per line. Therefore, two STPs should handle the BHR throughput of 100,000 messages per second. To ensure an acceptable CCSS reliability, we will employ two STP pairs. In case of a single STP failure, the remaining three STPs should have capacity to spare. In fact, the system should function even if two STPs fail. The COG analysis showed that these four STPs should be located at sites 96, 13, 63, and 55. To satisfy the standard SS7 model, the STPs at sites 96 and 13 were assumed to be a regional pair. The second regional pair was located at sites 63 and 55. The STP pairs were selected on the basis of two regions: eastern and western.

We will also assume two SCP pairs, collocated with the STP pairs. It was done to reduce cost of transmission facilities. We are assuming that four SCPs have the same capacity as the four STPs. If the same vendor who provides an STP with 255 line-termination capability, provides an SCP with a throughput of only 3000 transactions per second and not the desired throughput of at least 50,000 transactions per second, there is something wrong with the marketing and/or product planning departments. One can easily employ the methodology of Chapter 4 to increase the SCP nodal throughput by many folds by ordering the accesses to the disk subsystem of an SCP. If that is impossible, one can employ the so-called massive-multiprocessor architecture to increase the SCP throughput. In the worst case, one can always increase the number of SCPs collocated at each of the four STPs, assuming four STP sites are optimum for our SS7 network design. We will present a design process equally valid for other situations.

The EcoNets tool is ideal for designing *economical* network topologies. A CCSS network is not an economical network. It is a highly reliable network. It achieves its high reliability through the duplication of AL connections and STP/SCP pairs in the CCSS network. A major EcoNets redesign to handle all the previously mentioned four CCSS architectures would have been a major undertaking with little or no returns. Instead, we will use simple tricks to provide AL and trunk bundle redundancies over the two STPs in each region. The trick involves making two passes to compute the cost of ALs.

During the first pass, we consider only two switches at locations 96 and 63. This will force the SSPs to be connected to the nearest STP. The total number of AL terminations is 640 and the AL bundle cost is about $1.4 million per month. During the next pass, the two switches will be located at sites 13 and 55. The total number of AL terminations is again 640 and the AL bundle cost is about $1.374 million per month. Since each AL bundle was designed for a utilization of about 92 percent (according to the ratio of 48,000 to 56,000), using the two sets of AL bundles implies an effective AL utilization of about 46 percent. That is just about right value for satisfying the CCSS requirement of about 40-percent utilization. One can achieve any average AL bundle utilization by varying W_m. The cost of redundant AL bundles is about $2.774 million per month. The two AL topologies are illustrated in Figure 14.5 and 14.6 with cost summaries.

Similarly, another pass was made by considering four switch locations, four STP switches located at 96, 13, 63, and 55. We now get a fully connected STP backbone with bridge (B) and cross (C) type trunks assuming that four *specialized* SCPs are at each STP location. (Use of identical four SCPs would have eliminated the need for any trunk bundles. EcoNets creates a fully distributed BBNet for generality.) It shows that a single T1 is required for each of the six trunk bundles for BBTF=2. Since these trunks show an average utilization of about 80 percent (per MTKU design parameter), we will need two T1 trunks to satisfy the SS7 requirement of 40 percent utilization. One can set the SDF design parameter MTKU to 0.4 and make another model of the BBNet. That doubles the cost of our trunking network to about $142,000 per month. One can also modify the automatically created FTF0 file (see Chapter 11 for modeling a backbone network) to reflect any traffic load requirements caused by STP or SCP failures.

The only other subnetworks that have escaped our attention concern the four SCP nodes. We chose to select four *identical* SCPs collocated with STPs to save on transmission facilities. The first two SCPs are attached to STPs located at 96 and 13. The next two SCPs are attached to STPs located at 63 and 55. Although we don't need any trunk bundles for no-failure conditions, we will need trunk bundles to handle STP or SCP failures. We will now need the BBNet modeling/optimization methodology of Chapter 11. Using the automatically

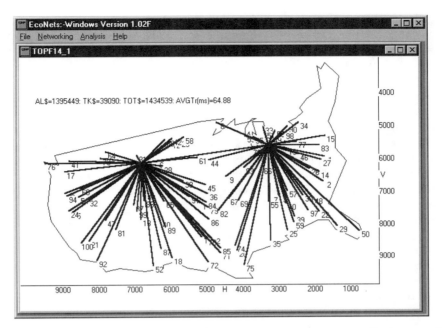

Figure 14.5 Topology of a CCSS network with two switches at 96 and 63.

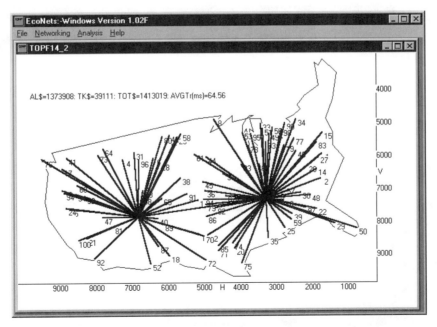

Figure 14.6 Topology of a CCSS network with two switches at 13 and 55.

created FTF0 file, we can now model the backbone network with modified inter-nodal traffic between an SCP to the regional STPs for the case of a single STP or SCP failure. See Figure 14.7 for an example trunk topology and Table 14.2 for the corresponding trunking network model.

Although we have analyzed two separate AL-subnetworks, and one final back-bone network, we haven't used EcoNets to plot the final network topology yet. Another trick comes to our aid. Each time the StarData or voice network topology is modeled, the software creates a transitory input file called FTFGN. By saving the corresponding two files and concatenating them using either the Windows' Notepad program or the FTF File Merge item of the File menu, we can easily create a combined FTFGN file. It will define all the connections, including the trunk bundles (which were computed for desired BBNet traffic flows). The FTFGN file can now be used to model the combined CCSS network topology using the GivenNet modeling item of the Networking menu. However, this will require the use of a LINK100 file with proper link-type values assigned to SSP and STP nodes. The creation of the LINK100 file is necessitated by the presence of two link types (2 and 3). The topology is shown in Figure 14.8 with the correct final monthly transmission cost of about $ 2.98M. Table 14.3 defines each connection of the combined network. One can repeat the process by employing 1.44Mbps high-speed links (HSLs) for both ALs and trunks (i.e., for ALT=3 and TKLT=3). The monthly cost of the entire CCSS (or SS7) network is reduced to about $2.07M, resulting in a monthly savings of about $0.91M (or about $11M

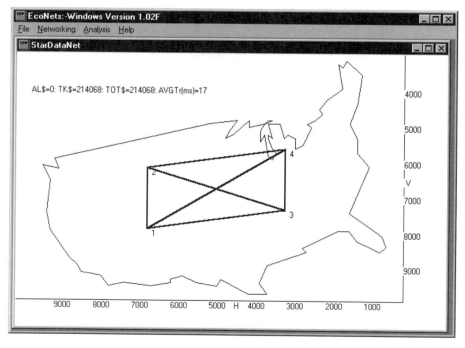

Figure 14.7 A four-node SS7 backbone network topology.

per year). Therefore, it is recommended to use HSL lines throughout the SS7 network. One can easily compute the transmission line costs for only two SCPs: one located at node 96 and another located at node 55. A simple exercise will show an additional penalty of $284K even after the equivalent monthly cost of SCP hardware is subtracted (assuming each SCP costs $2M, seven years for a life cycle, and an annual interest rate of 10 percent). Not many designers realize that deploying additional hardware can actually save money for a WAN backbone. What a neat trick!

Table 14.2 SS7 (CCSS) Trunking Modeling/Design Data

Trunk LINK/NODE Type = 3 :   BackBone Trunking Factor =2
MAX ALLOWED Data TK LOAD =0.8:   FDX FACTOR/Data =1
TK Multiplexing FACTOR =0: VoiceTKBlocking =0.1   Fftd=1
Max. Allowed Data Rate(bps)/AL=1440000:   Traffic Growth Factor= 1
TOTAL INTER/INTRA NODAL TRAFFIC=21330400

NO. OF SWITCHES=4   AT   1 2 3 4

ORIGINAL TRUNK TRAFFIC MATRIX AS FOLLOWS

0	3577600	3591200	3564000
0	0	3521600	3552000
0	0	0	3524000
0	0	0	0

MODIFIED TRUNK TRAFFIC MATRIX IS AS FOLLOWS

0	3577600	3591200	3564000
0	0	3521600	3552000
0	0	0	3524000
0	0	0	0

MODIFIED TRUNK MATRIX IS AS FOLLOWS

0	3	3	3
0	0	3	3
0	0	0	3
0	0	0	0

TRUNK COST MATRIX IS AS FOLLOWS

From	To	Miles	$/TK	TOT.$
1	2	541	8730	26190
1	3	1135	13037	39111

Continued

Table 14.2 *Continued*

From	To	Miles	$/TK	TOT.$
1	4	1332	14465	43395
2	3	1178	13349	40047
2	4	1135	13037	39111
3	4	542	8738	26214

*****SUMMARY OF SYSTEM COSTS*****
ACCESS LINK COST=0
TOTAL TRUNK COST=214068
TOTAL SYSTEM COST=214068
Cents per 1 MBITS=0.92

The response time requirements as set by Belcore are easily met for both network topologies studied. Belcore objectives demand 90 milliseconds and 50 milliseconds for cross-office delays (equivalent to one way delay between one STP to another when 25-character-long messages are employed) during periods of full loads. EcoNets computes the values of bid-to-start of message (BST) that includes a turnaround between the SSP and an SCP (equivalent to a Host in our example). The computed values of BST are 62 ms and 26 ms for the two topologies during periods of full load. Assuming half of these values are conservative numbers for cross-office delays, we easily meet the Belcore objectives.

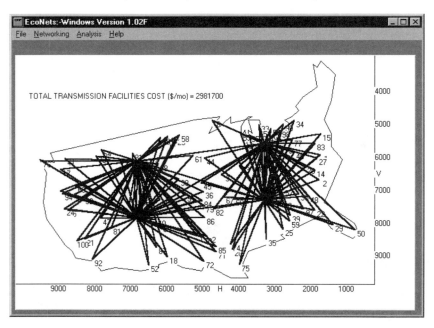

Figure 14.8 The final SS7 network topology with four STPs and four SCPs.

Table 14.3 SS7 Network Modeling/Design Data

********** APPLICABLE DESIGN PARAMETERS **********

FTD File = FTFGN4.txt:	VHD File = VHD100c:	LINK File = LINK100	
ALT / F_{lk} = 3 / 1:	Tariff No. = 3:	TGF = 1	

*********** ACCESS LINE CONNECTION DATA **********

NN	Sw	ALs	MIs	$/AL	$/ALB	
1	63	3	501	2225	6676	
2	63	9	587	2337	21036	
3	63	5	499	2222	11114	
4	96	3	138	1385	4156	
5	96	9	760	2562	23060	
6	96	3	128	1362	4086	
7	63	5	494	2211	11056	
8	63	7	505	2230	15615	
9	63	9	483	2185	19672	
10	63	5	582	2330	11654	
11	96	3	889	2729	8189	
12	96	9	335	1842	16582	
13	96	3	540	8730	26190	Normal Trunks
14	63	7	511	2238	15670	
15	63	5	499	2222	11114	
16	96	9	647	2415	21738	
17	96	9	672	2447	22030	
18	96	9	957	2818	25365	
19	63	5	161	1438	7194	
20	63	5	1059	2950	14754	
21	96	3	876	2713	8139	
22	63	5	778	2585	12928	
23	96	7	403	2000	14001	
24	96	5	783	2592	12960	

Continued

Table 14.3 *Continued*

NN	Sw	ALs	MIs	$/AL	$/ALB
25	63	9	841	2667	24008
26	63	3	423	2046	6139
27	63	5	479	2176	10882
28	96	7	185	1494	10461
29	63	9	984	2853	25681
30	63	7	567	2311	16179
31	96	7	86	1247	8729
32	96	7	584	2333	16334
33	63	3	185	1494	4483
34	63	9	340	1854	16686
35	63	7	919	2768	19382
36	96	3	654	2424	7273
37	96	5	406	2007	10036
38	96	9	405	2004	18044
39	63	7	727	2519	17635
40	96	5	581	2329	11647
41	96	7	632	2395	16771
42	96	5	363	1907	9537
43	96	3	296	1752	5256
44	63	7	544	2281	15970
45	63	9	672	2447	22030
46	63	7	271	1694	11858
47	96	9	632	2395	21562
48	63	9	638	2403	21633
49	63	7	84	1239	8678
50	63	3	1149	3067	9203
51	63	9	131	1369	12322
52	96	5	1005	2880	14403

NN	Sw	ALs	MIs	$/AL	$/ALB	
53	63	7	226	1589	11127	
54	96	3	640	2406	7218	
55	63	3	541	8737	26212	Normal Trunks
56	63	3	152	1417	4253	
57	63	9	467	2148	19338	
58	96	7	450	2109	14765	
59	63	7	779	2586	18108	
60	96	5	725	2516	12583	
61	96	9	495	2213	19923	
62	96	3	937	2792	8377	
63	55	1	541	8737	8737	Normal Trunks
64	96	7	331	1833	12832	
65	96	7	416	2030	14212	
66	63	9	221	1578	14202	
67	63	5	625	2386	11933	
68	96	7	586	2336	16352	
69	63	7	598	2351	16461	
70	96	5	929	2781	13909	
71	96	7	1114	3022	21157	
72	96	3	1126	3038	9114	
73	96	9	361	1902	17125	
74	63	3	1016	2895	8685	
75	63	3	1185	3114	9344	
76	96	3	842	2668	8006	
77	63	7	245	1633	11435	
78	63	7	159	1434	10039	
79	96	7	724	2515	17608	
80	96	9	307	1777	15997	
81	96	7	677	2454	17180	**Continued**

Table 14.3 *Continued*

NN	Sw	ALs	MIs	$/AL	$/ALB	
82	63	5	771	2576	12882	
83	63	9	439	2083	18753	
84	96	3	683	2462	7386	
85	96	9	1080	2978	26804	
86	96	3	800	2614	7842	
87	96	7	843	2670	18691	
88	96	9	361	1902	17125	
89	96	7	652	2421	16953	
90	63	9	243	1629	14661	
91	96	5	528	2260	11303	
92	96	9	1021	2901	26114	
93	63	9	304	1770	15935	
94	96	5	721	2511	12557	
95	63	5	157	1429	7147	
96	13	1	540	8730	8730	Normal Trunks
97	63	9	718	2507	22569	
98	63	7	181	1485	10396	
99	96	9	463	2139	19255	
100	96	9	934	2788	25096	
1	55	3	655	2425	7277	
2	55	9	515	2243	20194	
3	55	5	106	1311	6556	
4	13	3	565	2308	6926	
5	13	9	601	2355	21200	
6	13	3	576	2323	6969	
7	55	5	48	1103	5515	
8	55	7	883	2722	19055	
9	55	9	446	2100	18900	

Continued

NN	Sw	ALs	MIs	$/AL	$/ALB
10	55	5	157	1429	7147
11	13	3	564	2307	6922
12	13	9	797	2610	23493
14	55	7	527	2259	15815
15	55	5	804	2619	13097
16	13	9	548	2286	20580
17	13	9	809	2625	23633
18	13	9	467	2148	19338
19	55	5	532	2265	11329
20	55	5	555	2295	11478
21	13	3	499	2222	6668
22	55	5	454	2118	10592
23	13	7	835	2659	18618
24	13	5	636	2401	12005
25	55	9	329	1828	16457
26	55	3	481	2181	6543
27	55	5	611	2368	11842
28	13	7	546	2284	15988
29	55	9	650	2419	21773
30	55	7	307	1777	12442
31	13	7	617	2376	16634
32	13	7	507	2233	15633
33	55	3	721	2511	7534
34	55	9	805	2620	23587
35	55	7	378	1942	13595
36	13	3	638	2403	7211
37	13	5	152	1417	7089
38	13	9	522	2252	20275
39	55	7	274	1700	11906

Continued

Table 14.3 *Continued*

NN	Sw	ALs	MIs	$/AL	$/ALB	
40	13	5	172	1464	7321	
41	13	7	846	2674	18718	
42	13	5	829	2651	13259	
43	13	3	249	1642	4928	
44	55	7	657	2428	16998	
45	55	9	577	2324	20919	
46	55	7	521	2251	15761	
47	13	9	312	1789	16102	
48	55	9	390	1970	17730	
49	55	7	622	2382	16680	
50	55	3	848	2676	8030	
51	55	9	669	2443	21995	
52	13	5	472	2160	10801	
53	55	7	665	2438	17071	
54	13	3	589	2339	7019	
56	55	3	685	2464	7394	
57	55	9	195	1517	13659	
58	13	7	877	2714	19000	
59	55	7	302	1765	12361	
60	13	5	711	2498	12492	
61	13	9	774	2580	23224	
62	13	3	633	2397	7191	
63	55	3	541	8737	26212	Normal Trunks
64	13	7	722	2512	17590	
65	13	7	272	1696	11874	
66	55	9	332	1835	16519	
67	55	5	352	1881	9409	
68	13	7	614	2372	16607	

Continued

NN	Sw	ALs	MIs	$/AL	$/ALB		
69	55	7	268	1687	11809		
70	13	5	607	2363	11816		
71	13	7	774	2580	18063		
72	13	3	726	2518	7554		
73	13	9	689	2469	22229		
74	55	3	523	2254	6762		
75	55	3	659	2430	7292		
76	13	3	1002	2876	8630		
77	55	7	627	2389	16725		
78	55	7	523	2254	15779		
79	13	7	603	2358	16507		
80	13	9	797	2610	23493		
81	13	7	241	1624	11370		
82	55	5	454	2118	10592		
83	55	9	700	2484	22358		
84	13	3	597	2350	7051		
85	13	9	757	2558	23025		
86	13	3	596	2349	7047		
87	13	7	334	1840	12881		
88	13	9	182	1487	13387		
89	13	7	228	1594	11159		
90	55	9	742	2538	22849		
91	13	5	478	2174	10871		
92	13	9	559	2300	20708		
93	55	9	392	1974	17772		
94	13	5	687	2467	12336		
95	55	5	636	2401	12005		
96	13	3	540	8730	26190		
97	55	9	360	1900	17104	Normal Trunks	**Continued**

Table 14.3 *Continued*

NN	Sw	ALs	MIs	$/AL	$/ALB	
98	55	7	677	2454	17180	
99	13	9	86	1247	11224	
100	13	9	573	2319	20872	
96	63	3	1133	13029	39088	Backup Trunks
96	55	3	1176	13341	40023	Backup Trunks
13	55	3	1134	13036	39110	Backup Trunks
13	63	3	1331	14464	43394	Backup Trunks

TOT. # SW TERMINATIONS=1252:   TOTAL ALs COST/MONTH($)=2981700

14.4 Concluding Remarks

A basic methodology for modeling highly reliable CCSS networks is presented. A useful CCSS network was synthesized for 100 SSPs, four STPs, and four SCPs. A monthly cost of about $2.98M results in a very reliable CCSS network employing 56Kbps A-Links and high-speed B- and C-Links. A monthly cost of about $2.07M results for a very reliable CCSS network employing high-speed links everywhere. The use of high-speed links uniformly in the CCSS network is therefore recommended. It is also recommended that one should not attempt to save money by using only two SCPs. That false move based on *common sense* could cost another $280,000 per month.

Exercises

14.1 Using the CCSS network design methodology of Chapter 14, design a CCSS network using VHD100b input file, 100 SSPs, and a correct number of STP/SCP pairs.

14.2 Compute the actual cost penalty for using only two SCPs for two cases: (1) SCPs are not collocated with an STP and (2) SCPs are collocated with an STP.

14.3 Using the CCSS network design methodology of Chapter 14, design a network using VHD100c input file, three STP pairs, and three SCP pairs.

Bibliography

Kessler, G. 1990. *ISDN: Concepts, Facilities, and Services.* New York: McGraw-Hill.

15

The Overall Network Planning Process

15.1 Introductory Remarks

Most corporations have already invested large sums of money in their voice, data, and information processing networks. Along with their employees, these networks have become their strategic resources for maintaining competitive positions. The customer premise equipment (CPE) consists of voice terminals, dumb data terminals, PBXs, hosts, intelligent workstations (IWSs), minicomputers, personal computers (PCs), and local area networks (data LANs). Whereas voice requirements are growing at about 3 to 4 percent a year, the PCs, desktop workstations (WSs), and LANs are increasing at the rate of 20 to 30 percent. Furthermore, most corporations must be capable of introducing new products, services, and applications quickly and economically to stay abreast of the competition. This requirement is forcing them to consider corporationwide system integration to achieve large economies of scale.

Since most of the existing CPEs are most likely to be interconnected via proprietary networks, any migration to a new integrated corporate network is further complicated by the emergence of new standards for voice, data, and distributed DP. Standards like X.25, IEEE 802, OSI, ANSI's FDDI, ISDN, B-ISDN, Frame Relay, and SMDS for distributed data communications/processing are confusing the situation. Several new technologies like ATM switching are forcing these already known standards to change and helping to create several new standards. These developments provide several alternatives to corporations for solving their complex problems, but this also causes many uncertainties in the minds of corporate officers.

A major portion of a corporate network cost is still in the form of monthly lease expenses for communication facilities connecting their widely scattered locations. Such facilities consist of voice grade, DDS, and digital T1 and T3 lines. Many corporations also own microwave and optical fiber routes. Before 1984, there was only one carrier, AT&T. Now there are many common and bulk carriers with excess fiber capacity. This has increased the number of alternatives for designing one's private network(s). Another choice is to use virtual private network (VPN) or software-defined network (SDN) and VNet services. They are also seeing an end to the so-called leased-line tariffs. This creates even more uncertainties.

There are other trends that have been gathering momentum ever since the divestiture of 1984. At that time, an enterprise spent about 25 cents of each networking dollar on network management and control. Today, that portion for NMC has shot to over 50 cents of each networking dollar (Morency, Lippis, and Hindin, 1995). This drastic increase in NMC costs has forced many enterprises to shirk their responsibilities for their most strategic resource. This is easily reflected in the number of outsourcing contracts and embracing of virtual networking. This trend towards networking complexity is causing havoc within the corridors of enterprise power.

How to pierce this ever thickening wall of complexity and bring some sense of order in the planning process? The following sections describe several concepts underlying the ongoing planning process and the evolving technologies and standards that can be employed to face the new realities, opportunities, and challenges. The techniques for optimizing network topologies as described in the previous chapters form only a small but important task of this planning process. The discussion that follows will be combined with the contents of Appendix B dealing with evolving international and de facto standards to understand the overall planning process.

15.2 The Basic Planning Process

Each enterprise can understand the concepts and the solutions offered in this section only within the framework of the planning process. The planning process for corporate networking should be divided into three ongoing cycles:

1. Strategic planning cycle

2. Tactical planning cycle

3. Operational planning cycle

The strategic planning cycle deals with studying (1) the ever-changing, long-term objectives and goals of a corporation, (2) understanding the competition, and (3) ways to harness future developments in corporate networking to stay abreast of the competition. The strategic planning cycle should be an ongoing effort resulting in the publication of a white paper every six months. Some of these white papers, if accepted by the top management, should trigger a major tactical planning exercise.

The tactical planning cycle requires one to two years and it deals with the actual design and evaluation of alternative solutions. Consequently, it requires the use of intensive model making and network design techniques. The tactical planning cycle could result in meaningless outputs if good design tools are not available.

The operational planning cycle is an integral part of the ongoing process with a yearly check on the deployed network management and control (NMC) techniques against a backdrop of latest progress made in the NMC-based standards.

Based on the preceding discussions, one can therefore say that a typical planning process is an ongoing process that requires the services of a number of dedicated, focused, and experienced persons. The entire planning cycle is doomed to failure if a full commitment from the top management is not available.

To accomplish the aforementioned tasks, one needs a new type of corporate thinking. The traditional telecommunication and DP managers of most corporations are generally adept in only the operational aspects of the planning process. New organizational structures must be in place for guiding and awarding the work of new thinkers and workers adept in the strategic, tactical, and operational aspects of the planning process. In some cases, the corporation should also reward its planners for not only guiding its own destiny but also influencing international standards.

15.2.1 The Strategic Planning Process

The strategic planning group must constantly study the new technologies and the evolving standards for voice, data, distributed data processing (DDP), and multimedia applications in their enterprises. The relationships between the enterprise's long-term goals, technology, and standards must be clearly understood by the strategic planners before recommending a clear migration path leading toward the future enterprise network. One can illustrate the strategic planning process by studying the recent developments for any lessons to be learned.

Most of the standards for modern voice and data networks derive their origin from VLSI, digital switching, digital transmission, and fast packet switching technologies. The VLSI technology has made digital communication and distributed data processing (DDP) systems affordable to enterprises. The phenomenal increase in the sale of PCs, desktop workstations (WSs), data LANs and internetworking products in particular can also be attributed to VLSI technology.

During the early 1980s, strategic planners saw the need to integrate their emerging new data applications with those for voice. Although voice traffic accounted for 70 percent of the networking costs, it was increasing only at the rate of about 3 to 4 percent annually. The data traffic, on the other hand, began to grow dramatically. The strategic planners had three choices: (1) select a digital PBX to integrate their voice and data applications, (2) select a standalone packet switching system (PSS) based on the X.25 standard, or (3) go with stand-alone LANs to handle their need for data communications. The standalone LAN won the battle for three reasons: (1) LAN could handle bursty traffic at higher speeds (e.g., an Ethernet operating at 10Mbps) better than a digital PABX operating at 56Kbps or a PSS operating at the maximum speed of 56Kbps, (2) a data LAN provides PS connections and disconnections almost instantaneously as compared to a digital PABX taking almost a second to set up a circuit switched end-to-end path, and (3) whereas a data LAN allows full sharing of the bus or a ring, a PABX does not allow an efficient utilization of circuits for bursty data traffic.

But a LAN could not grow too large, either in the number of stations served or its footprint. Multiple LANs belonging to a workgroup or a department mushroomed. They also needed to communicate with one another. This was how the LAN internetworking technology emerged. Bridges operating over the OSI's second layer came to the rescue first. Routers operating over the third network layer provided a more flexible internetworking capability. Since routers were generally interconnected over point-to-point leased lines, a misutilized network topology emerged. While this upheaval was just beginning , the VLSI technology gave birth to Intelligent Hubs (IHs). It completely replaced the physical LAN with a logical LAN based on a star topology. Whereas an original LAN employed either an open bus (of an Ethernet) or a closed ring (of a Token Ring) topology, an IH always connects the WSs to itself via a star topology using the ubiquitous unshielded twisted pairs(UTPs). Furthermore, the IH completely duplicated the original IEEE 802 LAN protocols. In other words, each message must be transmitted to all the stations even if the message has a single destination address. The network interface card (NIC) attached to the WS accepts or rejects the message based on the address as was demanded by the IEEE 802 standards.

This is when a golden opportunity was lost to create a seamless, end-to-end, connection-oriented data communications technology. An IH could have been easily transformed into a new data PABX with a capacity to handle voice. *But the installed base of older LANs could not be ignored.* Instead, an IH was doomed to employ the connectionless PS technology of older LANs, thus preventing a seamless transition to all the existing and evolving broadband public networks. The vendors didn't see it. Neither did the user community and the nonexistent strategic planners of enterprises. The academic community is always ignored by the standards-making bodies. Meanwhile, the complexity of enterprise networks grew by leaps and bounds. Point-to-point broadband circuits ended up connecting the hordes of routers. During this period, NMC costs also skyrocketed. Most of the energy of vendor and user communities was spent in creating many additional but confusing standards just to fight self-ignited fires. T1-multiplexers tried to create some sharing of broadband WAN facilities for voice, data, and video applications. But they couldn't go very far since channelized data circuits were still misutilized by bursty traffic and they can't operate at the highest data rate capacity of the broadband facilities. In simple words, their networks were never scalable in terms of the bandwidth required and could not achieve low latency inherent in broadband facilities.

The original standards for data communication (as initiated by CCITT, now under the ITU umbrella, and later incorporated by ISO) arose from the underlying need for rapid data interchange in the highly developed countries and the availability of affordable X.25 packet switching products. This resulted in the lower four layers of ISO's Open Systems Interconnect (OSI) reference model as shown in Figure B 2.1 of Appendix B. The top three layers of the OSI reference model deal with DDP and generally reside in host processors located in separate locations. Figure B.4.9 illustrates the reference model for Local Area Networks (LANs) as originally defined by IEEE and now accepted by ISO. Later came the matured ISDN standard. The ANSI-initiated standard called the Fiber Distributed Data Interface (FDDI) is based on the optical fiber technology. Such a standard for Metropolitan Area Networks (MANs) employs a data rate of 100Mbps. Another data networking standard based on a subset of IEEE's 802.6 MAN is now called Switched Multi-Megabit Data Service (SMDS). Such an architecture is supposed to interconnect a large number of LANs for both MAN and WAN environs. All these standards suffered from many defects. Some were ill timed. Some were too costly. Also, these can handle only data. Most importantly, these standards could not satisfy enterprise's basic need for integrating their voice, data, and now the growing multimedia applications into a single network.

In the meantime, the need for bandwidth has been growing steadily in American enterprises. Many high-performance workstations already demand 50Mbps. Now the time has come again for the user and vendor communities to make an important choice about a new technology called Asynchronous Transfer Mode (ATM). It is built on the fast PS technology and employs short 53-byte cells (with 48 bytes of payload) to provide low latency to all applications like voice, data, and video. The ATM switching technology is now an integral part of the original broadband ISDN (B-ISDN) architecture and its evolving standards are being managed by the ATM forum with membership exceeding 700 vendors/users. ATM is now the only technology that handles LAN, MAN, and WAN environs in a seamless manner. However, there are many distracters in the consultant community. Their voluminous articles are very confusing to the user community. Whereas one writer predicts the early demise of ATM due to the extremely high per-termination costs, another article extols the ATM technology for its ability to offer the lowest cost per Mbps data rate. Some other articles attack the slow standards-making process and predict that ATM will not be a force until 2010. Then one reads about a few bold vendors (e.g., Nortel) that are taking orders for ATM switches providing a seamless integration of voice (based on variable bit rate option not yet standardized) and LAN data. The savings realizable are similar to those shown in Chapter 10 dealing with Integrated Broadband Networks. Some vendors are coming up with proprietary hardware early to achieve most of the promises of ATM today. They believe that the VLSI technology (e.g., application-specific ICs) can be used to retrofit their switches later when the standards mature. Meanwhile, their major thrust is to treat the installed base as simply a set of interfaces to ATM switches (via LAN emulation, multiprotocol over ATM, etc.). They also believe that the enterprise network will be simplified to such an extent that the proportion of NMC costs to the total networking costs will go down to the pre-1984 levels.

What about the DDP revolution? It still has still a long way to go. The OSI never took off as the ISO and the academic communities expected. Proprietary standards like Unix, Windows NT, and SNA still hold their grounds. The TCP/IP protocol—which OSI was groomed to replace—has grown to become the de facto standard, particularly due to sudden growth of the Internet. However, one cannot ignore the contributions of OSI in the area of multilayer DDP architectures. Even IBM's SNA architecture, originally introduced in 1974, ended up employing seven layers (Figure B.6.2 illustrates the similarities and differences between the layers of SNA and OSI reference models). In 1981, IBM unleashed another de facto standard called the PC-DOS. The sale of their PC just took off and the DDP revolution got a big boost. Dumb terminals finally got a backseat. But PCs cannot live by word processing and spreadsheets alone. They must also communicate with one another and with mainframes. LANs came to the rescue. Finally IBM had to shift gears. SNA has evolved to face the new realities of DDP by introducing Advanced Peer-to-Peer Networking (APPN). While DOS still survives, additional layers have been built to create the new de facto standards like Windows 3.1, Windows 95, and Windows NT.

As time progresses, IBM will also be forced to offer OSI interfaces. They already offer many such interfaces to European customers and a few for U.S. customers (e.g., NPSI for accessing X.25 network and X.400/EM). This does not imply that SNA will disappear. Instead, the OSI-based networks will provide vehicles for internetworking just as PTNs did for voice applications. Which network standards will be employed for strategic purposes will depend mainly on the existing investments in hardware and software products and the migration path a corporation intends to follow to meet its long-term goals. The tactical planning process can provide additional help in choosing the right technologies and standards in a manner as discussed next.

There will be cases when a large corporation must also influence the development of a new standard to handle its operations (e.g., General Motors asking for the MAP-based network architecture of Figure B.6.7).

15.2.2 The Tactical Planning Process

When the strategic planners develop a successful *white paper*, the enterprise may decide to actually define a set of viable solutions to achieve its long-term goals and objectives. The tactical planning group has the duty to provide concrete networking solutions. This is the group that actually quantifies the relationships between the products or services provided by the corporation and the performance of its communication and information processing networks. Based on these analyses, one can assess the applicability of several viable technologies and various levels of system integration for increasing the productivity of the worker and/or providing better products/services at a lower cost than before. Only the tactical planning group can assess the true benefits of a new technology (e.g., ATM) to an enterprise. Without their analyses, all statements about the potential benefits should end up as hot air. One should remember the Gulf War of early 1990s that was the subject of intense discussion by the so-called TV experts. The actual tactical war as managed by trained generals and fought by the dedicated soldiers proved most of these pundits wrong.

The tactical planning process consists of the following well-defined tasks or steps:

1. Model the current environments related to CPEs, applications/services, associated levels of performance provided to the users, and their relationships to productivity levels.

2. Model the new environments in terms of additional CPEs, applications, and desired levels of performance.

3. Develop a life cycle traffic growth model for each of the applications/services.

4. Assess the new technologies and associated standards available for applicability.

5. Develop viable designs for one or several integrated networks with life cycle cost and performance models. Chapters 2 through 14 of this book describe the design process for most environs.

6. Develop viable cutover models ensuring uninterrupted service to existing users and modify the life cycle cost models accordingly. This task must include facility planning.

7. Select the most desirable or least risky solution and sell it to the management.

The first step of the tactical planning process requires the longest period to complete, especially if it has never been attempted before. The second and third steps require a full cooperation from the top corporate managers. The fourth, fifth, and sixth steps are greatly facilitated by the availability of time-proven modeling, analysis, and design tools. The success of the seventh step is obviously dependent upon (1) a clear understanding of the problem, (2) depth of the task accomplished, and (3) faith in one's approach or solutions. The use of meaningful return-on-investment (ROI) models also helps. Quick and revolutionary solutions will never accomplish the goals of the corporation.

Effectiveness of this group is directly dependent upon the availability of time-proven tools that are based on algorithms and not on artificial intelligence. Many of the traditional tools required large mainframes for providing "very accurate" answers. Such accuracy was found to be needless since the tariff and traffic environments usually change while going from the network synthesis stage to the implementation stage. Furthermore, the network topologies are generally independent of the actual tariff employed. This was clearly shown in Chapter 8. The choice of mainframes was natural since only large computers could handle the huge amount of data associated with tariffs and traffic. However, the use of bypass facilities, availability of discounts to large users, and leasing of private lines from bulk carriers have destroyed the constancy of tariffs. It now appears that tariffs as we know may not exist anymore. The user must negotiate the rates individually. Simplified models of tariffs must be constructed for computing approximate costs of any networking alternative during the early stages of network planning. Such models allow the use of powerful desktop computers employing Intel's Pentium™ or Motorola's RISC CPUs. Graphs detailing network topologies and text detailing link/trunk costs can be stored as files for later retrieval, editing, and printout. Due to the interactive nature of the activity, one can now evaluate a large number of alternatives within days instead of months. Several technologies can be quickly evaluated for application. Several network topologies can be superimposed to synthesize integrated voice, data, and video networks. Chapters 5 through 14 illustrate this process in detail.

These days there is a movement that is tilting toward the use of expert systems for tactical network planning. Such an approach looks completely inappropriate in view of the availability of meaningful network design algorithms. Expert systems are only useful for highly complex situations like patient diagnosis or network fault prediction/location. Using expert systems for network planning is like using lawnmowers for shaving one's beard. Use of expert systems will also tend to encourage the use of mainframes again and thus slow down the tactical planning process.

Some employ computer simulations for designing networks. There are several problems with that approach. Simulation results are only valid when an optimum network topology system is fully known. And the network system must be in a state of statistical equilibrium before meaningful outputs can be derived. For a typical large network system, it takes a long time to reach such states. Simulation of large network systems also requires the use of a mainframe. Finally, most simulation studies cannot handle a real system operation without spending a huge amount of money and effort on the development of the simulation program and running that program on a mainframe. For example, the operation of an automatic call distribution system requires the scheduling of agents every half hour. This creates a large variety of system transients that are hard to simulate individually due to the lack of a statistical equilibrium. It is easier to learn about such subtleties from the ongoing operations of a real ACD system. That is the responsibility of the operational planning process discussed in the next section.

15.2.3 The Operational Planning Process

Ongoing network planning is one of the many services provided by a well-designed network management and control (NMC) system. There are nine separate but highly interrelated functions that must be addressed by the NMC system:

1. Configuration management dealing with network inventory, topology. and directory on a real-time basis, preferably using a relational database.

2. Fault management dealing with network status supervision, end-to-end testing, diagnostics, fault-reporting/trouble ticket creation/repairing, backup, and reconfiguring.

3. Performance management dealing with performance definitions, monitoring, trending, and thresholding on a real-time basis.

4. Accounts management dealing with planning of network related budgets, allocating costs to end users, and verifications of network system goals.

5. Security management dealing with establishing and maintaining criteria for network access management and reconfiguration control for any number of well-defined partitions.

6. Network planning dealing with periodic network optimizations, modeling of network contingencies for disaster recovery, and ongoing development of strategies to help the user to achieve short-term and long-term goals.

7. Operations management dealing with the efficient running of all operations centers, including the tasks of staffing, training, and controlling information flows.

8. Programmability dealing with the management and enhancement of existing software packages and development of new custom software packages.

9. Integrated system control dealing with a capability to manage and control a multivendor network transparently from any user-friendly console.

The objective of the operational planning group is to maintain a cost-effective private network despite the ongoing perturbations occurring within the corporate network(s) dealing with traffic, outages, and catastrophes. Since the mid-1980s, the field of NMC has been gaining the undivided attention of not only sophisticated large users and vendors but also ISO.

Large users have already discovered that a sophisticated NMC system is essential for operating a cost-effective, multivendor, multiapplication, private network. Unfortunately, each vendor started out with a proprietary NMC system good for only its subnetwork. Since its NMC system could not communicate with the NMC systems of other vendors, the multivendor network could not be operated effectively. This adds to the complexity of existing enterprise networks. To illustrate, the TCP/IP momentum created a very popular NMC standard called Simple Network Management Protocol (SNMP) in 1988. It represented a minimal but simple set of NMC capabilities. Most of the LAN and LAN internetworking vendors immediately embraced it. Its original version (called SNMPv1) has been greatly enhanced by the use of Remote Monitoring (RMON) hardware and later by the introduction of its second version during 1993 called SNMPv2. Although it now allows a distributed NMC capability, it still lacks a true hierarchical NMC architecture as required by enterprise networks.

In 1994, several vendors joined the Management Integration Consortium (MIC) to create a brand-new protocol. But many large players could not agree to anything. Users' long-term interests really suffered with the death of MIC in 1995. Now under the umbrella of Network Management Forum, many vendors have started to introduce products based on the concepts of network elements (NEs), elemental management systems (EMSs) for handling multivendor environments, and some services based on the evolving OSI message protocol. AT&T's Unified Network Management Architecture (UNMA™), DEC's Enterprise Network Management (ENM™), HP's Open View™, Cabletron's Spectrum® and IBM's Netview™ are major examples.

Since the completion of basic OSI framework, ISO has already defined the NMC functions performed at each OSI layer. ISO Draft International Standards (DIS) 7498-4 as drafted in October 1988 defines the OSI Management Framework. It contains the Common Management Information Protocol (CMIP) and Common Management Information Services (CMIS). CMIP is the basic protocol that defines the format of messages that will be passed within and between multivendor systems for implementing services. CMIS defines services dealing with how the standard messages are to be passed within and between multivendor systems or network elements. Since ISO has not yet dealt with the extent and type of services (i.e., with the content of NMC messages), a lot of work still needs to be accomplished. Several vendors and carriers who belong to the OSI/Network Management (OSI/NM) Forum are working to finish this work. Their objective is also to achieve multivendor network interoperability in the shortest possible time. Consequently, their NMC protocol will start with the X.25/IEEE 802.3 transport layers and the upper layers' functionality as expressed by ISO's CMIP/CMIS DIS.

During the next decade, one will start seeing NMC products conforming to the OSI standards. That will enable large sophisticated users to fully manage and control their networks. Interfaces to handle SNMP-based devices will be introduced. The real benefit of such a control will be a full understanding of the network behavior under varying operational environments. This in turn will encourage the development of expert systems (ESs) that will enable the NMC team to not only achieve self-healing at all levels of the network but also help provide the desired level of performance at the lowest possible cost. Since such an approach is fully automated, the cost of providing a full spectrum of NMC services will be drastically reduced as compared to the existing costs.

The introduction of the ATM technology in enterprise networks will finally usher in the truly hierarchical NMC capability. It will enable the collection and analysis of monitored NMC data on a hierarchical manner. One can then implement a true NMC based on the so-called "manager of managers" technique.

15.3 Final Conclusions

The book devoted its first six chapters to describing concepts, analytical tools, algorithms, and a software tool for the analysis and design of network topologies for all types of enterprise networks. Chapters 7 through 14 concentrated on providing a consistent design process for voice, automatic call distribution, data, integrated broadband, backbone, personal communication, and common channel signaling networks. Chapter 15 dealt with the overall planning process for creating meaningful network architecture for the enterprise. Chapter 15 also emphasized that each corporation must hire a new set of employees who focus on the strategic, tactical, and operational network planning processes as an ongoing activity. The corporations must learn to award these passionate planners for innovative thinking and bold initiatives. In some cases, it may be necessary to influence new network-related standards through regular participation in standards-making bodies. One can say with emphasis, however, that no single architecture or utopia will solve all problems. Each corporation will always require unique solutions for its unparalleled set of problems. Each corporation will also find a growing dependence on internetworking to achieve a cost-effective, multivendor solution. This in turn will require the synthesis of networks with increasingly mixed topologies. The network planners should find the tools and algorithms of this book useful for evaluating any hypothesis regarding a new technology, networking, and internetworking solutions. But they must never forget that the analysis and design tools provided in this book are only a small part of the overall network planning process. People behind the process are the most important part of the process. The worst thing that can happen to an enterprise is a complete abdication of its responsibility to control its strategic network resources. No one can ignore the calling of the new information age any longer.

Exercises

15.1   What are the main advantages of ATM-based networks?

15.2   How can one assess the true benefits of ATM technology to an enterprise?

Bibliography

Morency, J. N. Lippis, and E. Hindin. 1995. "The Cost of Network Complexity." *Network World*, July 31, 1995, pp. 44–45.

Derivation of Equations for an M/D/N Type Queue

Let $\qquad C = A / N = \rho = $ Average server utilization $\qquad$ (A.1)

and

$$\tau = (t / t_s - v)\ldots \qquad 0 <= \tau < 1 \qquad \text{(A.2)}$$

where
$v = $ Integer (t / t_s)
$N = $ the number of servers
$A = $ total traffic intensity handled by N servers in erlangs
$t = $ any real number for delay time
$t_s = $ fixed service time

One can define the Prob (delay > t) as follows:

$$\text{Prob (delay > t)} = \sum_{i=1}^{N-1} \sum_{w=1}^{\infty} V_i \, \text{Exp} \, [-A \, (w - \tau)] \, [R] \qquad \text{(A.3)}$$

where

$$V_i = \sum_{j=0}^{i} \mu_j \qquad \text{(A.4)}$$

$$R = \left\{[A*(w - \tau)]^{[N(v + w) + N - i - 1]}\right\} / [N(v + w) + N - i - 1]! , \quad (A.5)$$

$$\mu_j = \text{Exp}(-A) \left\{\sum_{i=0}^{N} \mu_i(A^j / j!) + \sum_{i=N+1}^{N+j} \mu_i [A^{(j-i+N)} / (j - i + N)!]\right\}. \quad (A.6)$$

The final values of μ_j as defined in Equation A.6 are computed by substituting the values of μ_j previously obtained again and again until the values of μ_j become stable from one iteration to another. This technique is only successful if "j" is selected large enough to satisfy the relationship $\mu_{j+1} = C*\mu_j$. Furthermore, one must select the first set of μ_j values as follows:

$$\mu_j = [(N*C)^j / j!] / \left\{[(N*C)^N / (1 - C)N!]\right.$$

$$\left. + [\sum_{i=0}^{N-1} (N*C)^i / i!]\right\} \dots\dots \text{ for } j \leq N \quad (A.7)$$

$$\mu_j = [(N^N*C^j) / N!] / \left\{[(N*C)^N / (1 - C)N!]\right.$$

$$\left. + [\sum_{i=0}^{N-1} (N*C)^i / i!]\right\} \dots\dots \text{ for } j > N \quad (A.8)$$

The published literature dealing with traffic theory does not yet provide simple expressions for the first two moments of the time spent in the M/D/N type queue in the form available for M/M/N type queue as shown in Equations 4.10a, b, and c. The author has simulated the M/D/N type queue for a large number of values pertaining to server load and N. He has discovered that the following normalized (for $T_s=1$) expressions agree very well with the simulation results.

$$\text{AVG}(T_q) = [1 + [P_o / 2(N - A)] \quad (A.9)$$

$$\text{VAR}(T_q) = [1 / (N - A)^2]*[P_o / 3 - P_o^2 / 12] \quad (A.10)$$

It is interesting to note that Equation A.9 reduces to Equation 4.5 and the square root of Equation A.10 reduces to Equation 4.6 for the single server (or N=1). The value of P_o for N=1 reduces to the average server utilization, C (we used ρ to represent server utilization in Chapters 1 through 4). The value of P_o in Equations A.9 and A.10 must be solved using Equation A.3 and the method described in this Appendix. One can use the curves of Figure A.1 to obtain the values of P_o for some useful values of C and N. It is also interesting to state that the values of P_o for the M/M/N queue are only about 6 percent higher than those shown in Figure A.1 for identical values of N and C. The reader can now use the analytical tools presented in this book to study the behavior of the worst

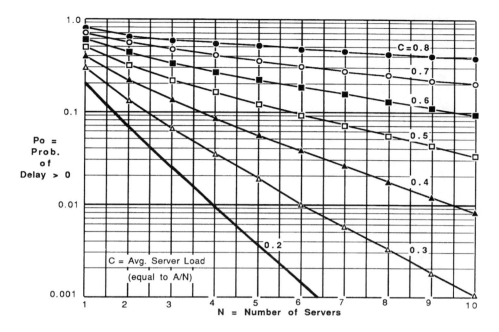

Figure A.1 Probability (delay > 0) or P₀ versus *N* and some useful values of *C*.

case (i.e., an M/M/N queue) and the best case (i.e., an M/D/N queue), and then interpolate the performance of all practical systems that generally fall in between the two extremes.

Note: The method just shown is attributed to J. L. Everett (*State probabilities in congestion problems characterized by constant holding times*, J. Operations Research, 1, 1953, pp. 279–285). His method which is now quite suitable for computations on the new desktop computers was first described in simple terms in an unpublished *Manual of Techniques for Traffic Analysis* prepared at the request of Mr. A. A. Collins, President of Collins Radio Company, in 1968. One of the author's colleagues at Collins, Dr. J. C. Shah in particular, is credited with researching the original papers dealing with M/D/N type queues. The letters M and D in M/D/N stand for the Markov (also Poisson or exponential) process related to arrivals, and D stands for deterministic service times.

B

Evolving International and De Facto Standards for Networking

B.1 Introductory Remarks

Public and private networks need standards for two main reasons: (1) to obtain a complete multivendor interoperability, and (2) to reduce the costs of hardware, software, and network management achieved through the mechanism of free enterprise. The telecommunications-related standards were originally initiated and finalized by the International Telecommunications Union (ITU) agency of the United Nations with full participation of the governments. But that is no longer the case. Many forums and task forces comprised of vendors are springing up in the United States and some other developed nations to complete new standards for network-related protocols with a greater rapidity than was possible through ITU. It is an entirely new game that becomes very exciting and anarchic at times. When these forums and task forces do complete a standard, it is then submitted to ITU for ratification. In older days, standards-making bodies were represented by people with research and academic backgrounds. The vendors and users were allowed only as nonvoting observers. Nowadays, the majority of the attendees at forums are vendors. The users are there only as observers (as part of the End-Users Network Roundtable, or ENR).

A networking standard developed by a forum/task force sometimes represents hardware/software already being marketed by a dominant vendor. In most other cases, the ratified standard gets retrofitted by most competing vendors. In any case, vendor representation at these forums is becoming essential but very costly to successful vendors. Therefore, the standards developed at these forums are generally more definitive as opposed to a grand but untried new entity. Such was the case with the Open Systems Interconnect (OSI) standard developed after a long period by International Standards Organization (ISO) of ITU. However, one must understand OSI before appreciating other standards since it has already influenced many other standards.

In the following two paragraphs, we will first describe the major international and national standards-making bodies before describing some well-known standards, for both public and private (de facto type) networks.

B.1.1 International Standards-Making Bodies

One of the many functions of the United Nations is to promote the development of international standards for telecommunications. To achieve this goal, the International Telecommunications Union (ITU) was created. Two committees known as CCITT and CCIR handle the development of standards for the telephone and telegraph (or the wire-related) and the radio systems. Both of these committees have recently been combined into one entity, ITU-T. We will continue to use the older terminology to stress the historical evolution of standards. This appendix will describe the activities of ITU-T and many popular forums, alliances, and task forces operating in the United States.

Each member nation sends its representatives to the CCITT-sponsored meetings held in Geneva and other cities at regular intervals. Once every three years, a meeting is held for the final approval of new standards under consideration. The national CCITT-sponsored subcommittees also meet in their own countries for finalization of national recommendations. There are many other standards-making bodies in technologically advanced countries (e.g., the United States) and regions (e.g., Common Market Countries of Europe) that coordinate with the CCITT activities. Many Bell Operating Companies (BOCs) and vendors also participate in the CCITT activities as observers. Users of the developed countries are not yet allowed to participate in the CCITT activities but they do participate in the activities of other national standards-making bodies.

The Consultative Committee for International Telephone and Telegraph (CCITT) and Consultative Committee for International Radio (CCIR) have already been mentioned. The International Standards Organization (ISO) represents the international computer and terminal manufacturers. The interests of the Electronic Industry Association (EIA) are self-evident. The American National Standards Institute (ANSI) oversees the interests of the U.S. manufacturers. The National Bureau of Standards and Institute of Electrical and Electronic Engineers (IEEE) also promote the development of standards for

American industries. On behalf of the BOCs and OCCs, the newly formed Exchange Carriers Standards Association (ECSA) is also interested in the development of standards that are suitable for their administrations. None of the aforementioned bodies operate in vacuum. They cooperate with one another all the time. Some of the standards may be named differently but equivalence exists between different nomenclatures employed by different industries of different countries. The relationships between the bodies during the standards-making process are illustrated in Figure B.1.1.

CCITT is composed of 18 study groups (SGs). Each study group (SG) has a responsibility for a particular area such as voice, data, or integrated services. These SGs are described as follows:

1. SGI influences standards dealing with telegraph and telematic services.

2. SGII handles telephone system operations.

3. SGIII handles tariff principles.

4. SGIV handles transmission maintenance of international circuits.

5. SGV handles protection against dangers and EM disturbances.

6. SGVI handles protection of and specifications of cable sheaths and poles.

7. SGVII handles public data networks.

8. SGVIII handles telegraphy and related DTEs.

9. SGIX handles telegraph transmission quality.

10. SGX handles telegraph switches.

11. SGXI handles telephone switching and signaling.

12. SGXII deals with telephone transmission performance.

13. SGXIII deals with automatic and semiautomatic public telephone networks.

14. SGXIV deals with facsimile telegraph transmission and equipment.

15. SGXV deals with transmission systems.

16. SGXVI deals with telephone networks.

17. SGXVII deals with data transmission over telephone and telex networks.

18. SGXVIII deals with digital networks such as ISDN.

The relationships between CCITT, ISO, and U.S. standards organizations are shown in Figures B.1.2. and B.1.3. The two major standards organizations in the United States, ANSI and ECSA jointly sponsor the T1 Technical Committees such as T1C1, T1D1, T1M1, T1Q1, T1X1, and T1Y1 to cover the standards related to Network/CPE, ISDN, NMC, Network Performance, Carrier-to-Carrier Interfaces, and other subjects such as SONET (as related to a synchronous optical fiber network transmission standard for frequencies starting at 50 Mbps). See Figure B.1.4 for an organization of T1 Steering Council.

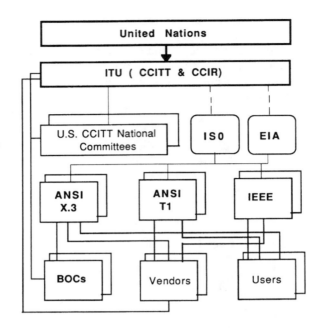

Figure B.1.1 An illustration of the standards-making process.

An interesting aspect of the standards-making process is the need for a complete consensus of all the members of the ITU before creating a new standard. As a consequence, the standards-making process is quite slow when compared to the way technology is changing in the developed nations. The disparity between

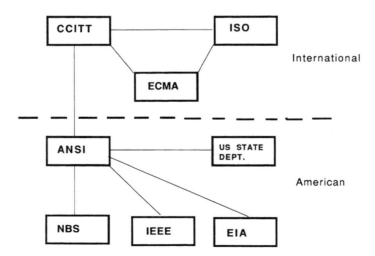

Figure B.1.2 Relationship between the international and U.S. standards-making bodies.

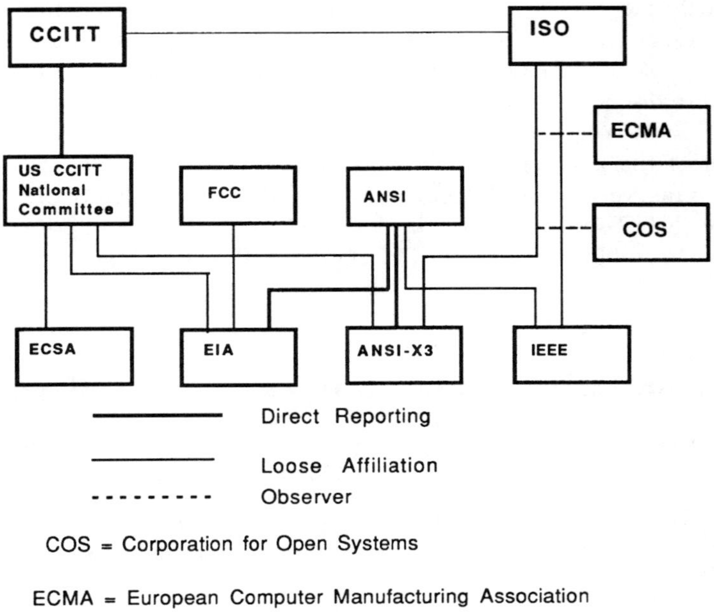

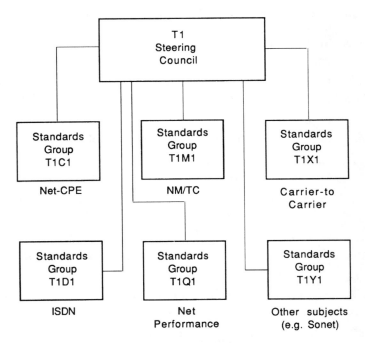

Figure B.1.3 Another view of the interplay between the
international and U.S. standards-making bodies.

Figure B.1.4 The structure of the ANSI T1 committee.

the developing nations and the relatively few developed nations (e.g., the United States, Sweden, Japan, and Germany) makes the process very slow. Furthermore, the presence of PT/T monopolies almost everywhere versus the free economies in developed nations also creates a disparity between the two interests. This gives rise to the creation of many de facto standards in developed nations. The making of the original T1 carrier structure as implemented by AT&T and IBM's Systems Network Architecture (SNA) within the United States could not wait for the CCITT-sponsored standards. Whereas the North American T1 standard has now been blessed with a CCITT standard, IBM's SNA will probably remain out of the CCITT fold forever.

B.1.2 Standards-Making Bodies of the United States

We will list the 10 most important standards organizations set by vendors. Their names and their activities are described as follows:

1. AIW (APPN Implementation Workshop) is focused on implementing APPN (A Peer-to-Peer Network) standards as related to IBM's SNA architecture.

2. ANSI (American National Standards Institute) is focused on LANs and WANs with the second highest membership of 1,400 companies, institutions, and government agencies.

3. ATM Forum is focused on fast packet switching (FPS) as exemplified by asynchronous transfer mode (ATM) switches with the third highest membership of 880 vendors and 150 user companies as part of the end-user network roundtable (ENR).

4. DMTF (Desktop Management Task Force) is focused on desktop WS and PC management.

5. Frame Relay Forum is focused on public and private networks based on Frame Relay fast packet switching technology.

6. GEA (Gigabit Ethernet Alliance) is focused on the development of an Ethernet LAN operating at a data rate of 1 Gbps.

7. IEEE (Institute of Electrical and Electronic Engineers) is interested in telecommunications networks and boasts the highest membership of 320,000 individual members from 147 countries.

8. IETF (Internet Engineering Task Force) is focused on Internet and associated technologies.

9. Network Management Forum is concerned with NMC capabilities of a public and private network.

10. SMDS Interest Group is focused on the Switched Multi-megabit Data Service as a subset of the IEEE 802.6 standard for metropolitan area networks (MANs). SMDS is now applicable for both MANs and WANs.

Some of the most well-known international and de facto standards are described in the following sections of this Appendix. Section B.2 describes how the ISO's Open System Interconnect (OSI) Reference model, is used to see how any existing standard compares to an international, multilayer model. The standards described in Sections B.3, B.4, B.5, and B.6 deal with voice, data, integrated voice and data, and distributed data processing (DDP), respectively.

B.2 Reference Model for Open System Interconnection

B.2.1 Introduction

The increasing use of Very Large Scale Integration (VLSI) techniques in the implementation of communication and data processing equipment has decreased the costs of hardware to such an extent that it is now possible to distribute the intelligence in an integrated fashion throughout the network in a very cost-effective manner. Each network node can be characterized by a collection of unique functions that define the NMC capability or the intelligence of a network system and, therefore, the associated network architecture as defined previously. Using the well-known technique of successive decomposition or iterative modeling, it should be easy to model any collection of communication and data processing functions handled by a network node. The international standards bodies like ISO and CCITT have provided just such a modeling technique. A collection of communication and data processing functions is generally known as the standard or the protocol. A protocol, on the other hand, defines the communication procedures and data encoding used to interconnect two compatible network nodes or systems. Most of the standards protocols (international and de facto types) can be represented in a consistent fashion. An obstacle to understanding the standards or protocols is their diversity and mundane complexity.

Reading about a protocol is never an exciting event. One way to deal with their diversity is to model a standard in terms of hardware/software layers. When two compatible network nodes communicate with one another to provide service, communication occurs between layers. Two adjacent layers within the same network node are separated by appropriate interfaces. Each interface creates a queue of tasks to be handled. Chapter 4 describes the methodology for analyzing such queues. Although these queues can cause delays, the multilayer approach for designing the software helps interoperability and simplicity in software design. Studying their software layers can help us understand the different protocols.

B.2.2 Open System Interconnection (OSI) Reference Model

In 1983, the International Standards Organization (ISO) created a standard (ISO 7498) for a reference model for distributing communication and data processing functions within a node. CCITT has also accepted this model as a basis for its evolving standards. A reference model is an attempt to represent the layers of communication and data processing functions within a network node and the required interprocess communication. See Figure B.2.1 for the seven-layered reference model for OSI designed to foster unfettered interoperability between hardware/software products manufactured by different vendors. The lower three layers—physical, data link, and network—represent the basic network service functions. These functions are distinct from the four higher layers—transport, session, presentation, and application. These four higher layers mainly deal with the data processing functions. Since this section is discussing only the OSI reference model, no attempt will be made to define the actual standards or protocols covered by each layer. The succeeding sections will deal with only the major standards for voice, data, integrated voice/data, and distributed DP environments respectively.

The physical layer deals with the following attributes:

1. Electrical (e.g., data rates and voltage levels)

2. Functional/procedural (e.g., definitions of data, timing, and control signals)

3. Mechanical environments (dealing with the type of connector and number of pins)

The data link control (DLC) layer performs the following functions:

1. Data link connection maintenance

2. Link activation and deactivation processes

3. Map data units to data link units

4. Multiplex data units to multiple physical connections

5. Delimit data link units

6. Error detection, recovery, and notification to end users

7. Network identifications (ID) and parameter exchange with peer DLC parties within the compatible network nodes

The network layer provides the following capabilities:

1. Providing network address and end-point ID

2. Multiplexing of network connection to DLC

3. Segmenting/blocking of datastreams

4. Selecting the appropriate service

5. Selecting the desired service quality

6. Error detection and recovery and their notification

7. Expediting data transfers

8. Connecting resets with loss of data and their notification

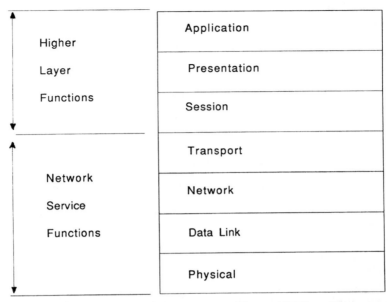

Figure B.2.1 ISO's Open System Interconnect (OSI) reference model.

The transport layer has three phases:

1. Call establishment phase dealing with transit set-up and network service, throughput, delay and error characteristics and data unit sizing

2. Data transfer mode dealing with blocking/segmenting and multiplexing of connections, end-to-end data flow control, data unit ID, connection ID, error detection and error recovery, and expedited data

3. Termination phase dealing with the cause of termination and the ID of termination connection

The session layer handles the following tasks:

1. Session connection

2. Session establishment and release/synchronization

3. Normal and expected data exchange

4. Quarantine service

5. Exception reporting

The presentation layer provides the following capabilities:

1. Data transformations required for information processing (IP)

2. Data formatting into either ASCII or binary or EBCDIC or numeric or Graphic form

3. Syntax selections

The application layer provides the following services necessary for any application:

1. Identifying participating users

2. Establishing authority to communicate

3. Testing of privacy mechanisms if required

4. Allocating costs or billing and testing quality-of-service

5. Recovering from error and allocating responsibilities

6. Selecting a dialog discipline

7. Transferring actual information between partners

A study of the various layers shows that the higher the layer, the more complex the software. It should also be obvious that the lower the layer, the more dependent it is on the physical hardware. Furthermore, it should be obvious that the application layer does not specify a given application—it represents only the interfaces provided for running the various applications/services handled by the network system. A network of compatible nodes represent an intelligent system quite similar to a single computer. Together, the interconnected network nodes provide a set of services to a distributed population of users while sharing distributed databases and computing resources. See Figure B.2.2 for an illustration of interlayer communications between interconnected nodes. In some quarters, the OSI reference model is referred to as the OSI architecture. This concept should not conflict with the one introduced in Section 1.7 if one considers the various layers as DP/DC entities and how these are related to one another. In the sections that follow, several popular standards (international or de facto type) and their similarities to the OSI reference model are discussed.

B.3 Standards Related to Voice Communications

B.3.1 Introductory Remarks

Most of the telephone networks in the developed countries like the United States, Canada, and Japan were developed before the international standards were finalized. The first attempt to interwork two voice networks was made after World War II when the gateways in New York and London were connected together to perform the necessary handshaking and to translate the signaling functions over the transatlantic cable. The networks in North America and European countries have evolved rather independently.

The CCITT Green Book describes most of the international standards dealing with voice communications that primarily employ the star topology at

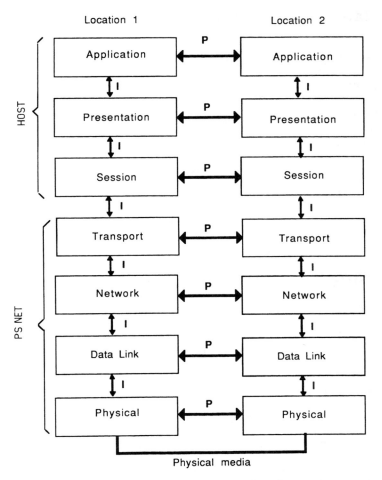

Figure B.2.2 An illustration of the interaction between two nodes based on the OSI reference model.

all levels of hierarchy. Therefore, most of the well-known standards will be divided by transmission and switching considerations only. For transmission, radio and wire line media will be considered. Both the analog and digital switches will be considered.

Whereas a pure data communication network standard can generally provide the OSI network layer services for the lowest three layers, namely physical, link, and network, most existing circuit switched (CS) voice networks provide only the lowest physical layer at this time. These networks must evolve to provide the higher layers of OSI services in order to handle other services on an integrated basis. A new set of standards that are the basis of new network architecture, Integrated Services Digital Network (ISDN), attempts to achieve that goal. ISDN is discussed in Section B.5.

B.3.2 Radio Transmission Systems

B.3.2.1 Analog Techniques

No important standards for this category

B.3.2.2 Digital Techniques: Some Standards

1. CCIR Rec. 382-3 for 2 GHZ Band, 90 Mbps rate and 1344 PCM channels
2. CCIR Rec. 383-2 for 4 GHZ Band, 90 Mbps rate and 1344 PCM channels
3. CCIR Rec. 384-3 for 6 GHZ band, 140 Mbps rate and 1344 PCM channels
4. CCIR Rec. 497-2 for 8 GHZ band, 70 Mbps rate and 960 PCM channels

B.3.3 Wireline Transmission Systems: Some Standards

1. CCITT Rec. G.120, G.121, G.123, G.133 define transmission performance
2. CCITT Rec. G.101, G.111, G.113, G114 define the transmission loss plan

B.3.3.1 Analog Techniques: Some Standards

1. CCITT Rec. G.232 for FDM Group with 12 Voice Channels @60–108 KHz
2. CCITT Rec. G.233 for FDM SGroup with 60 V. Channels @312–552 KHz
3. CCITT Rec. G.232 for FDM MGroup with 300 V. Channels @812–2044KHz
4. CCITT Rec. G.232 for FDM SMGroup with 900 V. Channels @8516–12388KHz

B.3.3.2 Digital Techniques: Some Standards

1. CCITT Rec. G.711 for μ-law PCM encoding of voice at 64Kbps
2. CCITT Rec. G.721 for ADPCM encoding of voice at 32Kbps
3. CCITT Rec. G.722 for Subband ADPCM encoding of teleconferencing at 48–64Kbps
4. ITU-T Rec. G.728 for LD-CELP encoding of voice at 16Kbps
5. CTIA-(NA) IS.54 for VSELP encoding of voice at 8Kbps

6. ITU-T Rec. G.723 for encoding of voice at 6.4Kbps

7. GSM-Europe, GSM/RPLPC encoding of voice at 13.2Kbps

8. ITU-ISO MPEG-1 for encoding 2-channel audio at 128–384Kbps

9. ITU-ISO MPEG-2 for encoding 5-channel audio at 320Kbps

10. ITU-ISO MPEG-1.2 for encoding addressable video at 1–8Mbps

11. CCITT Px64 for encoding videoconferencing at 64–1,536Kbps

12. FCC (NA) HDTV for encoding television signals at 17Mbps

13. CCITT Rec. G.733 for T1 @ 1.544Mbps with 24 PCM channels (N. America)

14. CCITT Rec. G.732 for CEPT @2.048 Mbps with 32 (30 data, 3 supv) PCM channels

15. CCITT Rec. G.743 for T2 @ 6.312Mbps with 96 PCM channels (N. America)

16. CCITT Rec. G.742 for CEPT @8.448 Mbps with 120 PCM channels

17. CCITT Rec. G.752 for T3 @ 44.736Mbps with 672 PCM channels (N. America)

18. CCITT Rec. G.751 for CEPT @34.368 Mbps with 480 PCM channels

19. CCITT Rec. (NA) for T4 @ 274.176Mbps with 4032 PCM channels (N. America)

20. CCITT Rec. G.751 for CEPT @139.264Mbps with 1920 PCM channels

B.3.4 Switching Systems

B.3.4.1 World Numbering Plan

This numbering plan allocates the country codes to each 15-digit international telephone number. Such a plan was necessary to allow international dialing.

B.3.4.2. CCITT Signaling System #5 for Intercontinental Telephone Cable

B.3.4.3 CCITT Signaling System #6 for Public Telephone Networks

Such a scheme was implemented within the Bell Systems and several other public telephone networks prior to its replacement by the CCITT Signaling System #7 for ISDN.

B.3.4.4 CCITT Signaling System #7 for Public Telephone Networks (for ISDN-Based Public Telephone Networks)

Some of major definitions of SS7 are found in ITU-T Document Q.600, Q.700, Q.800, and Q.900.

B.3.4.5 Q.22/Q.23 Recommendations for Push-Button Telephone Sets (Vol. GVI-1)

B.3.4.6 Q.35 Recommendations for Characteristics of Various Tones (e.g., Ringing, Busy, Congestion, and Warning): (Vol. GVI-1)

B.3.4.7 Q.40-45 Recommendations for the Characteristics of Transmission Plans: (Vol. GVI-1)

B.3.5 Network Planning and Traffic Engineering

1. CCITT Rec. E.520 for the use of Erlang-B formula for sizing trunk bundles
2. CCITT Rec. E.541 for the use of P.01 for the final choice route trunk

B.4 Standards Related to Data Communications

B.4.1 Introduction

Whereas the need for data communications for a distributed data processing (DDP) system is apparent, the need for data communications for a voice network is not that obvious. Practically all of the existing voice communications systems use some form of data communications for signaling and NMC functions. In many cases, signaling data are sent in-band on the same circuit that is employed for voice. In some parts of the system, special order-wire circuits are used to transport alarms and diagnostic data from one node to another. Most existing voice communications systems employ a proprietary protocol for data communications. An international standard for accessing a public data network (PDN) is described in the following subsections.

B.4.2 V. Series Standards for Analog Modems

Nowadays everybody is familiar with the modulator/demodulator (MODEM) devices employed to connect personal computers (PCs) with mainframes.

Sometimes modems are employed to synthesize multidrop links to allow several communication controllers (each controlling several CRT devices) to communicate with a mainframe. In most systems, modems constitute a large investment. For that reason, many intelligent PABXs and hosts employ modem pooling (or sharing) to reduce the costs of communication hardware. Modems are evolving into very intelligent devices, thanks to the VLSI technology. This intelligence is provided to achieve a desired amount of NMC functionality for a PDN.

See Table B.4.1 for a list of V. Series standards for analog modems or their maintenance and/or performance monitoring. Of course, some vendors sell modems that are based on proprietary protocols and used within a private data network.

Table B.4.1 Some Standards for Analog Transmission and Modems

Standard	Description
V.3	International Alphabet#5 for Information Exchange
V.15	Acoustic Couplers for Data Communications
V.16	Standards Description for Analog Modems
V.19	DTMF Modem for PTNs
V.20	Extended DTMF Modem for PTNs
V.21	300baud Modem for PTNs
V.22	1200bps Modem (FDX)
V22bis	300baud (Async), 1200/2400bps (Async or Synchronous)
V.23	600/1200bps Modem for PTNs
V.24	Definitions for DTE-DCE Information Exchange Circuits
V.26	120()/2400bps Modem for4-W Circuits
V.26bis	1200/2400bps Modem for PTNs
V.27	4800bps Modems for 4-W Circuits
V.27bis	4800bps Modem with Equalizer Standard for PTNs
V.27ter	2400/4800bps Modem for PTNs
V.28	Definition for the Unbalanced Interface
V,29	9600bps Modem for 4-W Leased Lines
V,32	9600bps Modem (FDX-Synchronous) for 2-W Leased/Dial-Up
V.35	48000bps Modem for 60-108KHz Group Circuits

Continued

Table B.4.1 *Continued*

Standard	Description
V.36	48000-72000bps Modem for 60-108KHz Group Circuits
V,37	96000-168000bps Modem
V.40	Parity Error Indicators
V.41	ErrorControllers
V.50	Transmission Quality
V.52/53	Bit Error Patterns/Limits

B.4.3 X.Series Standards for Wide Area Public Data Networks

Many assume that X.25 is the only international standard that is required to define a Public Data Network (PDN). Actually, a large collection of international standards is used to plan and implement a PDN. See Table B.4.2 for a list of the related X. Series standards required to define a PDN. The X.1 standard specifies all of the user classes of services provided by a PDN. See Table B.4.3 for a list of such services.

Table B.4.2   Some Standards for Public Data Networks

Standards	Description
X. I	International User Classes of PDN Services
X.2	International User Facilities in PDNs
X.3	Packet Assembly/Disassembly (PAD) Device
X.4	International Alphabet#5 for PDNs
X.20	Start/Stop DTE-DCE Interface in PDNs
X20bis	V.21 Modem for PDNs
X.21	Synchronous DTE-DCE Interface in PDNs
X.21bis	DTE-DCE Interface for Synchronous V.Series Modems
X.24	DTE-DCE Information Exchange Circuits (IECs)
X.25	DTE-DCE Interface for DTEs Operating in Packet Mode
X.26	Electrical Characteristics of Unbalanced Double Current IECs
X.27	Same as above for Balanced IECs
X.28	DTE-DCE Interface for StartlStop DTEs Accessing PADs

Continued

Standards	Description
X.29	Exchange of Control and User Data between DTE and PAD
X.32	Dial-Up PDN Service (< 9600bps)
X.60	Common Channel Signaling for Synchronous Data Service
X.70	Terminal and Transit Control Signaling for Start/Stop Data Service over International Circuits between Asyncronous PDNs
X.7 I	Same as above for Synchronous PDNs
X.75	Data Exchange on International Trunks
X.95	Network Parameters in PDNs
X.96	Call Progress Signals
X. 121	International Numbering Plan for PDNs

Figure B.4.1 illustrates a topology of a typical PDN based on the X. Series standards. Any packet-mode data terminal equipment (DTE) based on the X.25 standard can directly access a PDN node by interfacing with its associated data communication circuit-termination equipment (DCE). A nonpacket mode terminal (any one of existing terminals such as ASCII, BSC, and SDLC) must be

Table B.4.3   PDN Services Related to the X.1 Recommendation

Class-of-Service (COS)	Description of Services Offered
1. Asynchronous, Circuit Switched	1.1 @300baud, 11-Unit Code, Start/Stop
	1.2 @50-200baud, 7.5 to 11 Unit, Start/Stop
2. Synchronous, Circuit Switched	2.1 @600bps Using Data Terminals
	2.2 @2400bps Using Data Terminals
	2.3 @4800bps Using Data Terminals
	2.4 @9600bps Using Data Terminals
	2.5 @48000bps Using Data Terminals
3. Synchronous, Packet Switched	3.1 @2400bps with Packet Mode Terminals
	3.2 @4800bps with Packet Mode Terminals
	3.3 @9600bps with Packet Mode Terminals
	3.4 @48000bps with Packet Mode Terminals

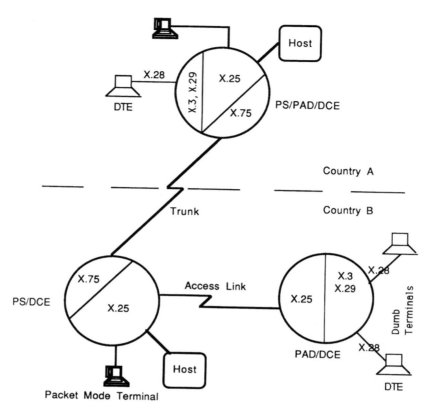

Figure B.4.1 A public data network topology based on some well-known standards.

serviced by another PDN node called the packet assembler and disassembler (PAD). To define the various interfaces and protocols between the PAD and PDN node, X.3, X.29, and X.28 were developed. A PAD is generally assumed to handle the X.25 protocol.

The three software layers for the X.25 protocol for accessing a PDN are illustrated in Figure B.4.2. These three layers are identical to the ones described for the ISO's OSI standard as described in Section B.2. The exact correspondence between the corresponding layers of the two standards resulted from the close association between the CCITT and ISO. Some of the well-known international and national (e.g., the CCITT and EIA recommendations) are specified as follows:

1. The first layer deals with the electrical, functional, and mechanical aspects of transmission:

   1.1 CCITT Rec. X.27 (or V.11) and EIA RS-422 specify the balance electrical interface with a data rate between 0 and 10Mbps. It is interoperable with RS-423 and X.26 (or V.10). Distance limitations are 4000 feet @100Kbps and 40 feet @ 10Mbps. CCITT V.28 and EIA RS-232C represent the first generation and CCITT X.26 (or V.10) and EIA-423 represent the second generation of standards for 0 to 300Kbps rate and shorter distances dealing with the unbalanced interfaces.

1.2 CCITT Rec. V.24 and EIA RS232C represent the first generation, and CCITT Rec. V.24 and EIA RS-449 represent the second generation of standards dealing with the functional definitions of data, timings, and control signals.

1.3 ISO 2110 and EIA RS-232C represent the first-generation standard for a 25-pin connector. ISO 4902 and EIA RS-449 represent the second generation standard for the 37-pin connector that is interoperable with RS-232C.

2. The second layer deals with the logical link control used for establishing, maintaining, and terminating a session between the DTE and DCE. It is identical to the HDLC protocol as specified in the X.25 standard for accessing a PDN. The IEEE 802.2 standard for LANs is also similar to the X.25 HDLC with some differences as shown in the following paragraphs. IBM's SNA employs a synchronous data link control (SDLC) protocol that is similar to HDLC. In all cases, each LLC protocol involves the use of a frame as shown in Figure B.4.8.

3. The third layer deals with the creation of packets with the calling and called node addresses, multiplexing packets from DLCs onto a network link, selection of service and its quality, error checking and recovery, and other related functions. Again, this is identical to the network layer of the CCITT X.25 protocol.

Figure B.4.3 illustrates the basic interfaces between the DTE, DCE, and the PDN at a very high level. Figure B.4.4 illustrates the physical and other standard layers of the interfaces between the DTE, DCE, and the PDN node. Figure B.4.5 shows two classes of DTE-to-DCE link access protocols (LAPs). The LAP A deals with the async-response-mode (ARM). The LAP B deals with the balanced mode of operation where either the DTE or DCE can initiate the process of data exchange.

The X.25 protocol employs both the virtual circuit (VC) and the datagram (or the connectionless) concepts for data interchange. For the VC approach, a logical VC (involving PDN nodes and links) is established before allowing any packet to move. This allows the sharing of the physical resources among many logical VCs. The connectionless approach will also allow the sharing of PDN's physical resources by using the dynamic routing technique described in Chapter 1. The VC and the connectionless concepts also allow the sharing of the physical link connecting a X.25-based host and the PDN among up to 4096 VCs. The resulting

OSI Layers

Application
Presentation
Session
Transport
Network
Data Link
Physical

X.25 Layers and Functions

NETWORK (X.25)
DATA LINK (HDLC)
Physical (X.27 or V.11, X.26 or V10, X.28, V.24, EIA RS-449, EIA RS-422, RS-232C, ISO 2110, ISO 4902 etc.)

Figure B.4.2 The X.25 layers and their relationships to the OSI reference model layers.

resource sharing is perhaps the most attractive feature of X.25 protocol. See Figure B.4.6 for an illustration of the VC operations related to the X.25 protocol.

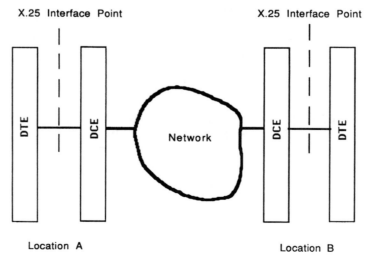

Figure B.4.3 An illustration of basic X.25 DTE-DCE interfaces.

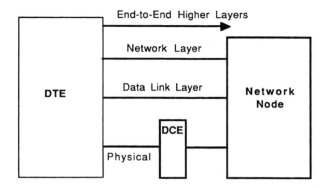

Figure B.4.4 An illustration of the X.25 layers involved in basic network interfaces.

Several types of data packets and data frames are required to make a PDN provide the data communication service. The structures of a call request packet, call accept packet, information packet, supervisory frame, unnumbered frame and the information frame are shown in Figures B.4.7 and B.4.8.

The X.25 protocol took a long time to standardize. It originally addressed the need to interchange data between dumb terminals over low-speed and error-prone lines. Immediately after the completion of the standard in 1977, intelligent WSs (IWSs) began appearing in the market and the common carriers started to offer highly reliable, digital facilities (e.g., DDS and T1). In 1988, CCITT published its recommendation I.122 describing what is now known as Frame Relay. It required only a slight modification of the X.25 protocol. It

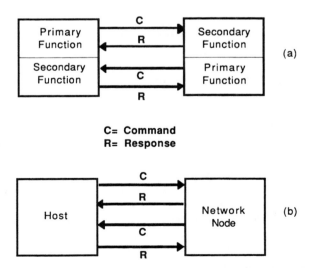

Figure B.4.5 Two classes of DTE-DCE link access procedures involved.

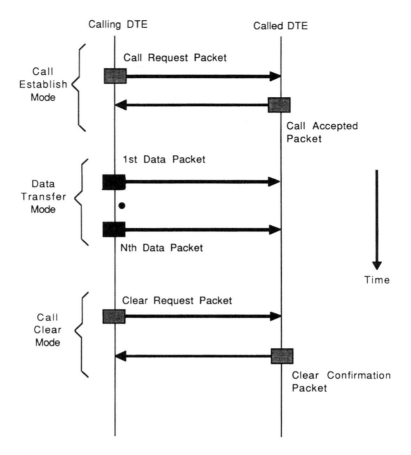

Figure B.4.6 An illustration of the X.25 data exchange operations in a virtual circuit mode.

employed the first two layers, physical and data link. If the originating IWS inserted a data link connection identifier (DLCI), then the frame could be delivered directly to the destination IWS without getting processed by the enroute Frame Relay switches (or modified X.25 switches). The need for the network layer was eliminated. The idea of bypassing enroute nodes and letting the end IWS perform the error control on the message is similar to that of fast packet switching discussed in Chapter 4 and Paragraph B.4.5.3 of this Appendix.

The final Frame Relay standard as specified by CCITT LAPD protocol (Q.921) and its subset CCITT Frame Relay protocol (Q.922) are the foundations of many public Frame Relay networks that provide fast, connection-oriented data communication on very high-speed digital lines (T1 and soon T3). ANSI document T1.606 and its Addendum defines FR service and congestion control. Due to its great similarity to the X.25 standard, the Frame Relay is now an ISDN bearer service designed to serve both LAN interconnection and host-based communications (e.g., SNA). See Section B.5 of this Appendix for a description of ISDN.

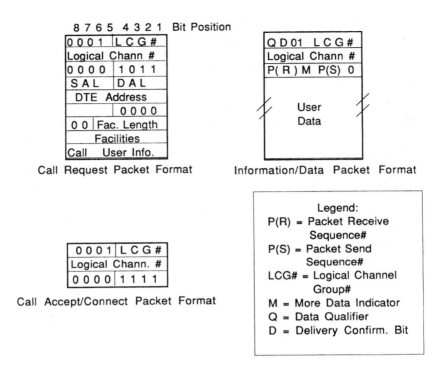

Figure B.4.7 An illustration of the various X.25 packet formats.

A Frame Relay system is not ideal for handling voice since it introduces variable delays caused by the handling of large but variable packet sizes. Only B-ISDN's ATM switches handle all applications with minimal constant latency. A discussion of that technology can be found in Paragraph B.5.5.

B.4.4 IEEE Standards for Local Area Networks (LANs) for Data Communication

In order to provide low-cost LANs to the computer community, IEEE Computer Society established a committee in 1980 to create LAN standards. Thus began the IEEE 802 project. Although the original goal was to establish a single standard, the effort ended up with several meaningful standards to face the reality of several evolving LAN architectures.

The work of the IEEE 802 committee is currently divided into the following subcommittees, each dealing with a specific LAN standard:

1. IEEE 802.1 dealing with Higher Layer Interface (HILI) standard

2. IEEE 802.2 dealing with the Logical Link Control (LLC) standard

3. IEEE 802.3 dealing with the Carrier Sense Multi-Access with Collision Detection (CSMA/CD) standard

4. IEEE 802.4 dealing with the token bus standard

5. IEEE 802.5 dealing with the token ring standard

6. IEEE 802.6 dealing with the Metropolitan Area Network (MAN) standard

Work has been completed on the 802.2, 802.3, 802.4, 802.5, and 802.6 and IEEE has approved related standards. Work on 802.1 is still progressing. It deals with higher layer interfaces for each of the other standards, internetworking, global addressing, and network management and control (NMC). Other study groups are considering other standards for LANs such as an IVDLAN (SG 802.9). The acceptance of the IEEE 802 standards has been remarkably quick and widespread. The National Bureau of Standards (NBS) which issues Federal Information Processing Standards (FIPS) for the U.S. government procurements, has issued an FIPS for the CSMA/CD and LLC standards. The others FIPS based

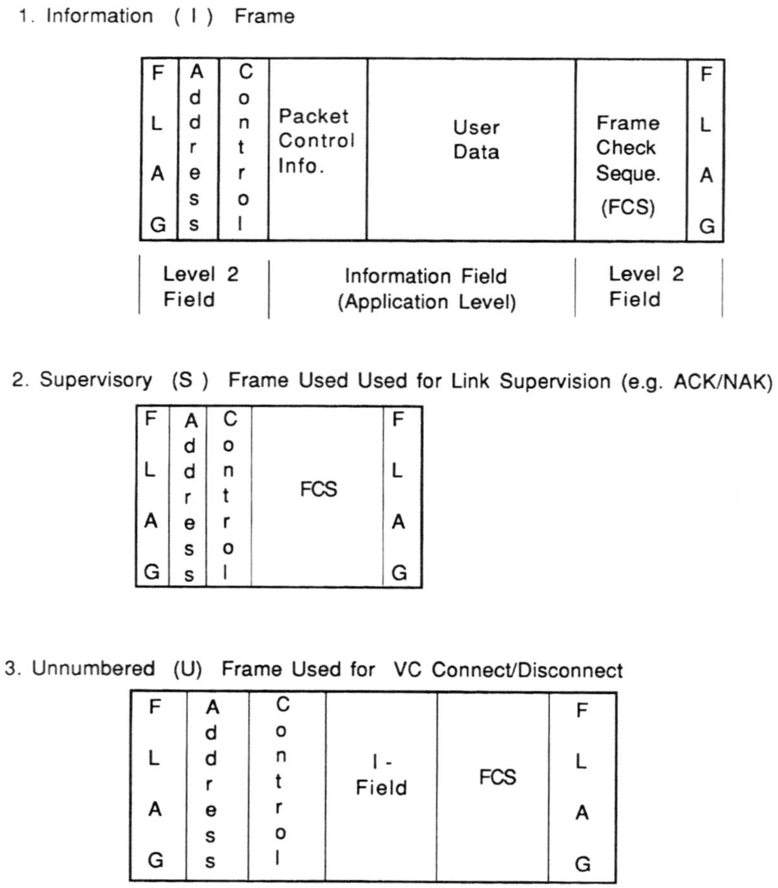

1. Information (I) Frame

2. Supervisory (S) Frame Used Used for Link Supervision (e.g. ACK/NAK)

3. Unnumbered (U) Frame Used for VC Connect/Disconnect

Figure B.4.8 An illustration of the various X.25 frame formats.

on IEEE 802 standards have been produced. The ISO has also adopted the IEEE 802 documents in their entirety. This event shows how the various standards organizations cooperate. Another influential standards organization, ECMA, is also adopted the IEEE 802 standards for a basis for its own LAN-related standards.

Each of the 802.3, 802.4, and 802.5 standards represents a unique LAN architecture and will employ the 802.1 and 802.2 standards to interface with the higher-layer protocols (e.g., transport, session, presentation, and application) and to handle the LLC layer, respectively. Whereas the higher layers are the same as those described for the OSI Reference Model (see Section B.2), the LLC for a LAN is somewhat different from the LLC for the X.25 protocol for WANs. The 802.2 LLC layer must support the multiaccess nature of the bus/ring media while sharing the link access with the medium access control (MAC) and providing a minimum amount of network layer functions. As a minimum, LLC must perform error control, flow control, and frame sequencing just as HDLC does. Since there are no intermediate switching nodes, a LAN does not need a formal network layer. The remaining parts of the network layer, if any, can be incorporated into the LLC layer. Most of the IEEE 802 standards are capable of handling only the datagram (or connectionless) services. IEEE 802.6 standard for a metropolitan area network as based on the DQDB (Distributed Queue Dual Bus) technology is also a connectionless PS protocol. A subset of 802.6 has now resulted in the Switched Multi-Megabit Data Service (SMDS) standard for use in both the MAN and WAN environs. Furthermore, all of the IEEE standards employ the asynchronous TDMultiplexing technique based on connectionless PS for sharing the common media.

See Figure B.4.9 for an illustration of a reference model for LAN standards and their comparison with the OSI reference model.

At the present time, the FDDI standard has been completed under the ANSI jurisdictions. ANSI Documents such as MAC (1987), PHY (1988), PMD (1989), and SMT (1994) represent the media access control, physical layer protocol, physical media dependency, and station management aspects of the FDDI standard. IEEE and ISO were mere observers. The FDDI standard has provided a capability to interconnect (1) very high-speed mainframes located in adjacent areas, and (2) low-speed LANs located within a metropolitan area. FDDI has a serious problem and it has to do with its very expensive interface cards. For that reason, FDDI never blossomed. FDDI has also been challenged by ATM's scalability for bandwidth and seamless switching fabric. ATM technology provides a better solution for implementing a LAN, MAN, and WAN. Strategic planners of an enterprise can now see the handwriting on the wall.

Since the media access control layer provides the unique character to the IEEE 802 standard, one should discuss the standard protocols associated with it.

B.4.4.1 CSMA/CD (802.3) Standard for the Bus Topology

For the CSMA/CD scheme, a station wishing to transmit first senses the presence of a carrier to see if another station is transmitting. If the medium is busy, the station will defer transmission of its packet/message until the carrier drops and a guard band time interval (also called the interframe gap) has elapsed. If the

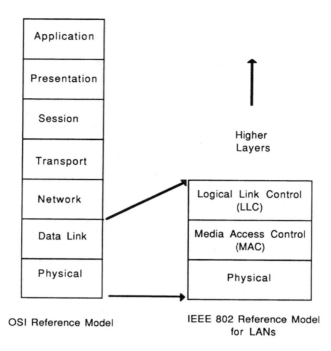

Figure B.4.9 Reference models for the IEEE 802 and OSI standards.

medium is idle, the station may transmit. During its transmission the station also tests for a collision with the transmission from another station. Such a collision is caused by the propagation-time separations between the stations. If there is a collision, the transmissions will be garbled and full of errors. If the station detects a collision, it will cease transmission and jam or reinforce the detected collision with a 48-bit pattern and then defer transmission attempts until a randomized retry time interval has passed. This standard is an evolved version of the original Ethernet product as developed by the Xerox Corporation.

B.4.4.2 Token Bus (802.4) Standard for the Bus Topology

According to this scheme, all stations on the bus receive the superposition of all other station transmissions as well as their own. Control is passed via a token (equivalent to a bit pattern). A station must receive a full message preamble before it can pass the token onward or transmit its own packet/message. This interval equals the time for energy to propagate from one end of the bus to the other end plus the time it takes for each station to perform the required signal processing. Such an interval is named as the interface delay.

B.4.4.3 Token Ring (802.5) Standard for the Ring Topology

In this scheme, each station receives from a single station and transmits to only a single station. Control is passed via a token that is equivalent to bit pattern. The performance of a token ring is determined mainly by the propagation time that defines the length of time to propagate energy from one station through all the other stations and back to itself. Each station interface unit is characterized by the time it takes to detect the token and decide whether to pass the token to the next station or transmit a packet/message. The interface time is therefore equal to one bit time (for a 1-bit token) only since the full message preamble can be handled by a subsequent serial-to-parallel conversion process.

B.4.4.4 IEEE 802.6 Standard for a Metropolitan Area Network

The only implementation of IEEE 802.6 standard is the Switched MultiMegabit Data Service (SMDS) architecture for implementing broadband data networks for both MAN and WAN environs. SMDS employs a *connectionless* protocol based on the mature IEEE 802.6 MAN specifications. However, SMDS is not a true MAN since it is only a subset of IEEE 802.6. It was originally developed by Bellcore for public deployment. It employs 3-level switches (that may or may not be 53-byte-cell-based) interconnected by T1 and T3 circuits. The SMDS Interface Protocol (SIP) allows the handling of two types of packets, one at the network level (up to 9.188 Kbytes) and one at the subscriber access level (53-byte cell). SMDS offers a series of Sustained Information Rates (SIR) ranging between 1.17 to 34 Mbps. It is being offered by most of the RBHCs and one IEC, MCI. Others are slow in offering. The hesitancy is caused by the large acceptance of connection-based Frame Relay and ATM networks.

B.4.5 Some De Facto Standards for Data Communications

A large number of private data networks employ proprietary protocols for LAN and WAN applications. Despite their widespread use, no attempt has been made to represent these in terms of a multilayer protocol in a manner similar to that for the OSI reference model. Only a brief description of these protocols is given.

B.4.5.1 Datapoint's ARCNET

About 8000 LANs currently utilize the Attached Resource Computer network (ARCnet) architecture. The topology of an ARCnet consists of hubs and resource interface modules (RIMs) as illustrated in Figure B.4.10 and it utilizes the token passing (of both the ring and bus variety) access scheme over a 2.5 Mbps bus for a large number (< 255) of intelligent devices and computers. Hubs and RIMs are interconnected via buses and all connected devices listen to transmissions. Each hub provides connections (via an internal ring) to RIMs serving PCs, file servers, and printers according to a physical star topology. Bridges and gateways are also available to interface with other ARCnets, and SNA/SDLC, HDLC or X.25 networks. Several higher-speed versions have also been marketed.

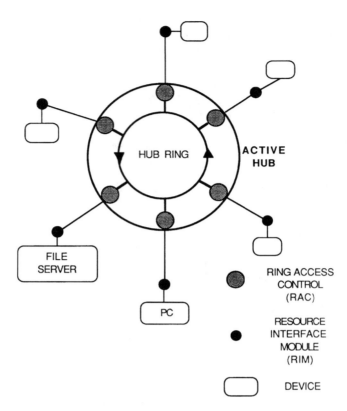

FILE SERVER

PC

RING ACCESS CONTROL (RAC)

RESOURCE INTERFACE MODULE (RIM)

DEVICE

Figure B.4.10 Mixed topology for Datapoint's ARCNet LAN architecture.

B.4.5.2 AT&T's ISN

After its divestiture in 1984, AT&T has been developing several data communication networking products for the BOC and the business applications. Two products, Datakit and Information Systems Network (ISN), are now available. Their unique data networking architecture is based on the concepts of shared virtual packet, shared high-capacity contention bus, and fast packet switching. Several LANs based on ISN can be interconnected to achieve a WAN. It employs the concept of VCs to establish end-to-end paths within the network. (See Figure 4.9 for an illustration of an ISN/Datakit-based LAN.) It consists of several input/output modules, a fast packet switch, a controller and high-capacity contention and broadcast buses. Most modules transmit on the contention bus and listen on the broadcast bus. The switch receives on the contention bus and transmits on the broadcast bus.

A concentrator is offered for interfacing with most of the existing terminal types. Gateways to X.25 PDN are also available. The high-speed packet switch can also service a multiple of 56-Kbps or T1 or fiber optical transmission lines while introducing very few end-to-end delays. Several of these PSs can be interconnected via high-speed trunk bundles according to a mesh topology to achieve a high capacity nationwide WAN. The resulting WAN should obey the mutistar, multicenter topology as illustrated in Figure B.4.11. The constraint to employ only the star topology usually results in high transmission costs with widely scattered CPEs. However, the transmission costs may be minimized when only pure LANs based on the ISN/Datakit architecture are interconnected to create a nationwide WAN. This implies that CPEs are located within a building or a campus allowing the use of privately owned cables.

B.4.5.3 Doelz's ESR/Virtual Packet Network

The Extended Slotted Ring architecture employs two types of nodes. A desktop box (called the Elite) provides packet handling and pure statistical packet multiplexing and demultiplexing (i.e., a PAD function) as illustrated in Figures 4.10 and 4.11. The large node (called the Esprit) provides HS packet switching functions for interchanging packets between a number of netlinks. Each Elite can service up to 16 different types of terminals. A fully shared transmission link (or netlink) can serve several Elites according to the DL topology. Several Esprits can be connected with one another according to a mesh topology to provide a WAN. See Figures B.4.12 and B.4.13 for an illustration of the extended slotted ring used in synthesizing a netlink and a typical directed netlink topology, respectively. The originating Elite node establishes a virtual circuit (VC) within the network before data packets can flow between the two ends. The ESR architecture packetizes every stream of data/message into very small proprietary packets and provides nodal bypass to achieve (1) protocol transparency and (2) very small end-to-end queuing delays. Gateways are provided to interface with X.25 Pans. This technology is now available from Ascom Timeplex.

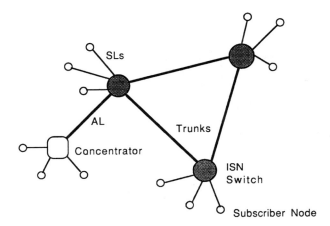

Figure B.4.11 Mixed topology for AT&T's ISN and Datakit architectures.

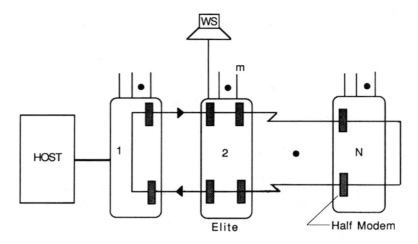

Figure B.4.12 An illustration of the extended slotted ring employed by a netlink in Doelz architecture.

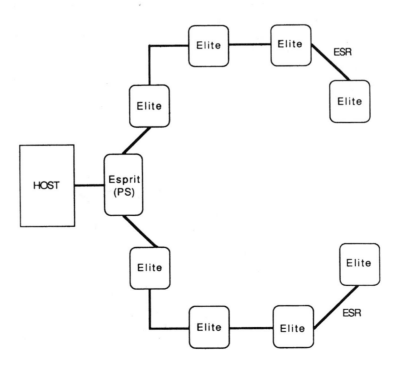

Figure B.4.13 A directed netlink topology employed in Doelz architecture.

B.5 Standards Related to Integrated Voice and Data Networks

B.5.1 Introduction

The evolving standards dealing with the concept of Integrated Services Digital Network (ISDN) are the subject of this section. ISDN framework was created to provide a direction for the evolution of existing public telephone networks in such a manner as to provide many additional services such as data, video, and leased variable bandwidth circuits. Most of the confusion regarding ISDN stems from the fact that ISDN is only a concept. Depending upon the set of ISDN-related standards, one can come up with a new variation in the ISDN concept. Basically, there are four sets of standards that deal with ISDN access interfaces: the basic-rate, primary-rate, the broadband ISDN, and common channel signaling scheme #7 (also SS7).

B.5.2 The Basic-Rate ISDN Access Interface

This interface involves the use of a digital subscriber loop (DSL) for accessing the ISDN central office. A DSL provides two B channels, each at 64Kbps, and one D channel rated at 16Kbps. Whereas a B channel can be utilized for the full-duplex (FDX) CS voice or data service, the D channel is employed for signaling and/or FDX PS and telemetry. See Figure B.5.1 for a detailed illustration of the ISDN interfaces on both sides of the network termination interface (NTI) unit. The ISDN service provider defines the U interface and the T interface is defined by the CPE. The U interface utilizes the existing 2-wire (or 1-pair) subscriber line with the restriction that its length must be in the usual range of 5–7 kilometers. The T interface utilizes a 4-wire connection that should not exceed 1 kilometer. The NTI unit provides line termination, layer-1 multiplexing/management, timing, power transfer, and interface termination functions. As shown in Figure B.5.2, additional ISDN terminals can be served through the use of a single NT2 unit using the star topology. In order to interface existing non-ISDN terminals (e.g., based on the RS-232C physical layer), a terminal adapter (TA) unit will be required. This results in an additional interface called the R interface.

B.5.3 The Primary-Rate ISDN Access Interface

This interface involves the use of a T1-based extended digital subscriber line (EDSL) for accessing the ISDN central office. An EDSL provides 23 B-channels, each at 64 Kbps, and 1 D-channel also rated at 64 Kbps. Whereas a B-channel can be used for the FDX CS voice or data service, the D-channel is utilized for FDX PS. The signaling functions are handled through the I.451 and SS7 standards and some NMC-related functions are accomplished through the use of an extended super frame (ESF) discussed earlier in Chapter 2. See Figure B.5.3 for an illustration of the associated T and S interfaces.

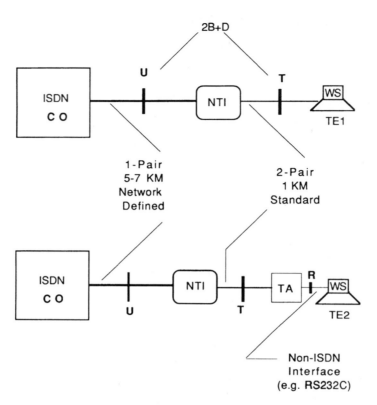

Figure B.5.1 Basic rate interface for a single user of ISDN.

B.5.4 OSI-Type Layers for ISDN and Related Standards

Study Group XI is concerned with the Q recommendations and SG XVIII is concerned with the I. recommendations for ISDN. Both the Q. and I. recommendations are concerned with the signaling and interfaces of first and the third ISDN layers. The link control layer is derived from the X.25 data link protocol. The applicable CCITT standards for the lower three layers are shown in Figure B.5.4. The first layer handles the physical and electrical interfaces. The applicable standards can be included in the set I.43X where X is any integer. This layer is concerned with only the raw bit transport. Separate standards for the basic, primary, and the broadband interfaces will exist. The CCITT recommendation Q.921 is for the ISDN link control. It is almost identical to the second layer of X.25 except that only the extended addressing mode is used and the address field defines the required service (e.g., voice, data, or NMC) provided by the ISDN access switch to the CPE. The Q.921 recommendation for the link control is also called the LAPD standard for ISDN. The third layer (network related) is defined by the Q.431 recommendation.

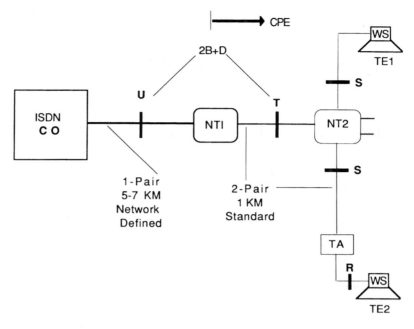

Figure B.5.2 Basic rate interface for several users of ISDN.

It specifies the OSI layer-3 functions for ISDN and the means for establishing, maintaining, and terminating network connections across an ISDN and between CPEs. Therefore, it controls the set-up of calls across an ISDN using structured

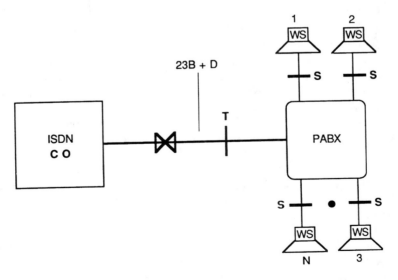

Figure B.5.3 Primary rate interface for an ISDN-compatible PBX.

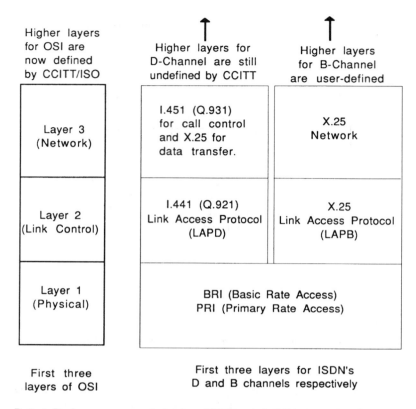

Figure B.5.4 Reference models for ISDN and OSI standards.

messages. The messages flow on out-of-band channels and consists of digital symbols. The message set can be easily extended to handle future growth of service.

The manner in which an ISDN provides voice and data related service on a wide area basis is illustrated in Figure B.5.5. Several ISDN COs are connected to an SS7-based mesh network to provide switched and nonswitched services using the separate public telephone and data networks. Figure B.5.6 illustrates the composition of an ISDN CO. Line termination (LT), control point (CP), PS, and CS modules are provided to interface with the CSS7, PTN, PDN, U, T, and primary rate signaling points. Figure B.5.7 illustrates the hierarchy and the topology of a typical SS7 subnet. Such a network is a replacement of AT&T's original CSS6 signaling network, which employed 2.4Kbps trunks and 180-bit messages. AT&T has already converted their old signaling network to SS7. The use of 56Kbps trunks and 256-bit message provides a tremendous improvement over the older SS6 network. The new AT&T's SS7 network consists of over 200 4ESSs and about 15 signal transfer points (STPs). It can handle up to 15 million call attempts per hour (compared to 2 million call attempts in SS6). Properly designed large private networks also employ a SS7 subnetwork for signaling and control and interworking with the public ISDN-based networks. Some medium-sized corporations may find the cost of a CSS7 subnetwork too high. In those cases, one can employ some hardware interfaces that allow the creation of a SS7 subnetwork using trunks operating at lower rates (e.g., 9.6Kbps). This approach will not only reduce the cost of the total network but also allow unfettered interworking with other private and public networks.

Many channel types have been defined for ISDN. The 64Kbps PS E-channel is recommended for use with the CCS7; the 384Kbps CS/PS H0-channel is recommended for HS facsimile or data or slow-motion video; and the 1.536Mbps CS H11- and the 1.920Mbps CS H12-channels are recommended for HS data and facsimile and full-motion video.

B.5.5 The Broadband ISDN (B-ISDN) Access Interface

The B-ISDN interface was originally developed in 1988 by CCITT and they chose ATM (asynchronous transfer mode) to implement B-ISDN. B-ISDN is a natural extension of the narrowband ISDN. Although it also provides ISDN-like definitions

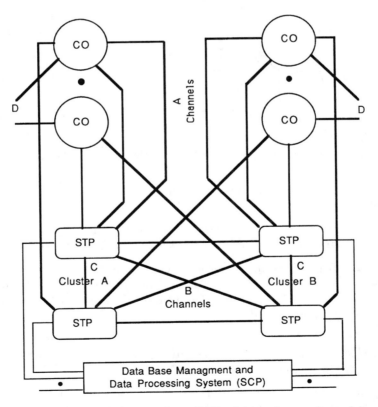

Figure B.5.5 Mixed topology of an SS7 network employed by ISDN.

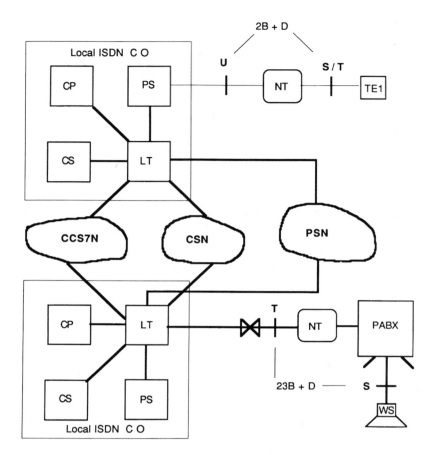

Figure B.5.6 Components of an ISDN central office.

for services and interfaces for public networks, it differs from its predecessor in the following ways:

1. It provides an increased amount of scalable bandwidth with an integrated, end-to-end grade-of-service management control.

2. It provides an identical switching interface to all applications such as voice, data, and video. An ATM CO differs drastically from a narrowband ISDN CO as shown in Figure B.5.6. There the CO consisted of separate CS and PS nodes.

The B-ISDN's protocol reference model was originally developed by ECSA's T1Y1 standards group. It is shown in Figure B.5.8. The B-ISDN reference model includes four layers:

1. The Physical Layer defines the electrical or optical interface. It provides dynamically configurable channels or packets at rates up to 150Mbps to ATM-savvy WSs. Higher rates for the WAN interface will be based on the

synchronous optical network (SONET) standard. This interface differs from the Basic and Primary interfaces in the fact that instead of using synchronous fixed-channel techniques, one can now provide a universal, extremely flexible optical transport mechanism not only for voice and data but also for services like high-density digital video. The physical layer also manages the transmission convergence/framing of incoming/outgoing data.

2. The ATM Layer performs the cell relaying and multiplexing functions. It also handles cell insertions/removals.

3. The ATM Adaptation Layer (AAL) consists of as many as five types of AALs for different traffic types. It also provides higher-layer information to ATM cell payload conversion.

4. The ATM Higher Layers (AHLs) handle all of the ATM control protocols and interfaces with end-user services and applications.

The ATM Forum was founded in 1992 to accelerate the process of ATM standards-making for the benefit of U.S. vendors. About 900 members (mostly

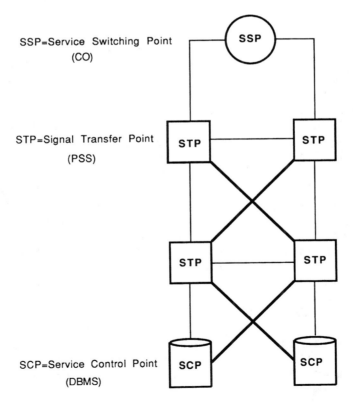

SSP=Service Switching Point
(CO)

STP=Signal Transfer Point
(PSS)

SCP=Service Control Point
(DBMS)

Figure B.5.7 Hierarchy and topology of a typical SS7 network.

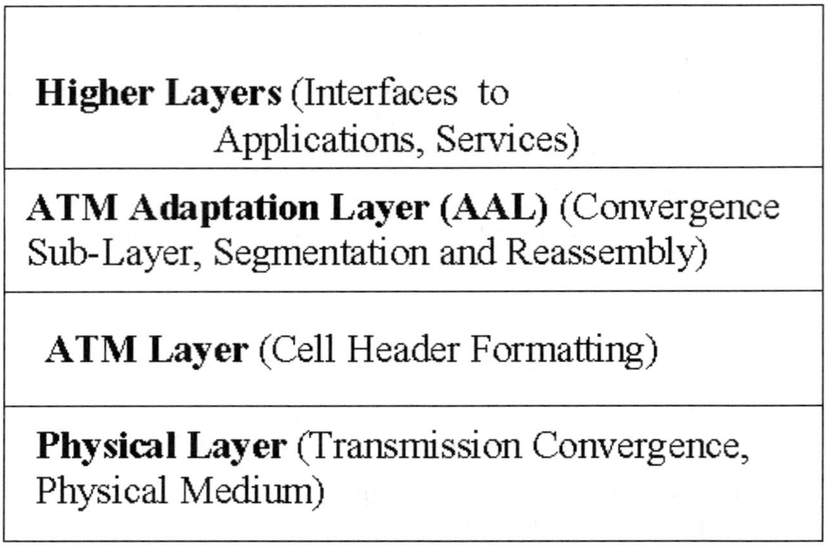

Higher Layers (Interfaces to Applications, Services)
ATM Adaptation Layer (AAL) (Convergence Sub-Layer, Segmentation and Reassembly)
ATM Layer (Cell Header Formatting)
Physical Layer (Transmission Convergence, Physical Medium)

Figure B.5.8 A B-ISDN model with lower three layers.

from the vendor community) constitute the Forum. They also listen to the wishes of end users via the End-User Network Roundtable (ENR). They have done a methodical job of defining and approving many recommendations and passing them to the International Standards Bodies mentioned earlier.

ATM technology derives its low latency from the small, fixed cell sizes employed for packet switching. Figure B.5.9 illustrates the components of the 5-byte header in each 53-byte cell. The price paid for small latency is in the form of additional fixed overhead. But this penalty will be eventually offset by the reduced WAN facilities achieved through traffic aggregation and statistical multiplexing capabilities inherent in ATM.

The class of services as defined by the ATM adaptation layer (AAL) can be illustrated by Figure B.5.10. It shows how ATM handles all applications with equal facility. Although some AAL protocols have not been defined yet, many vendors have announced products based on proprietary schemes derived through software. It is their belief that benefits of ATM along with the penalties of being late to market are too great to wait for the final standards to be completed. Figure B.5.10 also shows that only five types of AALs have been defined to cover most of the traffic types.

The ATM Forum has already defined most of the user-to-network and network-to-network interfaces for ATM. Many types of such interfaces are illustrated in Figure B.5.11. These standards have already accelerated the marketing of many ATM-based products for WAN applications.

See Figure B.5.12 for an illustration how the aforementioned AAL appears for a LAN-to-Host interconnection.

Legend: GFC = Generic Flow Control : VPI = Virtual Path ID
VCI = Virtual Circuit ID : PTI = Payload Indicator
RES = Reserved: CLP = Cell Loss Priority: HEC = Header Error Control

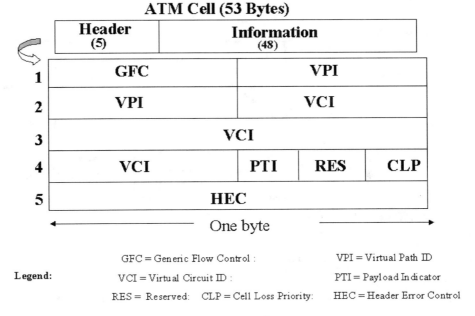

Figure B.5.9 ATM cell and cell header format.

Two additional standards dealing with LAN Emulation (LANE), which is already completed, and Multi-Priority Protocol over ATM (MPOA), which is yet to defined, will finally enable the implementation of a fully integrated digital

Class	A	B	C	D
Application	V/Vd	PV/Vd	D	SMDS
Conn. Mode	VC Oriented			N/VC
Bit Rate (BR) Mode	CBR	VBR		
AAL #	1	2	3,4,5	3,4

Legend: P=packetized, V=voice, D=data, Vd=video, VC=virtual circuit,
N/VC=no VC used, CBR=constant BR, VBR= variable BR

Figure B.5.10 B-ISDN calss of services.

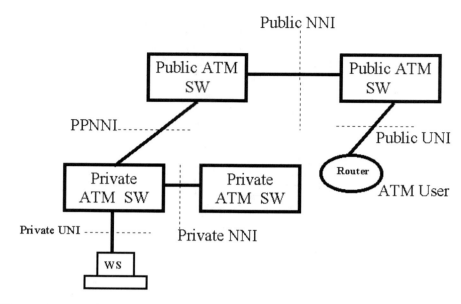

Figure B.5.11 User-to-network and network-to-network interfaces.

network (also called FIDN in Chapter 1) while preserving the investments in existing CPEs. Only then will all of the existing protocols function as interfaces to the FIDN yield sizable returns.

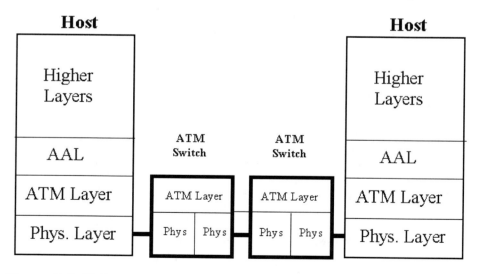

Figure B.5.12 Peer-to-peer communication via ATM network.

B.6 Standards Related to Distributed Data Processing

B.6.1 OSI-Related Standards

Since the acceptance of the OSI Reference Model standard ISO 7498 in 1983, the ISO subcommittee SC16 has been engaged in the arduous task of developing standards for the OSI protocols. CCITT, which cooperates with ISO, is also engaged in the development of relevant standards for PDNs. Both ISO and CCITT create standards that obey the OSI reference model and foster the interconnection between multivendor computer systems. Figure B.6.1 illustrates the two example sets of standards that have been accepted for some unique environments (e.g., LANs and WANs). Figure B.6.1 does not imply that the effort for developing standards is finished. Additional standards and protocols are still needed to satisfy the requirements of other meaningful environments. It should be interesting to observe that X.25 forms the basis of the first three WAN layers. ISO has added a few additional standards for LANs and connectionless (or datagram) protocol. Although the CCITT standards appear different from those developed under the auspices of ISO, there are many similarities. For example, the CCITT standard X.400 for electronic messaging requires only the basic activity subset of the ISO Session protocol ISO 8327 and as represented by the CCITT standard X.225. In general, the duties of ISO and CCITT are to handle all DDP environments and the PDN-based services, respectively. Eventually, these two subsets of standards will merge together to provide us with a single set of fully developed standards. The road to OSI has been a slow one, especially in the United States where TCP/IP, Unix, and Windows NT dominate. Due to the recent resurgence in Internet, one can now say that OSI will never penetrate the enterprise networks as was once envisioned.

B.6.2 IBM's System Network Architecture (SNA)

IBM first announced its proprietary SNA in 1974. It is a collection of hardware and software modules for distributing the computing resources and handling the needs for their large user base. It can be termed as a de facto standard due to its omnipresence in the United States. Since its inception, SNA has been continuously evolving. Its architecture can also be represented by seven layers. Since SNA was developed before the availability of the OSI reference model, there are several differences between the original SNA and OSI models. See Figure B.6.2 for a comparison between the two models. Each SNA node consists of at least a physical unit (PU). Some PUs are assigned the responsibility of a system service control point (SCCP). Some PUs also consist of a logical Unit (LU) through which an operator or an application program accesses the network. Each LU is associated with a network address and contains a service manager and one or more half-sessions. One half-session of an LU can be connected to another half-session of another LU in another SNA node to form a full session. On a logical level, the SNA network consists of interconnected LUs, resulting in a distributed network. Therefore, an SNA network can consist of intelligent workstations and a large

	ISO	CCITT
Application (Layer 7)	CASE 8650. FTAM 8571	DIRECTORY SVC. X.400 EM
Presentation (Layer 6)	PROTOCOL 8823 SYSTEM ASN.1	SYNTAX X.409
Session (Layer 5)	8327	X.225
Transport (Layer 4)	8073	X.224
Network (Layer 3)	X.25 8208/8878. CLNP 8473	X.25
Data Link (Layer 2)	X.25 LAPB 7776. LAN LLC 8802	X.25 (LAPB)
Physical (Layer 1)	DTE-DCE. LAN 8802	V.24(RS.232C) V.35, X.21

ASN.1 = Abstract Syntax Notation 1
CASE = Common Application Service
CLNP = Connectionless Network Protocol
EM   = Electronic Mail
FTAM = File Transfer and Access Management
LAPB = Link Access Protocol (Balanced)
LLC   = Logical Link control

Figure B.6.1 Notable equivalences between OSI and CCITT standards.

mainframe that controls a large SNA domain. Originally, an SNA-based network was always centrally controlled by the mainframe. However, since the recent introduction of the LU type 6.2, peer-to-peer communication can occur if the communicating programs are written on the LU6.2 platform. Interestingly, LU6.2 can be spread across several layers—a fact that violates the clean-separation-between-layers requirements of ISO. It was for this fact that ISO will not accept LU6.2 as an OSI application layer protocol. See Figure B.6.3 for some typical SNA network topologies. Since SNA supports BSC and SDLC, conventional multidrop topologies based on minimal spanning trees are very popular with the SNA networks. LU6.2 encourages the formation of clusters based on the mesh topology in the future. But that was the old SNA. Times have changed IBM, which has been recreated, from top to bottom. It is lean and hungry.

The original SNA model has been evolving ever since 1974. First APPC (advanced program-to-program communications) and then APPN (advanced peer-to-peer networking) came to make SNA a truly peer-to-peer model. SNA has

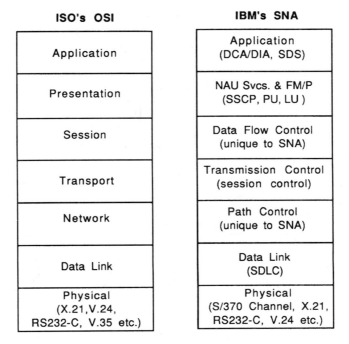

ISO's OSI	IBM's SNA
Application	Application (DCA/DIA, SDS)
Presentation	NAU Svcs. & FM/P (SSCP, PU, LU)
Session	Data Flow Control (unique to SNA)
Transport	Transmission Control (session control)
Network	Path Control (unique to SNA)
Data Link	Data Link (SDLC)
Physical (X.21,V.24, RS232-C, V.35 etc.)	Physical (S/370 Channel, X.21, RS232-C, V.24 etc.)

Notes:
DCA/DIA = Document content and interchange architecture
SDS or SNADS = SNA Distribution service
SSCP = System service control point (e.g. HP)
FM/P = File management and presentation

Figure B.6.2 Reference models for IBM's SNA and ISO's OSI standards.

also been providing interfaces to public/private X.25 networks, TCP/IP networks, and Frame Relay networks. IBM has discovered a gold mine in powerful routers. The old mainframes have been reincarnated. Lately, IBM has become committed to the ATM-to-the desktop strategy. Toward that goal, IBM has been marketing LAN switches and ATM-savvy switches. IBM is committed to preserve its market share and continue to serve the needs of entrenched SNA users for interworking.

B.6.3 DEC's Digital Network Architecture (DNA)

DNA, as originally conceived, is also a collection of hardware and software products and Digital Equipment Corporation first announced it in 1975. Whereas SNA is generally characterized by hierarchical control, DNA was developed to provide a truly distributed control. See Figure B.6.4 for the evolution of the various DNA layers and the applicable standards. One should note that the DNA Phase IV application layer handles the interfaces to the network management and control functions in an integrated fashion. At this time, DEC seems to have folded all of the OSI layers and the related standards into a product called DECnet/OSI while also

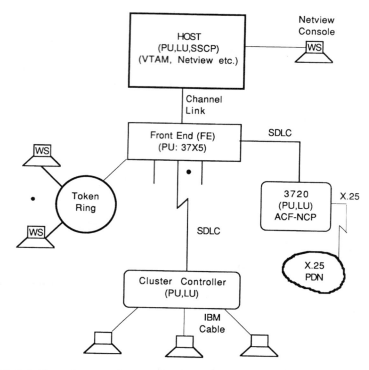

Figure B.6.3 Topology of a typical network based on SNA.

providing a proprietary alternative at each layer of the DNA protocol. Now one can run any application over any transport of one's choice. Many DECnet/OSI products are now available for full multivendor connectivity. DECnet/OSI attempts to preserve existing investments in network hardware and applications.

B.6.4 Manufacturing Automation Protocol (MAP) Architecture

MAP is a seven-layered protocol to help achieve factory automation by several firms, notably General Motors Corporation. The MAP architecture is primarily based on the OSI reference model applicable for the LAN environment. Due to the response time constraints, the Token Bus (IEEE 802.4) standard was chosen for the first physical layer. The applicable standards for each of the OSI layers are illustrated in Figure B.6.5. The bus topology applies for a MAP-based network.

B.6.5 Technical Office Protocol (TOP) Architecture

TOP is a seven-layered protocol to help achieve office automation and productivity increase among the technical workers of several firms spearheaded by the Boeing Corporation. The TOP architecture is primarily based on the OSI reference model applicable for the LAN environment. The office environments dic-

OSI Layers	DNA Layers	DNA Layered Functions	DNA/OSI Layers	DNA/OSI Layered Functions
Application	User	File Transfer, Remote Access,	Msg. Router (X.400 Gtwy) ----------------	OSI FTAM, CCITT X.400 Messaging
	Network Managmnt.	Down-Line System Load, Remote Command File Submission,	OSI Apps. DNA User DNA Mngt	
Presentation	Network Application	Virtual Terminals	OSI Prestns DNA Net Apps.	
Session	Session Control	Task-to-Task	OSI Session (OSAK) DNA Session Control	OSI Session
Transport	End-End Communications		OSI Transport (VOTS)	Task-to-Task
Network	Routing	Adaptive Routing	OSI Network	OSI Internet, Data Link Mgmnt. Adaptive Routing X.25
Data Link	Data Link	DDCMP (pt. to pt. & Multi Pt.) X.25 CSMA/CD	OSI DATA LINK	HDLC MAP 802.3 (see Appendix B)
Physical	Physical Link		OSI Physical	RS232, IEEE LAN 802.3 (see Appendix B)

Figure B.6.4 DEC's DNA layers/functions, as related to OSI layers.

tated the choice of the CSMA/CD (IEEE 802.3) standard for the lowest physical layer. The applicable standards for each of the OSI layer are illustrated in Figure B.6.6. The bus topology also applies for a TOP-based network.

B.6.6 Government Open System Interconnection Procurement (GOSIP) Architecture

The National Bureau of Standards (NBS) defined an OSI-based architecture for distributed data communication and processing for all U.S. government agencies during 1988. Thus far, the U.S. government agencies have been procuring data

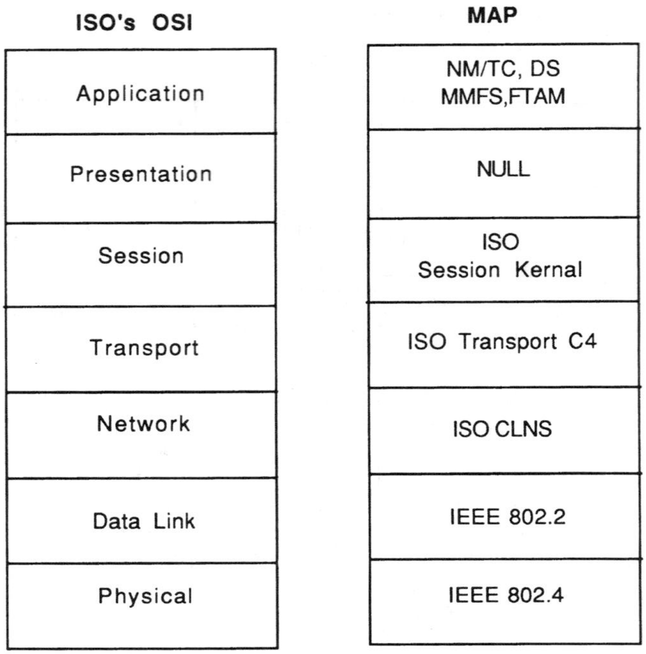

ISO's OSI		MAP
Application		NM/TC, DS MMFS,FTAM
Presentation		NULL
Session		ISO Session Kernal
Transport		ISO Transport C4
Network		ISO CLNS
Data Link		IEEE 802.2
Physical		IEEE 802.4

Notes:
NM/TC = Network Management and Technical Control
DS = Directory Service   : C4= Class4
MMFS = Manufacturing Message Format Standard of EIA
FTAM = File Transfer and Access Management
CLNS = Connectionless Network Service (also = Datagram)

Figure B.6.5 LAN layers for MAP.

communication products that satisfy the Transmission Control Protocol/ Internet Protocol (TCP/IP). Most vendors still manufacture products for the government based on TCP/IP. The transition to GOSIP is still doubtful considering the installed base of TCP/IP. See Figure B.6.7 for an illustration of the four-layered TCP and the two-layered IP protocols. See Figure B.6.8 for the standards related to the seven GOSIP layers for both the LAN and WAN environments. All the network topologies for LAN and WAN environments discussed in earlier chapters will apply to a GOSIP-based network. The recent developments in B-ISDN/ATM switching have cast additional doubts over the future of GOSIP.

B.6.7 Advanced Intelligent Network (AIN) Architecture

An advanced intelligent network (AIN) employs a common channel signaling network (also called SS7) to provide any service provider the capability of introducing new services and features quickly. Before the introduction of AIN, also called IN by PTTs of other countries, most of the services were realized through the software resident in switching service point (SSP). Such an approach was

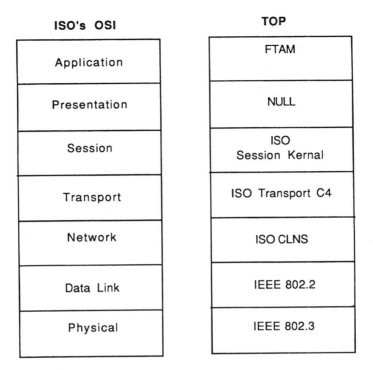

ISO's OSI	TOP
Application	FTAM
Presentation	NULL
Session	ISO Session Kernal
Transport	ISO Transport C4
Network	ISO CLNS
Data Link	IEEE 802.2
Physical	IEEE 802.3

Notes:
FTAM = File Transfer and Access Management
CLNS = Connectionless Network Service (also = Datagram)
C4= Class4

Figure B.6.6 LAN layers for TOP.

quite adequate in the past. But as the competition between the emerging new service providers intensifies, public networks must utilize the intelligence based in the SS7 network instead of increasing the complexity of their SSPs. The recent increase in the demand for nationwide PCS networks is bound to increase the use of AINs in the near future.

An AIN places most of the intelligence in the centralized database management (DBM) nodes called the SCP of SS7. See Section B.5 of this Appendix for a description of the SS7 architecture. The only thing missing in the SS7-related Figure B.5.5 is Intelligent Peripheral (IP) connected to CO or SSP. These IPs (with localized databases) reside at customers premises and are capable of triggering any previously agreed upon features or service.

The SS7 related standards are defined by Bellcore and their component protocols are defined by ANSI, ITU-T, and ISO. For example, the STP functions are defined in ANSI Message Transfer Part/Signaling Connection Control Part (MTP/SCCP) and its equivalent by ITU-T-TSS MTP/SCCP. ITU-T Q.600, Q.700, Q.800, and Q.900 specifications define SS7 architecture. The SS7 operations are also specified by ANSI documents T1.200, T1.600, and T1 S1.3.

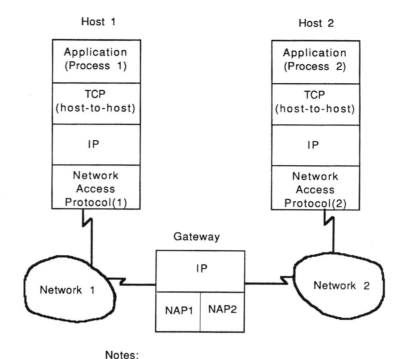

Notes:
TCP = Transmission Control Protocol
NAP = Network Access Protocol
IP = Internet Protocol

Figure B.6.7 A reference model for TCP/IP.

OSI Layers	GOSIP's Present and Future Protocols
Application	X.400 EM, FTAM, Virtual Terminal, Directory Service, Network Management, Association Control Svc. Element
Presentation	OSI Presentation Protocol
Session	OSI Session Protocol
Transport	Transport Protocol Class 4 & Connectionless Transport Protocol
Network	Connectionless Network Service + Dynamic Routing, X.25, ISDN, Connection Oriented Network Service + Internetwork Protocol
Data Link	X.25 HDLC LAP B, IEEE 802.3, 802.4, 802.5, FDDI MAC, ISDN,
Physical	BASEBAND, BROADBAND, Shielded Twisted Pair, FDDI Physical, Military Standard 114-A, RS-232

Figure B.6.8 Layers and protocols related to GOSIP.

Glossary

10Base-T 10Mbps Ethernet using twisted pairs

100Base-T 100Mbps Ethernet using twisted pairs

100VG 100Mbps Any LAN using voice-grade wiring scheme (also called 100VG-AnyLAN)

AAL ATM adaptation layer that converts higher layer packets into ATM cells

ABR available bit rate: traffic class defined by ATM Forum

ADPCM adaptive differential pulse code modulation, an ITU-T standard voice encoding scheme @ 32Kbps

ADSL asymmetric digital subscriber line (1.8 -8Mbps downstream, 16–640Kbps to S)

AIN Advanced Intelligent Network, a concept proposed by Bellcore

AIP ATM interface processor: ATM interface card in Cisco hardware

ALT access line type (it defines the link capacity, allowable data rate on the link, potential data rate per terminal, and the applicable tariff)

ALMF access link multiplexing factor (= #VG circuits/AL)

ALU arithmetic logic unit

AM amplitude modulation

AMA automatic message accounting

AMLC asynchronous multiline controller

AMPS advanced mobile phone service based on analog technology

ANI automatic number identification

ANSI American National Standards Institute

APPN Advance Peer-to-Peer Networking, as part of SNA offering

ARP address resolution protocol used by TCP/IP

ARPA Advanced Research Project Agency

ARS automatic route selection

ASK amplitude shift keying

ASYNC asynchronous mode of communication

ATA American Transportation Association

ATDM asynchronous time division multiplexing

ATDMA asynchronous time division multiple access

ATM asynchronous transfer mode

ATM Forum vendor consortium for ATM-related standards-making

ATP analysis type (for computing response times)

AWG American wire gauge

Balun BALanced to UNbalanced impedance matching device used as a cable to twisted-wire pair adapter in a premise wiring system

BBTF backbone trunking factor

BC bridged channel (for a multidrop link)

BCC blocked cell cleared

BCD blocked call delayed; also binary coded decimal

BCH blocked call held

BER bit error rate

BHR busy hour

BINS Barclay's integrated network system

BISDN Broadband ISDN

BLC bridged local channel

BLKL block length in bytes

BLU basic link unit

BNR Bell Northern Research

BOC Bell Operating Company

BOP bit-oriented protocol

BPS bits per second

BR block redundancy associated with each encoded block

BRI basic rate interface for ISDN, proving two B-channels @ 64Kbps each and one D-channel @ 16Kbps

BRA basic rate area

BSC binary synchronous communication

CAC customer administration center

CAS centralized attendant service

CATV cable television

CBR constant bit rate traffic class defined by ATM Forum

CCC copper cross connect

CCC clear channel capability

CCIS common channel interoffice signaling (AT&T's SS6)

CCITT Int. Telegraph and Telephone Consultive Committee (now simply ITU)

CCM customer coordination/control management

CCP communication and control processor

CCS common channel signaling (also equivalent to CCITT's SS#7)

CCS hundred call seconds

CCSA common control switching arrangement (AT&T nomeclature)

CCU communication control unit

CDMA code division multiple access

CEC Commission for European Communities

CIF cells in frame

CLLI city location ID for a CO or wiring center

CLNS connectionless network service

CLP cell-loss-priority bit used by an ATM switch during congestion

CLT central limit theorem (for statistical analysis)

CMST capacitated minimal spanning tree

CNCC customer network control center

CNCP Canadian National Canadian Pacific

CNDP communication network design program (based on EW Algorithm)

CNS communication network system

CO central office

COG center of gravity

COS class of service or communications operating system

CP customer premise or communications processing

CPE customer premise equipment (e.g., telephone, PABX)

CPU central processing unit

CRC cyclic redundancy code

CRT cathode ray tube (also used for a dumb computer terminal)

CS circuit switching (also circuit switched)

CSMA/CD carrier-sense-multiple-access with collision detection

CSMT Capacitated (or Constrained) Minimal Spanning Tree

CSU channel service unit that provides interface to leased T1 circuits (along with DSU)

CSV customer service vehicle

CV customer vehicle

DAMA demand-assignment multiple access

DAP device access protocol

DARPA Defense Advanced Research Project Agency

DB decibel

DBA database access

DBA dynamic bandwidth allocator

DBM database management

DCA distributed communications architecture

DCE data circuit-termination equipment or data communication equipment

DCP distributed communications processor

DCPSK differentially coherent phase shift keying

DCS distributed computing system

DCS data circuit switches

DDBM distributed database management

DDCMP digital data communication message protocol

DDD direct distance dialing

DDP distributed data processing

DDS dataphone digital system

DENOM denominator

DF disk file

DF dividing factor as used in ROI analysis

DL directed link associated with Doelz networks

DLA delayed acknowledgment

DM delta modulation

DMA direct memory access

DMTF desktop management task force for standards making

DNA distributed network architecture of DEC

DNA Digital (Equipment Corp.) Network Architecture

DNIC data network identification code

DP data processing

DS directory service

DS-0 a single 56/64Kbps channel of a DS-1 digital facility

DS-1 digital signal level-1 (T1) rated at 1.444Mbps and 24 channels at 56Kbps each

DS-3 digital signal level-3 (T3) rated at 44.736Mbps

DSA digital service area

DSDDS Dataphone switched digital data system

DSI digital speech interpolation (e.g., TASI)

DSL digital subscriber line (also see xDSL)

DSU data service unit that provides an interface to a leased T1 circuit along with a CSU

DTE data terminal equipment

DTMF dual-tone multifrequency

DTN data transporting network

DTS digital tandem switching

DUC data unit control

DUV data under voice

DXI data exchange interface between a router and a DSU or ATM facility

E1 European digital facility (similar to T1) rated at 2.048Mbps

E3 European digital facility (similar to T3) rated at 34Mbps

ECD expected call duration (SDF design parameter)

ECSA Exchange Carriers Standards Association

EDB enterprise database for all sites, CPEs, and network-related information

EEC European Economic Community

EFT electronic fund transfer

EIN European Informatic Network

EIS electronic information system (AT&T)

EM electromagnetic or electronic mail

EMA enterprise management architecture of DEC

ENR end-user network roundtable for getting user-community inputs for standards making

EO end office

EOSN end-of-service node (e.g., PBXs, PADs, data concentrators, key telephones sets, etc.)

EPSCS enhanced private switched communication system (earlier AT&T offering)

EPSS experimental packet switching service

ES expert system

ESF extended superframe

ESP enhanced service provider who uses SS7-based database services

ESR Extended Slotted Ring of Doelz Networks, Inc. (now part of Ascom Timeplex)

ESRA enroute station routing approach (a traditional approach for public telephone networks)

ESS electronic switching system

ETN electronic tandem network (AT&T offering for analog systems)

ETSI European Telecommunications Standards Institute

EVCC economical voice call cost related to a VPN

EW Esau-Williams network design algorithm

FADS force administration data system (ACD related)

FAX-PAK facsimile packet switching system

FCC Federal Communications Commission

FCC fiber cross connect (also Federal Communications Commission)

FCS frame check sequence based on CRC

FDDI fiber data distribution interface (ANSI-defined standards I and II)

FDM frequency division multiplexing

FDMA frequency division multiple access

FDX full duplex

FDXFACT full duplex factor regarding trunks

FE front end

FIDN fully integrated digital network

FIFO first-in-first-out

FM frequency modulation

FPS fast packet switching (e.g., ESR, Frame Relay, ATM)

Frame-RelayForum Frame Relay Forum for standards making

FSK frequency shift keying

FTAM file transfer and access management

FTP file transfer protocol for the IP layer of TCP/IP

FX foreign exchange

GEA Gigabit Ethernet Alliance for standards making

GOS grade of service

GPS global positioning system that employs satellite signals for defining an exact position

GSM global system for mobile communications based on TDM encoding scheme

HDLC high-level data link control

HDSL high bit rate DSL with full duplex T1 or E1data rates for up to 15,000 feet

HDX half duplex

HEC header error control

HEHO head-end-hop-off as a routing scheme for voice networks

HF high frequency

HLC high-level center

HMD half modem delay (ms)

HP host processor

HPR host port rate (bps)

HTT host think time in seconds

HU high usage circuit in a hierarchical network

IATA International Airline Transportation Association

IBC interbridge channel (for multidrop links)

IBDS integrated building distribution system

ICPB information characters per block

ICS Info Call Service

IEC interexchange carrier for inter-LATA services

IEC information exchange circuit in a PDN

IED Information exchange distribution

IEEE Institute of Electrical and Electronics Engineers

IES Information Exchange Service

IETF Internet Engineering Task Force

IGS Information Gram Service

IH intelligent hub that provides protocol emulation and/or switching

IMA input message acknowledgment

IML input (inquiry) message length in bytes

IMP interface message processor

IMS information management system (IBM offering)

IMT intermachine trunks

IN intelligent network that is also called advanced intelligent network (AIN)

INAP Infogram network access protocol

IOC interoffice (or POP) channel for inter-LATA use

IP intelligent peripheral that is an AIN element used to trigger requests for services

IPL initial program load

I-PNNI integrated private network-to network interface for ATM-based networks

IPSS international packet switching service

IR inquiry response

IR interest rate

IRC international record carriers; also interrate center

ISDN Integrated Services Digital Network

ISN information system network of AT&T (similar to Datakit)

ISO International Standards Organization

ISP Internet service provider

IXC interexchange carrier

ITU International Telecommunications Union

IVD integrated voice and data

IVDN integrated voice/data network

LAMBDA average message arrival rate per VMPT circuit per second

LAN local area network (usually the data variety)

LANE LAN emulation

LAP link access procedure

LAPB link access procedure, balanced (for X.25)

LAPD link access procedure-D channel (for ISDN)

LATA Local Access and Transport Area

LC local channel as a part of the total end-to-end circuit

LCM line concentrator module

LD long distance

LEC local exchange carrier for intra-LATA services

LIFO last-in-first-out (queuing system)

LL local loop

LNA local network architecture

LSI large-scale integration

LSO local service office

LTU line termination unit

LU logical unit

MA multiple access

MAN Metropolitan Area Network

Map Manufacturing Automation Protocol

MAU media attachment unit (equivalent to an Ethernet transceiver)

MATDM message-asynchronous time division multiplexing

MCI Microwave Communications, Inc.

MCMD multicenter, multidrop

MCMS multicenter, multistar

MD multidrop (also see CMST or MST)

MDF main distribution frame (equivalent to a major wiring hub)

MF MultiFrequency signaling scheme or a PCM MultiFrame or multiplexing factor (equals number of voice conversations carried by circuit)

MFJ modification of final judgment (resulting in AT&T divestiture)

MIB management information base for NMC database interchange (for SNMP and CMIP)

MIC Management Integration Consortium; it died after one year of activity amidst controversy

MIPS million instructions per second

MIS management information system

MJU multipoint junction union

MMFS manufacturing message format standard

MMOS message multiplex operating system

MPOA multiprotocol over ATM

MS message switching

MSK minimum shift keying

MST minimal spanning tree

MSU multiplex service unit

MT magnetic tape

MTBCF mean time between catastrophic failures

MTBEB mean time between error bursts

MTBF mean time between failures

MTBMF mean time between major failures

MTBNF mean time between nodal failures

MTBSF mean time between subsystem failures

MTKU maximum trunk utilization factor

MTS message telecommunications service

MTKU maximum trunk utilization factor for data networks

Multicast a LAN packet copied to a subset of stations (equivalent to a group message in MS)

MW MULDEM microwave multiplexer-demultiplexer

N-SMLC network synchronous multiline controller

NAK no acknowledgement

NAU network addressable unit

NBS National Bureau of Standards

NCC network control center

NCPM number of calls per month

NCU number of control units per virtual multipoint circuit

NIM network interface machine

NLT node-link type data for each location (it defines the link capacity, allowable link rate, and the applicable tariff)

NM/TC network management and technical control

NMC network management control (also network management center)

NM Forum Network Management Forum for standards making

NMP network management protocol (CCITT, ISO); also network management processor

NNI network to network interface as defined by ATM Forum

NNX network numbering exchange

NPA numbering plan area

NPL National Physical Laboratories of U.K.

NPSI NCP packet service interface (IBM)

NSC Network Systems Corporation

NSP network service protocol

NTI network trunk interface (ISDN)

NUM numerator

OAM operations, administration, and maintenance to be used by ATM cells for NMC functions

OCC other common carriers

ONA open network architecture allowing all ESPs equal access to RBOC networks

OS operating system

OSI open system interconnection (ISO's 7-layer framework for data interchange)

OSRA originating station routing approach

PABX private automatic branch exchange

PAD packet assembly/disassembly

PAM pulse amplitude modulation

PATDM packet-asynchronous time division multiplexing

PC primary center or personal computer

PCM pulse code modulation

PCR peak cell rate as defined by ATM Forum for NMC

PCN personal communications network for providing cellular/mobile services

PCS personal communications services provided on a PCN

PCTG programmable channel termination group

PDM pulse duration modulation

PDN public data network

PDS premise distribution system (also premise wiring system)

PDU protocol data unit (e.g., packet or frame or cell)

PLU primary logic unit

PM phase modulation

PMT packet-mode terminal

PNNI private network to (ATM public) network interface

POI point of interface (applies to IEC/OCC exchange)

POP point of presence (applies to IEC/OCC exchange)

POS point of sale

PPL Pluribus private line

PPM pulse position modulation

PPS private packet service

PPX private packet exchange

PRI primary rate interface for ISDN, providing 23 B-channels and single D-channel

PS packet switching (also packet switched)

PSE packet switching exchange

PSI Pluribus satellite interface message processor

PSK phase shift keying

PSS packet switching system

PSTN public switched telephone network (also PTN)

PSV public service vehicle

PTK propagation transmission constant

PTT Postal, Telephone, and Telegraph

PV principal value

PVC permanent virtual circuit

PWS premise wiring system (also premise distribution system)

QOS quality of service

RAC ring access control for Datapoint's ARCnet

RAM random access memory

RBOC Regional Bell Operating Company (BOC)

RC regional center

RFC request for comments; documents prepared by standards-related groups

RFP request for proposal

RHO regional hub office

RIM resource access module for Datapoint's ARCnet

RIP routing information protocol as employed by TCP/IP

RJE remote job entry

RJEP remote job entry protocol

RML response message length in bytes

RMON remote monitoring standard MIB defined in RFC 1271 for SNMP/NMC

ROE remote order entry

ROI return-on-investment

ROM read-only memory

RTIU remote terminal interface unit

RTS request to send (an EIA/TIA-232 control signal)

RU request unit

SAR segmentation and reassembly layer of ATM protocol

S/F store and forward

SBS Satellite Business Systems

SC sectional center

SCC specialized common carrier

SCE service creation environment of an AIN

SCMD single-center, multidrop

SCP service control point (or a DB management node)

SCSS single-center, single-star

SDF system design file

SDH synchronous digital hierarchy for the SONET standard

SDLC synchronous data link control

SDM space division multiplexing

SDN software defined network offering of AT&T (also see Vnet, VPN)

SDSL full-duplex T1 or E1 for up to 10,000 feet

SG study group (associated with CCITT)

SIO standard information outlet as defined by CCITT

SIP SMDS interface protocol

SIR sustained information rate

SITA Société Internationale de Télécommunications Aéronautiqes

SL subscriber line (or special line created before MD network is designed)

SMDR station message detail report

SMDS Switched Multi-Megabit Data Service

SMI structure of management information as specified in RFC 1155 for SNMP/NMC

SMLC synchronous multiline controller

SNA system network architecture as offered by IBM

SNAP standard network access protocol

SNMP simple network management protocol originally developed for Internet, now used by LANs

SNRC Successive Node Route Control

SOM start of message

SONET synchronous optical network; a set of standards

SOW statement of work

SPC stored program control

SPLS shared private line service

SPNS switched private network service

SQR square root

SSB single-side band

SSCP system service control point

SS7 signaling system No. 7 for common channel signaling network

STP signal transfer point (or an HS packet switch)

STS1 SONET electrical Synchronous Transport Signal at 51Mbps

STS3 SONET electrical Synchronous Transport Signal at 155Mbps

SWIFT Society for Worldwide Interbank Financial Telecommunications

TA terminal adapter (for ISDN)

TAC Telenet access controller

TASI time asynchronous speech interpolator

TC toll center

TCA time-consistent average

TCAM telecommunications access method (IBM)

TCO Telenet central office

TCP/IP transmission control protocol/internet protocol

TCPM total cost per minute of a voice call

TCTS Trans-Canada Telephone System

TDM time division multiplexing

TDMA time division multiple access

TDX time division exchange

TEHO tail-end-hop-off

TELCO telephone company

TGF traffic growth factor (an SDF design parameter)

THI terminal handler interface

TIH terminal interface handler

TIP terminal interface processor

TK trunk

TKMF trunk multiplexing factor = number of VG circuits carried on a trunk

TKLT trunk link type (an SDF design parameter)

TM transaction module

TMC total monthly cost of a network

TN transport network

TOD time-of-day

TOP technical office protocol

TP Telenet processor

TS time-sharing

TST time-space-time

TTTN tandem tie-trunk network

TTY teletypewriter

TWX teletypewriter exchange service

UBR undefined bit rate class of traffic as defined by ATM Forum

UITP universal information transport system

UNI user-network interface as defined by ATM Forum for both private and public networks

UNMA unified network management architecture (of AT&T)

UPR user port rate (bps)

UTP unshielded twisted pair as commonly used in LANs and 10BaseT Ethernets

VAN value-added network

VBR variable bit rate option of an ATM-based system

VC virtual circuit

VCL virtual-circuit-link

VCL virtual circuit and link data for each location

VCU video compression unit

VDSL variable speed DSL (13–52Mbps, 1.5–2.3Mbps) for 1000–4500 feet

VG voice grade

VGC voice grade circuit

VHD vertical (V) and horizontal (V) coordinate data of each location

Virtual LAN a virtual realization of a workgroup mapped over several physical LANs

VMPT virtual multipoint circuit associated with Doelz network

VNet virtual net offering of MCI

VNS virtual network services

VPN virtual private network offering of Sprint

VTAM virtual telecommunications access method (IBM)

VT virtual tributary for transporting subbase rate payloads in SONET

VTP virtual terminal protocol

WATS wide area telecommunications service

WC wiring center that serves the customer premise equipment

WPM words per minute

xDSL x-type digital subscriber line (x = A, R, H, S, V) allowing high-speed data rates

Index